CHRISTIANITY IN THE LIGHT OF SCIENCE

CHRISTIANITY IN THE LIGHT OF SCIENCE
Critically Examining the World's Largest Religion

EDITED BY **JOHN W. LOFTUS**

Prometheus Books
59 John Glenn Drive
Amherst, New York 14228

Published 2016 by Prometheus Books

Cover design by Nicole Sommer-Lecht
Cover design © Prometheus Books

Inquiries should be addressed to
Prometheus Books
59 John Glenn Drive
Amherst, New York 14228
VOICE: 716–691–0133
FAX: 716–691–0137
WWW.PROMETHEUSBOOKS.COM

20 19 18 17 16   5 4 3 2 1

Library of Congress Cataloging-in-Publication Data

Names: Loftus, John W., editor. | Stenger, Victor J., 1935-2014, dedicatee.
Title: Christianity in the light of science : critically examining the world's largest religion
    / edited by John W. Loftus.
Description: Amherst, New York : Prometheus Books, 2016. | Includes index.
Identifiers: LCCN 2016007392 (print) | LCCN 2016012665 (ebook) |
    ISBN 9781633881730 (paperback) | ISBN 9781633881747 (ebook)
Subjects: LCSH: Stenger, Victor J., 1935-2014. God. | Religion and science. | God—
    Proof. | Atheism. | BISAC: RELIGION / Atheism. | SCIENCE / Essays.
Classification: LCC BL240.3.S7383 C47 2016 (print) | LCC BL240.3.S7383 (ebook) |
    DDC 230—dc23
LC record available at http://lccn.loc.gov/2016007392

Printed in the United States of America

This book is dedicated to the memory of Victor J. Stenger (1935–2014).

Throughout history, arguments for and against the existence of God have been largely confined to philosophy and theology. In the meantime, science has sat on the sidelines and quietly watched this game of words march up and down the field. . . . In my 2003 book, **Has Science Found God?** *I critically examined the claims of scientific evidence for God and found them inadequate. In this present book, I will go much farther and argue that by this moment in time science has advanced sufficiently to be able to make a definitive statement on the existence or nonexistence of a God having the attributes that are traditionally associated with the Judeo-Christian-Islamic God.*
        —**Victor J. Stenger, from the preface to his 2007**
New York Times **bestseller,** *God: The Failed Hypothesis*

CONTENTS

Foreword   9

Introduction   13

PART 1. SCIENCE AND RELIGION

1. How to Think Like a Scientist: Why Every Christian
    Can and Should Embrace Good Thinking
        Guy P. Harrison   27

2. A Mind Is a Terrible Thing: How Evolved Cognitive Biases
    Lead to Religion (and Other Mental Errors)
        David Eller   47

3. What Science Tells Us about Religion:
    Or, Challenging Humanity to "Let It Go"
        Sharon Nichols   69

PART 2. SCIENCE AND CREATIONISM

4. Christianity and Cosmology
        Victor J. Stenger   97

5. Before the Big Bang
        Phil Halper and Ali Nayeri   119

6. Intelligent Design Isn't Science,
    and It Doesn't Even Try to Be Science
        Abby Hafer   141

PART 3. SCIENCE AND SALVATION

7. Saying Sayonara to Sin
    Robert M. Price and Edwin A. Suominen   169

8. The Soul Fallacy
    Julien Musolino   187

9. Free Will
    Jonathan Pearce   207

PART 4. SCIENCE AND THE BIBLE

10. Biblical Archaeology: Its Rise, Fall, and Rebirth
    as a Legitimate Science
    Robert R. Cargill   239

11. The Credibility of the Exodus
    Rebecca Bradley   253

12. Pious Fraud at Nazareth
    René Salm   275

PART 5. SCIENCE AND THE CHRIST

13. The Bethlehem Star
    Aaron Adair   295

14. If Prayer Fails, Why Do People Keep At It?
    Valerie Tarico   313

15. The Turin Shroud: A Postmortem
    Joe Nickell   335

About the Contributors   357

Notes   363

FOREWORD

I t is a solemn honor to have been asked to contribute the foreword to a book dedicated to the memory of Victor J. Stenger—a comrade in reason, defender of science, debater of dissemblers, and unmasker of voodoo in all its many disguises. It is an ineffable sadness, however, to hold in my hands what probably is the last bullet—shot into the future—from the gun with which Vic stalked the metaphysiological parabeasties that haunt our civilization. The wonderful book that lies here before the reader contains his last essay, his last attempt to communicate with a world that he tried so long to lure to a life of reason. Readers cannot help but be inspired by the final words of his chapter, "Christianity and Cosmology": ". . . we are special, at least on Earth and the solar system. Even though magical thinking and hubris may still destroy us, we can hope that our unique abilities will lead us to a better future." Amen.

Perceptive readers will notice that the *fantôme gris* of Christian apologist William Lane Craig drifts darkly through many of the chapters of this book, and it lurks unseen in all the rest. It is sobering to realize how many different experts have been needed to refute his multifarious and nefarious claims. At each appearance, however, his gray ghost is bleached into nonexistence by the light each author casts upon his shade. Again and again, Craig's "God" is seen to be a god of the gaps. Only beyond the frontiers of science can Craig's deity find a place to hide. As an example: Craig's Kalam Cosmological Argument can be seen to depend either (1) upon a claim that is scientifically meaningless (because there is no way to test it) or (2) upon a claim that may someday prove possible to test but can't yet be answered by science one way or the other.

In the first case, if the *ultimate* origin of time and space was a Big Bang (for which we have much evidence), then to argue that there was something *before* that (a god of some sort) is scientifically meaningless: there is no way even in our imagination to test such a claim. That places it outside the realm of reason, as well as outside the realm of science. This is for the simple reason that an observer would have to exist *before* the beginning of time—and that is philosophically incoherent.

In the second case, since Craig depends upon the Big Bang being a *singularity*—an ultimate beginning point for our universe, instead of being merely an extremely hot, momentary phase of compact energy—he simultaneously depends upon an issue that is susceptible to testing. As I write, the Advanced Laser Interferometer Gravitational-Wave Observatory (LIGO) has just announced that its twin detectors have heard the gravitational "ringing" produced by the collision of two black holes. They have found strong evidence for gravitational waves, adding yet another crucial piece of evidence in favor of Einstein's General Theory of Relativity. Some cosmologists have argued that this would also be evidence in favor of a "multiverse" in which the Big Bang was just a local event. It is unclear if the discovery might also be evidence for a "Big Bounce" instead of a Big Bang—the most recent expansion phase in a universe oscillating between Big Bounces and Big Crunches.

Craig is probably representative of the most successful latter-day defenders of the Christian faith, and so it is heartening to see just how precarious his positions have here proven to be. Those who carefully read *all* the arguments in this book—from "Before the Big Bang" to the demise of Nazareth, the "home town of Jesus"—will realize that *if any one of them be true*, Craig's war is lost. He has to be a twelve-ball juggler, keeping all his glass bubbles moving slickly through the air while trying to prevent them from falling to the floor or being intercepted by someone able to discover their vacuous content.

It will be interesting to see how Craig will try to stabilize his jugglery when hit by the blast of René Salm's discovery ("Pious Fraud at Nazareth") that Nazareth, the alleged hometown of "Jesus of Nazareth," was not inhabited at the time when the Holy Family should have been living there. No land of Oz, no Wizard of Oz. No little town of Nazareth . . .

As noted, if any one of Craig's arguments should fail, his whole case must collapse, due to the *ad hoc* way in which all his arguments must be cobbled together. By contrast, finding even fairly large errors in any particular scientific argument would not be of great overall effect, due to the consilience of all the scientific lines of evidence. All the roads of science are leading to the same place. This is largely because the sciences are seeking to explain the unknown in terms of the known—or at least, the better known. Apologists, by contrast, must commit the fallacy that medieval logicians termed *ignotum per ignotius*—explaining the unknown in terms of the even more unknown. A completely unknown—indeed, probably unknowable—supernatural cause (Craig's god of the gaps) is invoked to explain the existence of the universe, the origin of life, the origin of *Homo sapiens*, and the nature of moral behavior. (Space does not permit discussion of what quantum physics has done to the concept of "cause"!)

Craig's position is like that of King Canute: enthroned in solemn state upon the shore of the beleaguered Isle of Faith, ludicrously bidding the tidal wave of science and scholarly research to stay and to encroach no further onto his sacred, sacrosanct terrain. Stop, cosmology! Halt, physics! Stay back, evolution! Cease, critical studies of the Bible! Back off, brain physiology! With the publication of this book, the tide has risen to the seat of his throne. He's going to have to find dry linens.

Nearly all the subjects that I have fought over, debated, and written about over nearly sixty years as an atheist activist are covered in this book. Gratifyingly, *all* of them are advanced in this book far beyond where I left off, and all are brought up to date—in some cases way beyond anything I might have predicted when I was actively debating these subjects.

This is a full-service manual of counter-apologetics. All who seek to do their utmost to defend science and reason from destruction by the darkling forces of ignorance and superstition have, I believe, a *moral* obligation to read and master *all* the chapters of this book. They must not leave any gap in their argumental armor through which even a minor god might penetrate.

These are parlous times. The moment is desperate, yet hopeful. The dragon of faith is dying, but the thrashing of its tail while in the throes of death can still do much damage. Armageddon could yet prove to be the only prophecy of the Bible ever to come true. Readers of this book must do everything they can possibly do to prevent a first-ever success of a biblical prophecy.

—Frank Zindler, February 14, 2016

Frank Zindler is Professor Emeritus of biology and geology at SUNY–Johnstown, and a member of the Jesus Project.
A Life Celebration: Frank Zindler
http://blog.edsuom.com/2014/04/a-life-celebration-frank-zindler.html

INTRODUCTION

This new anthology is the fourth one in a series of books I've edited. My first three are named after *New York Times*–bestselling books by the so-called new atheists. This present one honors the late Victor J. Stenger, and his book *God: The Failed Hypothesis—How Science Shows That God Does Not Exist* (Prometheus Books, 2007). Although the title of this volume does not echo Stenger's title directly, I still consider it to be part of that same series.

In this present volume, Christianity and her many sects are critically examined by the results of science. Like my other anthologies, this one contains new essays written by superior authors on many important issues relevant to its theme. I am honored and very grateful to have such well-qualified authors to write chapters for it, as with the previous books. I am particularly grateful for David Eller and Valerie Tarico, who were able to write chapters for all of them to date.

The chapters in this book reflect Stenger's commitment to using the tools of science to critically examine the God hypothesis, and for communicating the results to the rest of us. He started out as a research scientist in the field of particle physics and ended up publishing twelve books on physics, quantum mechanics, cosmology, atheism, and the relationship of science to faith. I am particularly pleased Vic reached out to the educated masses, not just the elite scholars. His career was an excellent example of Karl Marx's aphorism: "The point is not merely to understand the world, but to change it."

Editing this book has brought me more joy than the others. Vic was a friend of mine, even though we never met. Vic was the most approachable new atheist who responded to my emails. He also wrote blurbs for my books and chapters for them. Vic deeply cared about the same kinds of issues I do. When I told him I was considering editing this work, he submitted a chapter for it in advance. It's his last known essay.

In this anthology, we're critically examining Christianity (or Christianities) in light of science. This has been a major theme of mine, first seen in *Why I Became an Atheist*, where I devoted no less than three full chapters to science and faith (6, 13, 15). Science also plays a significant role in my anthologies. In *The Christian Delusion* there are six chapters largely dealing with it (2, 3, 5, 6, 9, 15). In *The End of Christianity* there is whole part of the book titled "Science Puts an End to Christianity," with four chapters (11–14). In *Christianity Is Not Great* there are nine chapters on science and faith, with one part of the book dealing with it (2–4, 10–15). Any important subject not treated in this present volume is probably discussed in them.

Even as much as I've dealt with science and the Christian faith, there are still many faith-based claims unaddressed. That's because they don't deserve any consideration, like the supposed locations of the Garden of Eden, the resting place of Noah's ark, the Tower of Babel, the pillar of salt supposed to be Lot's wife, or where Moses parted the Red Sea. There are a multitude of these types of under-evidenced claims in the Bible, such as a talking snake and donkey, ancient people living nine hundred years, a bush that was on fire but didn't burn up, a man's hand that turned leprous then was restored to health, shoes that didn't wear out for forty years in a wilderness, a mass of millions being fed "manna" for forty years from heaven who were led by a pillar of cloud by day and a pillar of fire by night, the sun standing still in the sky, an axe head floating on water, a great fish swallowing a man, and so on. In the New Testament, there isn't credible evidence for the miracle claims we read there, especially involving a shadow or a cloak or clumps of clay or a healing pool of water, or a man who was bitten by a poisonous snake and shook it off unharmed. The reason none of these claims are taken seriously is because there has to be some sort of credible evidence to begin with. If there was a scientific court to address these faith-based claims, any independent reasonable judge would simply dismiss them with prejudice due to the woeful lack of evidence.

SCIENCE AND LIBERAL CHRISTIANITY

To their credit, many Christians don't believe these faith-based claims because of the lack of any credible evidence. Not only that, but after nearly two thousand years they're also rejecting some of the major beliefs of their predecessors for the same reason. Over sixty-five years ago, New Testament scholar Rudolf Bultmann conceded that several key Christian beliefs could no longer be defended in a scientifically literate world:

> The cosmology of the N.T. [New Testament] is essentially mythical in character. The world is viewed as a three-storied structure, with the earth in the center, the heaven above, and the underworld beneath. Heaven is the abode of God and of celestial beings—angels. The underworld is hell, the place of torment. Man is not in control of his life. Evil spirits may take possession of him. Satan may inspire him with evil thoughts. It is simply the cosmology of a pre-scientific age. To modern man . . . the mythical view of the world is obsolete. It is no longer possible for anyone seriously to hold the N.T. view of the world. We no longer believe in the three-storied universe. No one who is old enough to think for himself supposes that God lives in a local heaven. There is no longer any heaven in the traditional sense. The same applies to hell in the sense of a mythical underworld beneath our feet. And if this is so . . . we can no longer look for the return of the Son of Man on the clouds of heaven. It is impossible to use the electric light and the wireless and to avail ourselves of modern medical and surgical discoveries, and at the same time to believe in the N.T. world of spirits and miracles.[1]

Christians are not within their epistemic rights to believe these things. There isn't any credible evidence to warrant belief in these and other claims within the Christian tradition, just as there isn't any for the other beliefs I mentioned. They're all based on texts found in their ancient superstitious prescientific Bible. It's just that sophisticated theologians have been able to convince believers that mere argumentation can substitute for sufficient evidence. Now for some mixed metaphors: They have woven a tangled web of words to tailor the data so it fits their preconceived Procrustean bed of faith. The main purpose of this book is to honestly and rigorously expose their dog and pony show for what it is. No evidence of a rabbit up this sleeve. No evidence of a rabbit up that sleeve. Presto! A rabbit is pulled out of a hat!

One of the earliest books I read when beginning to doubt my Christian faith was written by Bultmann's protégé Uta Ranke-Heinemann, who became a Catholic theologian. Her book was titled *Putting Away Childish Things*, and had a subtitle that was both provocative and revealing: *The Virgin Birth, the Empty Tomb, and Other Fairy Tales You Don't Need to Believe to Have a Living Faith*.[2] Her book received blurbs of high praise by liberals Karen Armstrong, Elisabeth Schüssler Fiorenza, and Bishop John Shelby Spong. These liberals could accept most of the evidence presented in this present anthology and still believe. But none of them takes seriously what Ranke-Heinemann actually said, not even Ranke-Heinemann herself. For after she convincingly showed that the nativity tales in the Gospel of Luke contain too many historical inaccuracies, discrepancies, and incorrect dates to think the story of the baby Jesus could be true, she said:

> If we wish to continue seeing Luke's accounts of angelic messages and so forth, as historical events, we'd have to take a large leap of faith: We'd have to assume that while on verifiable matters of historical fact Luke tells all sorts of fairy tales but on supernatural matters—which by definition can never be checked—he simply reports the facts. By his arbitrary treatment of history, Luke has shown himself to be an unhistorical reporter—a teller of fairy tales.[3]

Ranke-Heinemann ends up endorsing the irrationalities of a leap of faith herself, since her own words can be repurposed against her:

> If we wish to continue being Christians we'd have to take a large leap of faith: We'd have to assume that while on verifiable matters of historical and scientific fact the church told all sorts of fairy tales but on supernatural matters like the existence of God, angels, and/or veridical religious experiences—which by definition can never be checked—the church simply reports the facts. By her arbitrary treatment of history, the church has shown herself to be an unhistorical reporter—a teller of fairy tales.

Ranke-Heinemann and her liberal friends cannot have it both ways. If they reject several prescientific claims of the early church because they're faith-based claims without the requisite evidence, then they should reject all faith-based claims. Hopefully this anthology can help Christians like them be consistent. For sooner or later one or more of the chapters in this anthology will expose their own fairy-tale faith. Eventually all faith-based claims run aground, into the cold hard truths of scientific evidence. So Guy P. Harrison is right on the mark when he wrote, "To date, there is no known verifiable scientific evidence that confirms any Christian supernatural claim." (Endnote 1 of his chapter.)

TWO DECLARATIONS OF VICTORY
BY CHRISTIAN APOLOGISTS

Christian apologists will futilely try to scare their progeny with the boogeyman of scientism here, declaring their faith victorious because they claim its critics think science can test everything, which is a self-refuting claim. But this is a non-issue because it's only a self-refuting claim if we say we're certain science can test everything. For if we leave room for reasonable doubt then it's not self-defeating to say science can test most everything or *almost* everything. When it comes to matters of fact like the nature of nature, its workings, and even its origins, science is the best and only way to gain objective knowledge about them. It isn't just that there is no *better* alternative. It's rather that *there is no other alternative!*

If faith is suggested as an alternative to science then let's be clear about it. Faith has no method and solves no problems. Therefore it doesn't gain us any objective knowledge about matters of fact. Saint Anselm of Canterbury said, "Faith seeks understanding." No one in the history of Christianity ever said "Understanding seeks faith." Faith is best viewed as a cognitive bias that misjudges the probabilities. If faith is defined as trust, then we should not trust faith.

Regardless of this nonissue, many religious beliefs can indeed be empirically tested. These are the types of beliefs we examine in this book.

So there is nothing inside its pages where the boogeyman of scientism can be used to reject the evidence presented. Even some God concepts can be tested empirically. For instance, in part 3 of a previous book of mine, *How to Defend The Christian Faith*, I tested the concept of a perfectly good, all-knowing, all-powerful God against the empirical evidence of the massive amount of intense suffering experienced in the fourteenth century by the Black Death plague, and found such a God concept was extremely unlikely to the point of refutation, despite the gerrymanderings of God's defenders.

Christian apologists will also futilely try to scare their progeny with the boogeyman of methodological naturalism, declaring their faith victorious because they claim its critics have a pretheoretical commitment to naturalism, which *de facto* excludes all supernatural explanations and answers. This is also a non-issue, which is fallaciously used as an excuse not to trust any of our conclusions. As I wrote about elsewhere in *How to Defend the Christian Faith*[4] it is not the case that by using a fair method, one that has produced objective knowledge in every discipline of learning, that supernatural results are excluded. Clinical studies on prayer could still obtain positive supernatural results, no matter what pretheoretical commitment scientists have.

Just the same, let this book be a way to test (or examine) that faith-based supposition of theirs. To support their contention, they should be able to find conclusions in this book that were determined in advance by such a method. What we're saying is the evidence is against what Christians believe. You won't see any Procrustean bed we're trying to force the evidence to fit. I defy anyone to read this book and say otherwise. Methodological naturalism did not bias any conclusions, which makes it a non-issue. We're just trying to honestly deal with the evidence. It's the evidence that convinced most of us to walk away from Christianity, despite having had a pretheoretical commitment to faith.

BRIEFLY INTRODUCING THE CHAPTERS

In part 1 of this book, the authors critically examine Christianity as the religion it is, from the perspective of science. In chapter 1, Guy P. Harrison persuasively argues that Christian people of faith should think like scientists. Then he teaches them how to do it. While doing so might not automatically lead them to reject their faith, he also says, "Thinking threatens Christianity; it's as simple as that." In chapter 2, David Eller argues that the origins of religious faith are rooted in our evolved brains, which tolerates a whole host of known cognitive biases, and, as such, should make believers into doubters. In chapter 3, Sharon Nichols challenges people of faith to abandon their religion altogether.

In part 2, the authors present the cosmological and evolutionary evidence that does not support creationism. In chapter 4, Victor J. Stenger shares a brief history of ideas about cosmology and how Christianity failed by incorrectly understanding the cosmos. If it failed on this issue, there should be no reason to accept its prescientific claim that God created the universe either. In chapter 5, Phil Halper and Ali Nayeri show that modern science does not support the idea of an absolute beginning to the universe, contrary to Christian philosophers like William Lane Craig. In chapter 6, Abby Hafer shows from the actual data that intelligent design is not science but rather creationism in disguise. This data was discovered by her without any pretheoretical commitment to naturalism.

In part 3 are four chapters that critically examine some crucial beliefs of Christianity. In chapter 7, Robert M. Price and Edwin A. Suominen argue that there was no Adam and Eve, and as a result no original sin and no need for a savior from sin. In chapter 8, Julien Musolino argues there is overwhelming evidence against the belief in souls, while, in chapter 9, Jonathan Pearce argues that human beings do not have free will. The upshot of their arguments is that any type of Christianity requiring salvation from sin by free-willed choices is ruled out.

In part 4 are chapters that present archaeological evidence disputing the claim that the Bible is based on history. In chapter 10, Robert Cargill shows that by employing modern science to interpret archaeological findings, rather than interpreting the archaeological findings through the lens of the Bible, biblical archaeology has been reborn. As a result, with more reliable findings since the second half of the twentieth century, many aspects of the Judeo-Christian faiths have been legitimately challenged and dismissed. In chapter 11, Rebecca Bradley provides an important Old Testament example with regard to the credibility of the Israelite exodus from Egypt led by Moses. In chapter 12, René Salm provides an important New Testament example with regard to the credibility of the claim that Jesus was from Nazareth.

In part 5, the chapters focus on three faith-based claims in the New Testament. In chapter 13, Aaron Adair examines attempts to use astronomy to verify the Bethlehem star and finds they all fail. In chapter 14, Valerie Tarico shows there is no evidence to think petitionary prayers are efficacious, so she asks why people still do it. Then Joe Nickell, in chapter 15, writes a postmortem on the Shroud of Turin, showing it to be a fake. Since it's a fake, there is no hard evidence that Jesus was raised from the dead. None. Yet, given what we learned in part 1 of this book, hard evidence is what we would need if we honestly wanted to know the truth.

DON'T BE CLOSE-MINDED TO SCIENCE

In this volume is found the evidence, the scientific evidence, the evidence that can convince open-minded people. Open-minded people will be open to the scientific evidence. Closed-minded people won't be open to it, but will instead try to denigrate or deny it. To help believers be open-minded to scientific evidence, I have argued quite extensively for the *Outsider Test for Faith*.[5] Professor Jerry Coyne, a scientist specializing in evolutionary genetics at the University of Chicago, says "the wisdom of this . . . quasi-scientific approach" is "unquestionable."[6] It asks believers to rationally test one's culturally adopted religious faith from the perspective of an outsider, a nonbeliever, with the same level of reasonable skepticism believers already use when examining the other religious faiths they reject.

The reason believers are not open-minded to science, in those areas where science conflicts with their faith, is because of confirmation bias. This bias is a strong tendency human beings have to search for data, or to interpret existing data, in ways that confirm their preconceptions. Michael Shermer calls it "the mother of all cognitive biases."[7] The outsider perspective nullifies this bias precisely because it's an outsider's perspective. It has nothing to confirm. When there is nothing to confirm, then confirmation bias doesn't get in the way of finding the truth. The outsider perspective is therefore the best and only way to objectively test religious faith. This perspective uniquely helps believers rid themselves of confirmation bias so they can honestly apply the same standards used in investigating other religious faiths across the boards. It gives them permission to accept the results of science rather than deny them. For that's what they do whenever there's a conflict between science and the other religious faiths they reject. They accept science, since that's the only intellectually honest thing to do.

Some scientists are denying there is any conflict between science and religion, following Stephen Jay Gould, who argued that science and religion are two different areas of inquiry and the two don't overlap. First there is the "magisterium" of science, which "tries to document the factual character of the natural world, and to develop theories that coordinate and explain these facts." Then there is the "magisterium" of religion, which "operates in the equally important, but utterly different, realm of human purposes, meanings, and values."[8] This non-overlapping magisteria of science and religion is known as the NOMA principle.

Certainly a lot can be written by way of criticizing NOMA. For our purposes, let this book be a test case with regards to it. Since science shows the world's largest religion is not factually based, and therefore false (What else are we to conclude?), then it isn't reasonable to believe human purposes, meanings, and values are to be derived from that supernatural source either (via divination, prophecy, revelation, inspiration, or illumination). They cannot be derived supernaturally when the religion itself is shown to be false. So any human purposes, meanings, and values were already there prior to its beginning, or subsequently learned by trial and error apart from it. In fact, we can critically examine and subsequently reject Christianity because of the suffering it has caused down through the centuries (barring sufficient evidence on its behalf). This is something which my anthology, *Christianity Is Not Great: How Faith Fails* (2014) does. Given that NOMA doesn't save the world's largest religion, the only recourse left for believers is to reject the science.

How can they reject science, you ask? Sean B. Carroll, a professor of molecular biology and genetics at the University of Wisconsin–Madison, tells us. An article he read in the journal *Pediatrics*, titled "Chiropractors and Vaccination: A Historical Perspective" sparked his thoughts.[9] The article traced the roots of anti-vaccination among chiropractors to its founder, Daniel David Palmer, in 1865, and highlighted this same attitude among practitioners in the last few decades. In the 1950s, for instance, many chiropractors denied the science of polio vaccinations, but they were proven wrong when polio was eradicated in the United States because of vaccines. The article went on to offer the six arguments chiropractors have used to deny the science of vaccinations for a century.

After reading it, Carroll saw a general pattern among science deniers. Carroll said, "I could superimpose those arguments entirely upon what I had been reading from the antievolution forces." The six chiropractic anti-vaccination arguments can be seen as "a general manual of science denialism," he said. He saw that "to deny a piece of science there was sort of common playbook, a common set of tactics." In fact, "You could throw any argument at me about evolution, climate change, etc., and it would be in one of these six bins." Here they are:

1. Use anything to cast some small measure of doubt on the science, no matter how small, disregarding that the probabilities are very high that the science is correct.
2. Question the motives of scientists, saying they are motivated by profit or some other underhanded reason.
3. Magnify any disagreements between scientists by citing gadflies as authorities who represent a tiny minority.
4. Scare the hell out of people by exaggerating the potential harms or risks in accepting the science.
5. Appeal to the value of personal freedom by claiming no one should be compelled to accept the science.

6. Object that the science repudiates some key point of philosophy or theology, which Carroll says is one of the most important tactics of science denialists. On this he quotes creationist Henry Morris, who said, "When science and the Bible differ, science has obviously misinterpreted its data."[10]

These are the arguments and tactics of close-minded people who seek to reject science. None of these tactics actually do anything of the sort. They are all efforts in argument substitution, where someone substitutes an argument when the evidence shows otherwise. But they aren't really even legitimate arguments. They are informal fallacies, where rhetoric itself substitutes as argumentation. They're rhetorical bluffs, or rhetoric without substance. For instance, in denying what I call the evolutionary paradigm (or theory, or fact), many believers object that if evolution is true there can be no morals. Whether that's the case or not is being debated, of course, but the issue of morality has nothing to do with the objective overwhelming evidence for evolution. Either the evidence is there or it's not, and it is. So don't accept the tactics of science deniers if you want to know the truth. Follow the evidence instead.

Yes, follow the evidence, but also follow the money. If you want to know why so many perceived authorities still question the results of evolutionary science, then follow the money, just like you should follow the evidence. For instance, the John Templeton Foundation aims to reject science in the interests of religion by offering academics millions of dollars to find reasons to believe. So anything financially supported by the Templeton Prize should be subjected to intense scrutiny. Lately, the Templeton Foundation has even funded climate change denialists.[11] This, too, has to do with religion, for if there is a god in control of the environment then climate change is nothing to worry about. But then, there shouldn't have been anything to worry about with polio either. In any case, if you reject the Templeton Foundation's antiscience view of climate change, you should clearly see its antiscience view with regard to creationism too.

Part 1

SCIENCE AND RELIGION

Chapter 1

HOW TO THINK LIKE A SCIENTIST

Why Every Christian Can and Should Embrace Good Thinking

Guy P. Harrison

The world's most numerically popular religion has a problem. Christianity's most important claims are not supported by good evidence and logical explanations. If this belief system had some significant degree of scientific confirmation going for it, it is likely that everyone alive today would be a devout Christian. Certainly all rivals would have wilted and died long ago, their hollow claims frozen out by the shadow of an obviously real Christian god. After two thousand years of Christianity, however, what we have is a religion that has been unable to convince even half the world's population that its claims are true.

What is required for someone to embrace an extraordinary claim that lacks supporting evidence? What kind of sacrifice is necessary to make such a leap? Specifically, do Christians give up some degree of their ability to reason in order to believe the unbelievable? They do, unfortunately. This may be putting it bluntly but there is no way around it. This religion—like many others, of course—stands in direct opposition to the full blossoming of a human mind. Thinking threatens Christianity; it's as simple as that. Most forms of Christianity discourage or outright forbid followers to vigorously and consistently think. Within this belief system, questioning, doubting, and mind-changing are commonly seen as bad. Meanwhile, believing simply because it feels good psychologically or a story says to believe is considered good. Fortunately, Christians and all others who may be intellectually off course can decide to make changes at any point in life. It is not overly difficult to do so and, as I will explain, doesn't even require a Christian to stop being a Christian. It's never too late to start thinking like a scientist.

Without the support of science or scientific thinking,[1] Christianity's success and survival depends almost entirely upon exposure during the intellectually and emotionally vulnerable years of childhood, along with reinforcement or coercion from the surrounding society. This sad state is not unique to Christianity. Family indoctrination and cultural influence are the key crutches of all major religions. Very few believers come to their convictions after diligently researching a variety of contradictory religions, objectively comparing claims, and also considering what science has revealed about the human brain's propensity to be fooled by others and to fool itself. But these are Christianity's issues to deal with. I have little interest here in criticizing this ancient belief system's survival methods. I am far more concerned with the challenge of helping individual Christians think better in their real-world lives.

NO CHRISTIAN LEFT BEHIND

Good Thinking is my umbrella term for understanding, appreciating, caring for, and using the human brain in ways that enable one to better avoid lies, mistakes, and delusions, instead of repeatedly running toward them with open arms. Good Thinking reduces one's error rate over a lifetime. It includes thinking like a scientist in daily life. It also requires us to acknowledge our vulnerabilities and tendencies toward misperceiving reality and believing in nonsense. It is in everyone's best interest to realize and remember that the default human way is to believe first and ask questions never. We become better thinkers by recognizing how poor we are at thinking. Good Thinking is not some anti-Jesus agenda or an attack on anything else specifically, other than on lies, delusions, and cognitive errors. Who in their right mind would oppose it? For the sake of clarity, I repeat: Good Thinking is not necessarily opposed to Christianity. I would like to believe that virtually everyone, including Christians, can agree that it's only smart to be on the alert for fraud and bogus beliefs. Few people, I hope, openly claim to be in favor of *bad* thinking. No Christian should reflexively reject Good Thinking. Those who feel it is somehow negative or unnecessary likely have been misled by unthinking or unscrupulous sources and would do well to reconsider.

Religions are not all bad. Of course there are good things about Christianity. I understand that for many people it feels great and inspires them in positive ways. Little or no harm is done when a man kneels before a cross and whispers a wish into the air around him. No one is hurt when a joyful woman weeps before the Stone of Unction in Jerusalem. But we all are diminished and cheated when so many versions of Christianity encourage or demand people to think with less force and clarity than they are capable. The championing of this pathetic and passive posture dims our world. It robs humankind of much potential progress in science, medicine, exploration, education, security, and cooperation. In its worst expressions, Christianity would have us not only stagnate but move backward. Too little attention is given to what is stolen by this deference to blind allegiance. Not only is every moment spent looking up at an empty sky a lost moment down here in the trenches of reality, it also distracts us from dreams of what we might achieve if we took better advantage of our limitless potential to reason and create.

The belief system that rests upon the biography, personality, and supposed desires of Jesus/God the Father/the Holy Spirit endures despite a profoundly strange and wholly unproven premise. Nevertheless, more than two billion people today apparently find it not only compelling but convincing. No doubt the central story would be dismissed as unbelievable by most current Christians had they heard it for the first time in adulthood. The basic points that one is asked to accept as real follow in the next paragraph. As you read them, ask yourself how such claims without evidence could ever win over minds without circumventing or shutting down one's ability to reason.

An all-powerful god who knew the future when he created humans later decides to slaughter nearly every man, woman, child, infant, puppy, kitten, ant, etc. in a global flood. The reason the humans he made were so bad and worthy of horrible deaths is because they were "sinful" or "fallen."

This state was a result of Adam, the first human, having bitten into a piece of fruit that had been designated off limits. Thanks to Adam, every human since has left the womb flawed and condemned. A single act of mastication-rebellion puts every newborn on death row. Never mind that the idea of punishment for an inherited criminal conviction is incompatible with all modern concepts of justice. Despite "our" crime, however, this god still loves us. For this reason he came to Earth in human form some two thousand years ago. This was so that he could be tortured and executed as a human sacrifice, which apparently was necessary for him to be able to forgive us. This event is commonly described as the ultimate act of love and sacrifice: God the Father gave up his only son for us. But it wasn't quite as bad as it might have been because the dead Jesus promptly lived again and ascended back to heaven to rejoin God the Father—who is also Jesus, by the way. (Jesus, God the Father, and the Holy Spirit are all one, according to the Holy Trinity doctrine.) The crucial point of the story is that this god sacrificed his son/himself—albeit only temporarily—so that he would be able to excuse us for Adam's crime against him, which we had nothing to do with. No one seems to know why a god who makes all the rules and answers to no one couldn't just pardon us and skip the barbaric crucifixion event entirely. But what matters is that the Christian god loves us and provided everyone with a loophole to escape his punishment for something we didn't do. Make sense?

Many atheists enjoy endlessly mocking and dismantling this bizarre tale, as well as other claims, ideas, and behaviors that various versions of Christianity promote, including: miracles, faith healing, symbolic and actual cannibalism,[2] opposition to human rights, doomsday predictions, and a six-thousand-year-old Earth. These are all fair game, of course. I never discourage poking holes in big claims that come with little or no evidence. I do believe in prioritizing, however, and skeptics of Christianity might do well to keep in mind that such things are mere sideshows next to the most serious problem this religion presents. Christianity's overarching opposition to Good Thinking is the crucial challenge, and there is only so much time in a day. This is where the most damage is done, and it should be the focus.

The reason so many people accept and defend the many minor supernatural claims within Christianity is because they have already fallen so deep into the game. They failed to apply Good Thinking initially and bought the lie that it's bad or unnecessary to demand evidence for unusual and important claims. Therefore it seems inefficient to spend a substantial amount of time combatting belief in demons and angels, for example. Better to educate and encourage Christians to think critically, to doubt, and question everything as a general way of life. Inspire them to work harder, embrace science, and accept that not every question has an answer that is immediately available. Good Thinking is the all-purpose cure and preventative. Illuminate one fallacy or cognitive error for a Christian, and one helps her think for a day. Show her how to watch out for them on her own, however, and one helps her think for a lifetime.

Christians promoting faith over skepticism and belief over science is nothing new, of course. The story of "Doubting Thomas," the apostle who wouldn't accept the resurrection claim without evidence, has been held up for more than a thousand years as exhibit A in the Christian case against science and reason. It is not difficult to imagine why Christianity so often opposes Good Thinking. Many leaders and the more vocal Christians seem to sense that atheism is the final destination of those who think well. This conflict between reason and religion is not to be taken lightly. Discouraging or hindering the optimal use of a human brain to ferret out lies and mistakes is one of the worst things one person can do to another. Immeasurable misery and loss stems from poor thinking. In the twenty-first century, humankind needs to strive for maximum brain power, not less. Religious terrorism, for example, is seen by many today as a problem that nothing but war or nicer US foreign policy can solve, depending on which expert you ask. But Good Thinking is probably the only long-term solution. People who question, doubt, and think freely are much less likely to kill innocents and sacrifice their own lives in the service of extraordinary claims that can be shredded easily by analysis and skepticism. They also are much less likely to follow leaders who are uninformed enough to declare that the earth is six thousand years old and a supernatural doomsday is imminent—without providing evidence, of course. If we want a world without religious terrorism then we must first raise a generation of children who are capable of thinking their way through lies and delusions.

HUMAN 2.0

Good Thinking is not a stance or set of skills that come naturally to us. To the contrary, we all are born to be sloppy and inconsistent thinkers. This is normal. It is who we are. Perhaps we can think of it as the secular version of original sin. Unfairly, we inherit the cognitive limitations, biases, and misleading subconscious shortcuts of our ancestors. Similar to the story of Adam, Australopithecines reach out from millions of years ago to burden us today. The brain you host right now is virtually the same as the brain our human ancestors were packing in prehistoric Africa more than 100,000 years ago. Our modern brains are great, no doubt, but still not the best match for the concrete, plastic, high-tech, socially intricate societies so many of us now inhabit. We are something like aliens who have been tossed into a confusing new world. We never evolved to be scientific thinking machines that instinctively apply critical thinking and skepticism when confronted with important experiences, claims, or choices. Our brains are still busy looking for deep meaning in the wind and worrying about big cats eating us. We have to compensate for this with effort and purpose.

Christianity, like many other religions and like supernatural/paranormal claims in general, exploits a variety of standard mental vulnerabilities. We have a strong tendency to visually or intellectually connect dots, for example, even when those dots have no meaningful relationship. Our brains are also uncomfortable with unanswered questions, so we often make up answers and then defend them by twisting our perceptions and thoughts. Hopes and fears can weigh heavy when we make important decisions, often without our awareness. Regardless of whatever intellectual gifts or educational achievements one may boast, Good Thinking is necessary if one is a human being. Sorry, no exemption for Christians. Good Thinking is the one thing capable of pushing back against the avalanche of sensory inputs, impulses, and flimsy assumptions that so often derail our thinking. We all need it because the brain is strange and rascally in the ways it goes about its business. Good Thinking is our only hope for living a life that is at least somewhat free from the lies, delusions, and madness that soak the world around us.

The good news is that this way of approaching life is within reach of us all. As I detail in my book *Good Thinking: What You Need to Know to Be Smarter, Safer, Wealthier, and Wiser*, this is much more about will and skill than social status, innate intelligence, or cultural location.[3] Unfortunately, for various reasons, most people stumble through life making one avoidable bad decision after another. Don't believe me? Look around at our world. Belief in various forms of nonsense is near universal. Here in the twenty-first century, belief in outlandish, unsupported claims such as astrology, psychics, ghosts, UFOs, and worse flourish to some degree in every society. Irrational thinking is universal. I have traveled through six continents and interviewed or chatted with people who were rich, poor, powerful, powerless, educated, and illiterate. No social stratum is spared from this persistent plague, no individual completely safe. This is bigger than a divide between the religious and nonreligious. Many atheists fail to consistently apply Good Thinking. The French, for example, are mostly nonbelievers when it comes to gods but can't seem to get enough of homeopathy and other forms of medical quackery. There are atheists in America who believe in the most absurd conspiracy theories. And, of course, most atheists worldwide make many irrational decisions in the routines of their daily lives, just like most religious people. Just because one manages to figure out that Mohammed probably didn't really ride that white winged animal up to heaven, Joseph Smith's seer stones most likely weren't legit, and it's doubtful that Jesus walked on water doesn't mean everything else automatically falls into place. Good Thinking is needed always and everywhere. Fortunately, anyone can have it. No doctoral degree, membership fees, or decoder ring required.

It is a common misconception that innate intelligence, or at least advanced classroom education, makes one immune, or at least less vulnerable to scams and irrational beliefs. Not necessarily. It depends on how that intelligence is applied and what one learns in those classrooms. There are numerous examples of this. The late Edgar Mitchell was highly educated and made it all the way to the Moon on Apollo 14. But that same brain that did so well academically and professionally also harbored beliefs in ESP, UFOs, and many other unlikely claims. Brain surgeon turned political hopeful Ben Carson is about as elite as one can be in terms of formal education, yet he consistently demonstrates horrendous shortcomings in critical thinking. Carson has publically promoted medical quackery and declared his belief in numerous unsupported claims that any sharp-thinking high schooler could see through. Carson has stated, for example, that evolution is a lie inspired by Satan and the pyramids at Giza were built by the Old Testament figure Joseph to serve as grain silos. He also holds the strange position that microevolution is real but macroevolution is not. This seems a bit like believing in campfires while rejecting the existence of forest fires. He also thinks that human moral behavior proves the existence of his particular god. Apparently Yale-educated Carson knows nothing about bonobo chimps who consistently exhibit what can only be fairly described as moral behavior. Or maybe he does and consequently believes there is a little hairy god up in bonobo heaven who is responsible for their acts of sharing and kindness. Ben Carson, a Christian, is not dumb, yet blatantly dumb ideas inhabit his well-educated and most likely gifted brain. No Christian, regardless of education or IQ score, should make the mistake of assuming that they can afford to wave off Good Thinking. If one is a human being, one needs it. Christians who insist on thinking themselves too bright or too credentialed to ever be a sucker for nonsense and fraud might recall Proverbs 16:18: "Pride goeth before destruction, and a haughty spirit before a fall."

GOOD THINKING EXAMINED

So what is this Good Thinking stuff really all about? How exactly does one get it and apply it? And how can skeptics who hope to bring more light to the world convince Christians to want it, both for their benefit and everyone else's? As stated earlier, Good Thinking is my umbrella term for understanding, appreciating, maintaining, and using a human brain well. It is practical and useful in virtually all areas of life. Every moment in every society, far too many people waste time on half-baked claims, buy junk products, join bogus organizations, and trust their health to attractively packaged lies over medical science. In the United States, for example, three of every four adults—many of whom are Christians—believe in one or more supernatural/paranormal ideas, such as astrology, spirit channeling, and ghosts.[4] Nearly one-fourth of adult Canadians believe that mediums really do carry on two-way conversations with dead people.[5] Astrological silliness is rampant in India and throughout Asia. Belief in magic thrives in Africa and the Caribbean. Trillions of dollars are squandered globally every year on medical fraud, bogus investments, and exploitive organizations. Countless hours are sacrificed daily by good people for no positive gain. Clearly humankind needs more Good Thinking, including Christians. They can't sidestep this mess. For all their talk about God's protection and the promise of an afterlife, many of them needlessly suffer the results of their own poor thinking. Remember that Christians have to make do with fallible human brains as they wade through this swamp of lies and delusions daily, just like the rest of us. Marketers con them. Politicians lie to them. Friends and family manipulate them. And in some cases their own church leaders abuse or rob them. All Christians need Good Thinking—even if they can't bring themselves to apply it directly toward their religious beliefs.

The following are the minimal requirements for Good Thinking. These seven points can significantly improve the way in which anyone operates in life and approaches important decisions. Collectively they represent an invaluable opportunity for self-improvement and an enhanced defense against bad ideas.

Know Basic Brain Structure and Function

Have no fear, an advanced degree in neuroscience is not required. It is immensely helpful to achieve at least a simple understanding of the major components of the brain, where they are and what they do. Most people never bother, or they fail to even think about this, but this is the one organ that is central to our lives. It guides us in everything we do. "You" begin and end with your brain. Know it! At the very least have a passing familiarity with the neocortex, amygdala, hippocampus, cerebellum, and so on. Learn something about the way neurons connect, pass on information, and form networks. Discover how vision and memory work. (Most people have no idea.)[6]

Imagine yourself as a racecar driver. How well would you be able to perform in races if you understood how the steering wheel worked but nothing else? How likely is it that you would ever approach your potential as a competitive driver if you knew nothing about gears, axels, fuel pumps, and tire pressure? Appreciating one's brain and making good use of it begin with getting to know this amazing three-pound electrochemical blob that built civilizations and flew us to the Moon.

Appreciate How Brains Evolved

Despite the impressive power and creativity of the human brain, it is not a perfectly designed thinking organ—far from it. The indifferent, unplanned twists and turns of our evolutionary past produced a brain with significant limitations and unexpected quirks, many of which we are not naturally aware of. Never forget that our brain took shape in prehistoric Africa, a time and place where finding food, selecting mates, and avoiding being eaten by predators were overwhelming priorities. Many sensory and cognitive processes that served well enough to propel us forward one more generation back then cause us problems today. Ignorance about this—coupled with the endless barrage of modern experiences, choices, and demands—routinely lead people to make bad decisions and cling to false beliefs. Nothing wrong with being in awe of the wonderful human brain, but it is necessary to also acknowledge its past in order to use it well in the present.

**Know How Nutrition, Physical Activity,
and Lifelong Learning Impact the Brain**

Food matters. Eating leafy green vegetables, nuts, fruits, and high-quality protein help keep the brain safe and working well for as long as possible.

Excessive calories and too much added sugar can harm the brain in the short term and long term. Physical activity is also crucial to brain performance and longevity. Our most important organ is trapped inside a body that evolved to be standing and mobile during most waking hours. Sedentary lifestyles hinder the brain's ability to perform well and withstand disease. Consistent exercise—as little as twenty minutes, five or six times per week—can literally grow your brain. Exercise stimulates the brain to make new neurons. This is important because more neurons generally equate to a healthier, sharper, and more disease-resistant brain. Finally, life-long learning is vital. Learn another language, start juggling, try playing a musical instrument, or take up a new skill sport. Fresh mental challenges help build a brain that works well and lasts long.

Be Conscious of the Subconscious Mind

Brains never sleep. Night and day they monitor and regulate body functions, assess sensory input, and come up with answers (rational or otherwise) to our problems. Most of this activity occurs without our awareness or consent. One may go too far to suggest that the conscious mind is led around by the subconscious mind like a blindfolded zombie on puppet strings—but it's uncomfortably close to that. Second guess your important decisions in the present and review all past conclusions because there is a silent, often irrational, majority partner with a big say in all you do.

Be Alert to the Brain's Many Hidden Biases and Shortcuts

Within all that previously mentioned subconscious activity there are many normal mental processes that can undermine rational thinking. It is crucial that one be aware of this and know to review and challenge the answers and impulses that emerge from behind the curtain. One who has no idea what hindsight bias is, for example, is less likely to learn from past mistakes. That person will instead be coddled and misled by the revisionist historian that dwells somewhere in the subconscious mind. Anyone who does not understand confirmation bias and motivated reasoning is condemned to squander time and energy building up lopsided defenses of beliefs and positions, regardless if they are valid or not. Know also that the brain automatically serves up quick and compelling suggestions or commands, even when it has little or no good information to go on. Those who fail to understand this are sitting ducks, waiting to be exploited by some thick-haired TV preacher with his hand out, a politician spouting scary soundbites, and salespersons with highly symmetrical faces. The subconscious means well, it's trying to help, and it does most of the time, but sometimes it leads us to terrible places and then works to make us feel correct and confident about how we got there. This is why it is so important to think like a scientist in daily life. We must consciously analyze, test, and confirm because we can't fully trust our brain's natural ways.

Be Humble

If a skeptic is arrogant, she or he isn't doing it right. Hubris about one's knowledge and thinking increases the chance of error. Humility makes it easier to both avoid and let go of irrational beliefs not worthy of taking up valuable real estate in our heads. Knowing that our natural and normal cognitive processes are not totally reliable and admitting that we can be wrong about things we see, hear, feel, remember, and think seems simple and obvious, but most people forget it at key moments. Those who resist Good Thinking and stick to a belief because they feel there is no need for honest doubt and analysis to challenge it are essentially declaring themselves to be perfect human beings who are incapable of making mental mistakes. Anyone who can admit to falling short of intellectual perfection should have no problem admitting to the value of Good Thinking. Let's be honest, we are all effectively nuts. Each one of us has at least one foot planted in fantasyland at all times. Good Thinking not only confronts but humanizes this problem. For it is only human to believe the unbelievable. Therefore we must be on guard against our natural and normal thought processes. Living a more rational life requires us to use our brain to protect ourselves from our brain. Anthropologist David Eller does an excellent job of exploring this strange struggle further in the next chapter of this anthology.

Be Courageous and Mature

Good Thinking requires us to not only doubt and question everything important but also to accept the inevitable absence of answers sometimes. This may feel frustrating, awkward, or even embarrassing, but we should not pretend to know things we do not. Saying "I don't know" does not have to mean defeat. It can be the starting point from which we go in pursuit of the ideas and evidence that may give us the answer. An "I don't know" declaration is also honest. Filling in a blank with a made-up answer is guessing at best, lying at worst.

We also must be brave enough and strong enough to accept answers that contradict our most cherished hopes and beliefs. Good Thinking requires one to be a grownup about this and admit that there is a difference between wanting something to be true and knowing that it is. Hope can be a wonderful thing, right up until the moment we turn it into a lie.

Note that Christianity, Jesus, Satan, prayer, or the apocalypse never once came up in the proceeding seven points. But Good Thinking is precisely the best way to challenge and diminish the appeal of such things. This is the ultimate light switch, one that can lead a believer to finally see beyond stories, emotions, and blind traditions. A Christian who has embraced these seven pillars of Good Thinking is a Christian who is likely to be on her way to the exit soon—but only because she decides it makes sense. This is the way a religious house of cards within a human mind finally comes tumbling down. The skeptic rarely if ever blows it up in a dramatic moment with the breath of heated words. The atheist does not destroy it in one rush with the force of a good argument. The believers must do the real work for themselves and getting them to adopt Good Thinking in general is the surest route to that achievement.

HOW TO THINK LIKE A SCIENTIST
EVERY DAY OF YOUR LIFE

Among the pillars of Good Thinking is the call to "think like a scientist." Too many people view science as something only certain kinds of people engage in or have access to. But science is not an exclusive club or some pinnacle of academic achievement that is beyond most of the population. To the contrary, the scientific process is a human creation for all humans. Science is a means to finding things out, testing, exploring, confirming, and correcting. View it as a tool, one you are a part owner of. Get past the cliché of some egghead in a white lab coat holding a test tube. It's much bigger than that. The words "scientist" and "science" also connect to the image of a lone *Homo erectus* imagining and experimenting before finally producing a beautiful and functional Acheulean hand axe, or an urban child wondering where a passing ant lives and how it spends its day, then following it to find out. Science is not the periodic table of elements. It's more than a college major or job. Science is a way of thinking, the best way we have to figure out what is real and what is not. There is no patent on the scientific process, no property rights or border restrictions. Why not use it?

Fortunately, "thinking like a scientist" is not nearly as strange or difficult as some might assume. The simple and accessible steps of the scientific process follow below. Anyone can and should use them whenever possible, toward becoming a more rational being. This won't make you invulnerable to fraud and false beliefs. But it can increase the odds in your favor when it comes to making important decisions and attempting to distinguish reality from fiction.

1. Research, Observe, Gather Information

This may seem like an obvious requirement for good decision making but very few people actually do it consistently. Instead they choose to rely on shortcuts such as a gut feeling, good story, tradition, or the word of some authority figure. But this first step is crucial. Often, all of the necessary information to make a sensible decision is readily available, and one only has to make the effort to look. This does not mean asking only someone who advocates or profits from the subject in question. Nor does it mean relying exclusively on personal memories to determine if something seems like a good idea, because human memory is notoriously unreliable. And it certainly does not mean trusting your subconscious often-irrational and bias-filled brain to nudge or shove you in the right direction. Thinking like a scientist means gathering information from diverse and quality sources.

Imagine yourself buying a new car. Relying on the salesperson to provide you with all the information about the car would be risky, right? He or she is biased due to economic self-interest and is likely to feed you a slanted presentation. Likewise, you have to watch out for the many biases within yourself. For example, what if you lean toward one particular car on the lot because many years ago in childhood you rode to Disney Land in a similar model of the same color and have positive but ultimately irrelevant memories tied to it? What if you decide to buy a car because you feel that the asking price is a great deal? It might seem so only because it is framed between an overpriced luxury model and a bare-bones cheap model. These are the moments Good Thinking needs to kick in. Think like a scientist.

2. Develop an Idea or Hypothesis That Might Support or Refute the Idea/Product/Claim

Once you have the idea, carefully think it over. Trim it. Get to the crucial core of what matters. If it's a car purchase, list the reasons you have narrowed your choice to one. Why this model? Nothing wrong with feeling good about a new car, of course, but don't lie to yourself about fuel efficiency ratings and safety features if the only thing motivating you is the car's shape and color. Your idea may be: "This model car is the best because . . ." Drill down in pursuit of the real reasons your brain has landed on this car model, and then challenge these reasons to see if they hold up.

3. Design and Conduct an Experiment

Once again, try to expose error and falsehood. Anyone can do experiments. No safety goggles required in most cases. With our car example, a simple test drive can be an experiment. The testing of ideas and claims gets at the core of what makes science so special. There is a critical difference between believing, wanting, and hoping that something is real and showing that it is. This is the fundamental divide between science and religion. Science *shows*, while religions simply make claims and leave it at that. Always remember that one good experiment can trump a million words.

Many Christians accuse science of being unreliable and nothing so special because it's "just another religion." Besides being an odd way to vilify something ("Science is just as lame as my religion."), this claim is wrong. Science offers no safe harbor for empty claims that can't be validated, at least not for long. Without good evidence, successful experiments, and accurate predictions, big ideas don't endure. The concept of "faith," for example, may impress many Christians but it is utterly worthless in science. Coming to a firm conclusion based on nothing more than a feeling, ancient stories, or trust in an authority figure is not Good Thinking. How different would it be if one was supposed to accept scientific theories on faith? Imagine Newton, Copernicus, or Galileo saying: "I didn't bother with math or astronomical observations, but have faith when I tell you that this is how the heavens go." Anton van Leeuwenhoek: "I didn't build any microscopes or observe any germs, but have faith, they're really there." What if Darwin had looked around at our world's biodiversity and simply declared, "Nature did it"? Faith certainly wouldn't be enough to get me to accept more than three billion years of evolution. Science shows, primarily through experimentation.

4. Evaluate Your Results, Repeat the Same Experiment, or Conduct Different Experiments

When it comes to thinking like a scientist, one's work is never done. Forget stone tablets and eternal answers, in science nothing is necessarily permanent. Everything is understood to be open to correction and improvement if better evidence demands it. Never forget this crucial point. View all facts as fleeting, all knowledge tentative. This may seem like it would be a wishy-washy or even disorienting way to live, but it is just the opposite. It can be very grounding to honestly accept the imperfection and incompleteness of present human knowledge. Being ready and able to change one's mind also means having sufficient confidence in yourself as a growing, learning, and thinking being. Striving to be smarter tomorrow than you were yesterday is an important part of Good Thinking. Approach truth and reality with the understanding that you will never quite get there, and take solace in the fact that you are moving in the right direction.

5. Share Your Conclusions with Smart People

Science says the more brains, the better. Ask others to look for errors in your reasoning. One reason science works better than guessing, wishing, or praying is because it's so often a communal effort that transcends national borders and cultural flavors. Though not safe from corrupting influences, science still is much less beholden to authority, tradition, and power than religion and politics, for example. One impoverished, unknown grad student with a superior evidence-backed idea can obliterate the work of a hundred rich and famous Nobel Prize winners. Careers are built by exposing error and clawing us closer to truth.

Politicians who change their minds are labeled "flip-floppers" and smeared as indecisive and weak. Religion is even worse in this regard because change is often viewed as a violation of eternal divine will—new evidence to the contrary be damned. This is why religions constantly clash with science. Reality and truth are the goals of the scientific process—regardless of what reality and truth may turn out to be. Most religions, however, are fixed on enduring laws and stories—regardless of what reality and truth may turn out to be.

The standard journey of a scientific discovery, experiment, or theory includes the publication of one or more articles in a science journal. Although it can be very difficult to get past the gatekeepers and onto the pages of a respected journal, that's not the finish line. Once published, anyone any-where can scrutinize, challenge, and, if able, falsify the claims, data, meth-odology, or conclusions. Self-correction is not science's weakness, as many religious people assert. Just the opposite, it is a key reason for the great-ness of science. To think like a scientist in our lives, we need to mimic this process. We must expose our ideas, conclusions, and beliefs to challenge, and then change our minds if it becomes clear that we are wrong.

Let's return to our car example to get a real-world idea of why this concept of sharing results and revising is so important. After doing your research, conducting a few experiments (test drives), and trying your best to fend off irrational subconscious biases, you decide which model to buy. But don't sign that contract just yet. First make the effort to share your decision with others, especially those who know a lot about cars, if pos-sible. Listen to their feedback with a mind pried open by Good Thinking, and be willing to reverse course if it makes sense to do so.

WHY EVERY CHRISTIAN NEEDS GOOD THINKING

For some, thinking like a scientist and the other aspects of Good Thinking may seem like an odd way to approach daily life, but it makes sense and it works. The list of undertakings that this approach can assist with is endless. We all can utilize the proven power of science and reason in one way or another. It comes in handy when shopping for food or clothes, choosing a college, picking a career path, deciding whether or not to stick it out with a business partner or love interest, making healthcare decisions, deciding whether or not to join ISIS, buying a car, and much more. It is my view that there is no good reason not to aim Good Thinking directly at one's reli-gious beliefs. But I understand that not everyone feels they can or want to do this. It's okay; this doesn't have to be a deal breaker. If a Christian feels that Jesus is out of bounds, that Christian can still use Good Thinking to be less gullible and irrational in the office, at the mall, in the voting booth, etc.

There are "earthly" problems and dangers lurking within church walls that Christians ought to be concerned about. These have nothing to do with the supernatural, so they should be fair game for all. It is undeniable that many Christians have been economically exploited, sexually abused, and pushed to make disastrous personal decisions by incompetent, deranged, or corrupt preachers and priests. In many cases, pain and damage could have been avoided with the application of Good Thinking—without ever challenging any cherished supernatural beliefs. "Some is better than none," may be a helpful way for reluctant Christians to view Good Thinking.

A motivated atheist or skeptic who wants to elevate reason among Christians might feel that promoting Good Thinking in general rather than going after specific Christian claims is a weak strategy. To the contrary, this is the most powerful approach of all, the one most likely to deliver help and earn results. Zeroing in on prayer, miracles, heaven, hell, the born-again experience, the Noah's Ark story, and so on tends to play out like a game of whack-a-mole. I'm not suggesting that these things should not be challenged—of course they should, as is done in this anthology. But there will always be a hundred more wild claims ready to pop up in place of one just knocked down by a well-reasoned argument. Individual bricks matter, but less so than the foundation upon which they rest. Poor reasoning skills and ignorance about the human brain's odd and unexpected ways are the root problem. The best way to make a positive and lasting impact on Christians is to sell them on the concept of Good Thinking for their own good. Do that and there is an excellent chance that they will figure things out on their own. And this is how it must be. After all, no atheist or skeptic can think for a Christian. The best one can do is explain, show, and inspire. Ultimately it is the Christian's choice to think or not.

Good Thinking is not the summit. It is a path. It is not another religion, nor is it meant to be a replacement for Christianity. Good Thinking simply is a way to get ourselves on track toward making the best of our brains. Through it we shed many of the distractions and delusions that don't deserve our embrace or allegiance. It makes us confront our deepest flaws while also focusing on the vast potential of every brain. We humans may be far from perfect in our thinking, but this does not mean that we cannot strive for perfection. Good Thinking is the attitude and behavior that offers us the chance to wake up, grow up, and live the fullest life possible.

Chapter 2

A MIND IS A TERRIBLE THING

How Evolved Cognitive Biases
Lead to Religion (and Other Mental Errors)

David Eller

Instead of merely describing religions, or attempting to prove if—and which—religion is true, scientists in many different fields have recently converged on the project of *explaining* religion. That is, rather than asking "what is religion?" science has begun to ask *"why* is religion?" As Scott Atran and Joseph Henrich put it, psychologists, anthropologists, biologists, neuroscientists, and many others have joined forces to determine "why supernatural beliefs, devotions, and rituals are both universal and variable across cultures."[1] The key, they all agree, is not to be found in the religions themselves, which are so diverse as to have little in common; certainly the answer is not in "gods," since not all religions even contain gods. Rather, the key is something that all humans have in common, the human brain and mind.

Since religions are undeniably human thought systems, attention reasonably focuses on the organ that makes human thought, including religious thought, possible. And as a natural object, the human brain is also a product of evolution, gradually forging the organ into its modern human form from pre-human precursors, and settling certain pan-human capabilities and habits. However, as all evidence powerfully suggests, and as this chapter will explore, the human brain and mind is a highly imperfect device, prone to all sorts of biases, errors, breakdowns, and outright delusions, among which is religious belief.

In other words, as many scholars have asserted, religion is "natural" to the human brain/mind, but this is not necessarily a good thing, and it certainly does not mean that religion is true or good. What it means, in the words of anthropologist Pascal Boyer, is that "mental processes create religion,"[2] that the way the brain/mind works tends naturally to spit out religion—and that the way it works *badly* makes religion inherently suspect and unreliable.

COGNITIVE SCIENCE OF RELIGION: HOW BRAINS MAKE RELIGION

This convergence of scientific disciplines on the question of how mental processes create religion has been called the cognitive science of religion (CSR), and over the past two or three decades it has advanced sufficiently to have its own professional organization (International Association for the Cognitive Science of Religion) and its own academic journal (*Journal for the Cognitive Science of Religion*). According to IACSR, the cognitive science of religion is multidisciplinary, bringing together insights and methods from natural sciences, social sciences, and humanities to understand why religion exists at all and why it has the specific features that it does. On the other hand, CSR and the professional organization do not attempt "to validate religious or spiritual doctrines through cognitive science,"[3] nor do they explicitly try to refute those doctrines, although, as we will see, the ultimate effect is to demonstrate that religious ideas are unsound and undependable.

As science, the bedrock concept underlying CSR is evolution, the fact that over millions of years human cognitive abilities and tendencies evolved in response to the environment, both physical and social. The first thing to note about neurological and cognitive evolution (and this point will figure prominently below) is that the human brain developed from pre-human brains *and retained many of the components and functions of those reptilian and mammalian brains*. Not the least of these primitive characteristics are emotions and instincts, along with or for the purpose of quick, even unconscious, life-saving action. To be sure, a layer of conscious analytical, potentially logical or rational, thought was later added in the prefrontal cortex, but much human brain activity continues to occur at those lower levels.

As early as the 1960s, anthropologist Clifford Geertz insisted that evolution had shaped the human mind to be not only wonderfully intricate and "plastic" (that is, capable of change in response to experience) but also remarkably "incomplete," desperately requiring social interaction and cultural input.[4] Thus, significantly, the evolved brain/mind had two profound weaknesses: it hard-wired in habits that were adaptive in earlier environments—and even in earlier species—but not necessarily useful today, and it became susceptible to shared and transmitted ideas (culture), whether or not those ideas are true.

Proponents of CSR commonly refer to the natural evolved tendencies of human thought as "intuitions," the term suggesting that we think *with* them but seldom think *about* them. They are our handy, automatic, and mostly unreflective ways of understanding the world. In 1980, in what was perhaps the first foray of cognitive science into religion, Stewart Guthrie posited that the most fundamental of evolved cognitive intuitions is anthropomorphism—that is, the tendency to see the world as human-like. That essay, and a subsequent book entitled *Faces in the Clouds: A New Theory of Religion*, argued that because much of our experience is ambiguous and uncertain, we must interpret or decode that experience, and the most obvious and available decoder is human nature itself, so "human-like models frequently are chosen to interpret ambiguous phenomena."[5] By "human-like models," Guthrie meant primarily the very brain/mind that humans had evolved; in other words, the most basic of all evolved intuitions is that other beings and phenomena—perhaps even trees and rocks, perhaps even wind and rain, perhaps even war and death—have human-like minds too. This is, of course, consistent with observations of what we like to call "religion," as in Irving Hallowell's study of Ojibwa (Native American) thought: noticing that Ojibwa people used person-words to speak about "inanimate" objects, he asked if, for example, all stones are alive and conscious, and one informant answered, "No! But *some* are."[6]

Hallowell went on to contend that many if not all of the things that Western civilization considers inanimate and impersonal were for the Ojibwa "not a natural object in our sense at all. . . . [T]he sun is a 'person' of the other-than-human class"; in fact, "any concept of *impersonal* 'natural' forces is totally foreign to Ojibwa thought."[7] Equally, "advanced" monotheisms continue to put forth a god who is a "person" and who is distinctly human-like in personality (e.g., prone to love, anger, jealousy, etc.) and sometimes even in body.

THE EVOLVED COMPOSITE NATURE OF MIND

Since Guthrie's pioneering declaration that "gods consist of attributing humanity to the world,"[8] CSR has made significant strides and become more sophisticated in its language and concepts. Anthropomorphism has developed into the more technical notions of "agency" and "intentionality." Agency refers to the capacity to be an actor, to choose one's own actions, rather than being a passive object to be acted upon. As I write this chapter on my computer, I am the agent, and the keyboard is a mere object of my action. Intentionality is the related idea of having one's own will, desires, goals, purposes—in short, one's own intent. I have intentionality; my keyboard does not.

Herein lies the connection to Guthrie and between evolution, brains, and religion. Humans have evolved brains that allow them to have intentions and agency. Humans are by no means the only beings that are intentional agents; unless one closes one's eyes, it is obvious that cats and dogs, chimps and chickens have intentions of their own (things that they like and dislike, that they approach or avoid) and can and do act as agents. But there are two crucial questions that humans have struggled to answer: Do other beings—including other people—have a mind like mine, and if so, which ones?

Guthrie called the tendency to attribute mind to other beings a "good bet": it is a bet because we really do not know, and it is a good bet because if we are wrong there is no harm (the shaking tree branch over there was just a random event) but if we are right then there could be great gain (the shaking tree branch over there belies a hidden bear or human enemy—or spirit). Scholars from philosophers to brain physiologists call this problem the "theory of mind" (ToM for short), the intuitive notion that other people, other animals, and, who knows, some stones have minds like mine.

Implicit in Guthrie's argument was an idea made explicit and central by Pascal Boyer in his *Religion Explained*. Following Justin Barrett, Boyer called the evolved and generally useful intuition to attribute mind to any and every thing "agency detection" or even, in Barrett's terms, *agency hyperdetection* because our minds are compulsive agency-attributors. Barrett subsequently actually claimed that we have an evolved "agency detection device" or ADD (a pithy acronym) as part of our cognitive apparatus, one that "suffers from some hyperactivity, making it prone to find agents around us, including supernatural ones, given fairly modest evidence of their presence"[9]—or, we might add, no evidence at all. In a word, the human mind promiscuously projects other minds wherever there is the slightest encouragement to.

A hyperactive agency detection device, or HADD, is an evolved prerequisite for religion, just as intuitive agency attribution is a requirement for life in social groups; it is essential to suppose that other people act for reasons and that those reasons are knowable. Other minds, including "spiritual" or "supernatural" minds, are merely additional potentially knowable agentive beings. But anthropomorphism and agency detection do not explain why human cognition produces *religion* as opposed to other kinds of mundane other-than-human persons. For that, we must understand more about the evolved human cognitive machinery.

Boyer reasoned that agency detection is only one of many intuitive tendencies of the brain/mind. That is, it is absolutely crucial to realize that evolution did not give us, or any species, a single monolithic brain but one with many different parts and processes operating simultaneously, which somehow yield coherent (although not necessarily accurate) experience. CSR researchers like Boyer, and even more so Scott Atran, dub these diverse operations "modules," Atran identifying four different sets or classes of modules:

1. Perceptual modules, for such specialized purposes as "facial recognition, color perception, identification of object boundaries, and morphosyntax" [i.e., structures in spoken language]—the first of these accounting for the predisposition to see human faces everywhere
2. Primary emotion modules, connecting the mind to the body in such states as fear, surprise, anger, disgust, happiness, or sadness
3. Secondary affect modules, enabling higher-level emotional reactions like anxiety, love, grief, guilt, pride, and so on
4. Conceptual modules, which include ideas and beliefs but also more elemental categories or concepts out of which ideas and beliefs are constructed.[10]

Boyer was particularly interested in this last class of modules, which supply what he called "templates" through which we organize experience. A template "is like a blank form with specific fields or boxes (e.g., 'animal'), and the concept is the specific way that form is filled out (e.g., 'walrus')."[11] With a fairly limited number of generic templates, some maybe innate (like perhaps "space" and "time") and others learned (like "animal" or "plant"), we categorize and make sense out of the world.

Now, for Boyer, "religion" creeps in precisely where an idea or claim violates the basic expectations and assumptions of our categories, but only in certain ways. Boyer and others, like Barrett, have called these exceptions "minimally counterintuitive" concepts, but I would suggest that they are not so much *minimally* as *optimally* counterintuitive. For example, a minimal violation of intuition would be a human who can fly, which is not inherently religious; an *optimal* violation would be a human who comes back from the dead or who can do miracles. *That* sort of behavior requires some special explanation. And a violation that goes too far, like Boyer's example of the being that only exists on Tuesdays, would likely be rejected as nonsense.

So, given our evolved capacities and our lived experience, certain ideas hit the sweet spot of not-too-hot, not-too-cold, but just right for the imagination. Some of these ideas end up being labeled "religion," although it is not apparent, and CSR practitioners have not figured out, why some are and some aren't. A vampire is an optimally counterintuitive idea, as is a Martian or Santa Claus, but no one calls them "religious." Atran tried to resolve the problem with his definition of religion, namely, "(1) a community's costly and hard-to-fake commitment (2) to a counterfactual and counterintuitive world of supernatural agents (3) who master people's anxieties, such as death and deception."[12] But this does not quite get the job done: first, as Hallowell found, a society might lack the notion of "supernatural," and second, supernatural agents do not always reduce anxieties but may increase them (i.e., belief in witches or demons, or a god for that matter, can make believers more afraid). In passing, let us not ignore the fact that Atran's "counterfactual" is a polite way of saying *false*.

The intensely important implication of the entire CSR project is that, as Atran phrased it, religious ideas and behaviors "involve the very same cognitive and affective structures as nonreligious beliefs and practices—and no others—but in (more or less) systematically distinctive ways."[13] Boyer concurred when he postulated that "the building of religious concepts requires mental systems and capacities that are there anyway, religious concepts or not."[14] Boyer continued:

> Religious concepts, as I have said, invariably recruit the resources of mental systems that would be there, religion or no. This is why religion is a *likely* thing. That is, given our minds' evolved dispositions, the way we live in groups, the way we communicate with other people and the way we produce inferences, it is very likely that we will find in any human group some religious representations of the form described in this book, whose surface details are specific to a particular group.[15]

Note, very importantly, that the particular local form of religious ideas and beliefs need not—and often does not—include "gods," such that much of the work of CSR on "gods" is misplaced and evinces its Christian origin and influence. The question is intentional other-than-human agents, not necessarily "gods."

At any rate, the CSR explanation of religion in terms of already-existing and not-distinctly-religious cognitive processes and mental modules is a direct challenge to the presumption, rampant among scholars of religion as well as believers of religion, that religion requires and deserves some special treatment—that there is something "different" about religion. Indeed, Boyer affirmed that "this notion of religion as a special domain is not just unfounded but in fact rather ethnocentric."[16] Even the eminent champion of religion, William James, had to admit as much in his *Varieties of Religious Experience*: there are no religious experiences that are distinct from other experiences, "no one elementary religious emotion, but only a common storehouse of emotions upon which religious objects may draw," just as there is "no one specific and essential kind of religious object, and no one specific and essential kind of religious act."[17] He further had to surrender to the fact that what we call religious feelings or objects or actions "are each and all of them special cases of kinds of human experience of much wider scope. Religious melancholy, whatever peculiarities it may have qua religious, is at any rate melancholy. Religious happiness is happiness. Religious trance is trance."[18] James's famous book was even subtitled "A Study in Human Nature," not "A Study in the Supernatural."

This means, contrary to some scholars and defenders of religion, that there is no need to claim that religion itself "evolved" nor that there is a special "god spot" or religious module in the brain. Admittedly, there are two competing versions of CSR, one of which proposes that religion has been evolutionarily adaptive, based on its contribution to social integration (holding society together), group selection (perpetuating the group, even if individuals suffer and die), prosocial behavior (roughly "morality"), and encouraging commitment to the group and its beliefs and values (the so-called "costly signaling" function of religion, which asks members to perform arbitrary, expensive, and quite frankly often absurd and self-destructive things as a sort of test of commitment). But then, religions can be as divisive as integrative, can get entire groups killed off, does not always promote beneficial behavior (at least not beneficial for the individuals, like sacrificial victims or homosexuals, who are oppressed if not destroyed by it), and is often too costly to sustain.

Most practitioners and observers of CSR would agree, I think, that the dominant stream argues that religion is not directly selected by evolution at all. What has been evolutionarily selected and stimulated by evolution are the underlying mental and social capabilities and tendencies, *which incidentally give rise to religion as well as to other thought- and action-systems*. "CSR's central idea is the *religion as by-product thesis* according to which religious beliefs and practices are informed by our non-religious cognitive systems working in different domains," explains Aku Visala.[19] Psychologist Lee Kirkpatrick expressed it most colorfully when he compared religion to cheesecake. Following a suggestion by linguist Steven Pinker, Kirkpatrick correctly insisted that there is no gene for cheesecake, nor any evolved need for cheesecake (eating it is definitely not adaptive!):

> We do not have an evolved mechanism dedicated to preparing or seeking cheesecake; instead, the invention of cheesecake capitalized on those preexisting mechanisms that evolved for noncheesecake purposes [such as desire for fatty and sweet foods]. Cheesecake is exquisitely designed (by humans, of course, not by natural selection) to maximally titillate these evolved taste-preference mechanisms. *In short, I think religion is a kind of socio-emotional-cognitive cheesecake.*[20]

In the words of Atran and Henrich, "ordinary cognition produces extraordinary agents,"[21] and religion is nothing exceptional or special. In the immortal words of Friedrich Nietzsche, religion is human, all too human.

DON'T TRUST YOUR BRAIN:
COMMON AND WELL-UNDERSTOOD BIASES OF THOUGHT

In summary, we can embrace Visala's characterization of the "Standard Model of CSR," according to which "religious beliefs arise and persist mainly for the following reasons":

1. They are *counterintuitive* in ways that make them optimally suited for recall and transmission (Minimal Counterintuitiveness, MCI hypothesis).
2. They emerge and are supported by cognitive mechanisms that

generate beliefs about agents and agency (Hypersensitive Agency
Detection Device, HADD hypothesis).

3. They typically represent the religious entities as minded agents
   who, because of their counterintuitive character, stand to benefit us
   in our attempt to maintain stable relationships in large interacting
   groups.

4. They are also inference-rich and thus allow us to generate narratives about them that enhance their memorability, make them
   attractive as objects of ritual, and increase our affective reactions
   toward them.[22]

The entire project of the scientific study of religion and cognition
converges on the understanding that our evolved brain and its supported
mind feature ways of understanding and organizing experience that make
us susceptible to ideas that we conventionally call "religion." It also
stresses that our brain/mind is at least overzealous in a number of critical
ways, as in detecting "other minds" and more fundamentally in absorbing
transmitted cultural ideas, and is quite frankly prone to error.

However, I do not think that CSR has followed the implications of the
brain-makes-religion viewpoint to their inevitable and ugly conclusion. It
is simply a fact, and a very well-established one, that the human brain/mind
is vulnerable to many, many errors of thought beyond hyper-enthusiastic
agency detection and inordinate attention to counterintuitive ideas. Gary
Marcus argues fundamentally that the brain/mind that we have inherited
from millennia of evolution is a deeply flawed machine for knowing the
world and making judgments. He calls the brain a "kluge," a general term
for a poorly assembled collection of parts that, according to Dictionary.
com, "while inelegant, inefficient, clumsy, or patched together, succeeds
in solving a specific problem or performing a particular task." And, as
Marcus declares, "that which is clumsy is rarely reliable."[23]

This perspective raises the question of how and why the human brain/
mind evolved in the first place. As for how, the answer, as Marcus and
all brain scientists recognize, is that it developed from earlier and more
primitive brains and, rather than starting all over again to design a better

brain, built on what was already available. Consequently, scientists can see that the primitive reptilian brain still resides at the core of the modern human brain, overlaid with a somewhat less primitive mammalian brain, which itself is overlaid with the modern neocortex or thought- and judgment-center. The lower or more primitive parts of the human brain are not particularly adapted for rational thought, and they preserve all manner of emotional and instinctual responses that may actually obstruct reasoning. Worse yet, the various layers are not even especially well integrated, making for odd (mal)functions in the human brain/mind.

Further, evolution for the most part prepares organisms for action more than contemplation. Both scientific and popular writers (the latter epitomized by Malcolm Gladwell) have emphasized that at least two parallel cognition systems co-exist in the human brain/mind. One, the more primitive one or what Marcus calls the "ancestral" or "reflexive" system, "seems to do its thing rapidly and automatically, with or without our conscious awareness"; it is "clearly older, found in some form in virtually every multicellular organism," yet still today it "underlies many of our everyday actions."[24] Some folks, like Gladwell in his popular and laudatory (and aptly named) *Blink: The Power of Thinking without Thinking*,[25] praise intuitive thought and response, and, like any kluge, the ancestral/reflexive system is sufficiently successful at the tasks for which it evolved, like evading predators and making other snap decisions. It is not, though, particularly good or successful at more complex problems requiring careful analytical thought, like assessing the veracity of religious claims.

Marcus, Gladwell, and everyone else agree that parallel to and/or on top of the old reptilian/mammalian brain is a more sophisticated system that Marcus calls "deliberative." An unlikely evolutionary leap in cognition, the deliberative system "consciously considers the logic of our goals and choices" and "is a lot newer"; "it deliberates, it considers, it chews over the facts—and tries (sometimes successfully, sometimes not) to reason with them."[26] Depending literally on different parts of the brain, Marcus underscores that "there is no guarantee that the deliberative system will deliberate in genuinely rational ways. Although this system can, in principle, be quite clever, it often settles for reasoning that is less than ideal."[27]

This is a really crucial point that even many supporters of evolution misunderstand or make unclear: "evolved" or "adaptive" does not necessarily mean optimal, let alone perfect. Evolution tweaks what it has at hand and usually ends up generating klugey results like the human brain and the entire human body, both of which could have been designed and engineered much more intelligently. The two brain systems operate in fundamentally different ways, are often at cross-purposes, and are each liable to their own set of errors. In particular, while the old brain is evolved to be fast and not terribly rational (for survival purposes), the new brain too has a variety of shortcuts and quick-fixes that are as likely to spawn falsity as fact.

Amos Tversky and Daniel Kahneman, in a 1974 paper, are widely credited with founding the "heuristics and biases" model of cognition and for coining the term "cognitive bias."[28] The fundamental problem, they asserted, is that human thought and decision-making very often operate under conditions of uncertainty: we do not know what is true, and we do not know what choices we should make. Pausing to deliberate every fact-claim and every decision would paralyze us and possibly threaten our well-being if not our lives. Thought—even higher-level "rational" thought—is characterized by, and to a large extent depends on, heuristics, "mental shortcut[s] that allow people to solve problems and make judgments quickly and efficiently. These rule-of-thumb strategies shorten decision-making time and allow people to function without constantly stopping to think about their next course of action."[29] Not surprisingly, Kendra Cherry adds that heuristics "are helpful in many situations, but they can also lead to biases" that undermine the whole effort to think rationally.

Tversky and Kahneman mentioned three common heuristic devices or methods—representativeness, availability, and adjustment and anchoring. Representativeness, as the name suggests, involves making judgments on the presumption that one thing or situation is representative of or similar to another. But precisely because it is a presumption, it is "insensitive" (in their words) to factors like sample size (for example, the dreaded "anecdotal evidence"), probability, and predictability. Because human brains are so dismal at statistics, people easily also mistake chance for meaningful pattern (or worse, for the intervention of a benevolent or malevolent mind); worse yet, Tversky and Kahneman discovered that this heuristic suffers from the "illusion of validity," in which "people often predict by selecting the outcome that is most representative of the input."[30]

Availability is simply the heuristic that latches onto the nearest, most familiar, or most common information or circumstances: decisions and beliefs are determined "by the ease with which instances or occurrences can be brought to mind."[31] Not surprisingly, judgment under the heuristic of availability favors what people have heard most recently and/or most frequently, as well as what other people think and say. Information that is uncommon, unfamiliar, or otherwise hard to process gets pushed aside. Finally, adjustment and anchoring refers to the process by which people begin their thinking with some information already in mind and then make gradual and incremental adjustments or changes in their thought, if any. Obviously, a preexisting idea or belief, especially one passionately held, is a strong anchor, from which individuals only slightly stray, rejecting information that challenges the anchor. One particularly famous and easy-to-manipulate form of anchoring is known as "priming," in which a stimulus or experience immediately prior to decision-making orients the mind in a specific way and overdetermines the outcome. For instance, if experimental subjects are exposed to a frightening thought or experience before given a task to perform, their performance will be colored by fear; if they are encouraged to think about spirits or gods before they begin, then spirits/gods serve as an anchor for subsequent thought and action, leading them to behave differently than if such beings had not been mentioned. Fear or threat is an especially powerful anchor: Marcus writes that the "more we are threatened, the more we tend to cling to the familiar"[32]—which makes us more prone to the representativeness and availability heuristics. Is it any wonder that so many religions feature fear and threat as key aspects of their doctrine and practice and even inspire members to feel afraid and threatened? Such tactics intentionally narrow the horizon of thought.

Having listed just three troubling heuristics, Tversky and Kahneman concluded that such strategies "are highly economical and usually effective, but they lead to systematic and predictable errors,"[33] including an array of what they called "cognitive biases." As that name suggests, cognitive biases tend to sway (bias) thought in one direction or another, often and alarmingly away from the truth, or at least what the facts should indicate. Such cognitive biases have been thoroughly identified and studied by psychologists, economists (who value rational decision-making), and other scholars. And there are so many, referring to different kinds of errors, that some experts have organized them into four types of categories—social biases, memory biases, decision-making biases, and probability biases.[34]

Social Biases

This set of biases emerges from thinking flaws related to how we relate socially to one another. A few examples include:

- In-group bias, the tendency to prefer and to overestimate the goodness and rightness of one's own group
- Halo effect, the assumption that good or bad qualities in one area of a person's life rub off on other or all parts
- False consensus error, the belief that other people, especially in the same group, all think the same way
- Herd instinct, the inclination to accept the opinions and examples of the majority
- Self-fulfilling prophecy, or acting in such a way as to produce the results that you expected to occur
- System justification effect, the tendency to prefer and preserve the status quo
- Fundamental attribution error, or the likelihood that individuals will exaggerate personality-based explanations and minimize the power of circumstances

Memory Biases

These biases emerge from inherent faults in how memories are stored, processed, and retrieved. It has been demonstrated experimentally again and again that memory is remarkably unreliable. Marcus called human memory "the mother of all kluges,"[35] highlighting the problem of "context-dependent memory" or the propensity to remember things better when the environment or situation of remembering is the same as the environment or situation of experience or learning. In other words, situations can "kick-start" memory, so that the more we are in certain situations, the better we remember things related to those situations. Among other memory biases are:

- Suggestibility, when ideas are placed in a person's memory after the fact, as by another person who asks leading questions or makes assertions about the past; this has been identified as a profound problem in the form of "implanted memories," which people will often swear that they remember as actually happening
- Consistency bias, or falsely remembering the past as if it was similar to the present
- Hindsight bias, which like the consistency bias reimagines the past through what we know today, making the past seem more obvious and predictable than it really was
- Egocentric bias, or remembering things in a way that serves the individual's interest, such as exaggerating the positive or conveniently forgetting unflattering things

Decision-Making Biases

Eric Fernandez lists an astounding forty-two biases relating to how we make choices and accept beliefs—a fact that in itself should make us suspicious of our rationality. A few of these fatal biases are:

- Confirmation bias, the common and crucial propensity to pay attention to information that supports what we already think, and of course to discount or ignore information that contradicts our beliefs
- Endowment effect, or the disposition to more easily or willingly accept an idea (or object) than to give it up
- Choice-supportive bias, combining decision-making and memory biases by remembering your past decisions as better than they really were

- Framing, a very important and frequent bias in which how information is presented or "framed" affects how one feels about it and what
conclusions one draws from it; language often plays a critical role
in framing, as in the classic case of the estate tax, which opponents
renamed a "death tax," with the result that the same individuals were
in favor of it when it was called "estate tax" but against it when it
was framed as a "death tax"
- Illusion of control, or the false sense that the individual has more
control over events than they actually do
- Irrational escalation, or the inclination to use past rational decisions
to make or justify further irrational ones
- Mere exposure effect, or the obvious tendency to prefer things that
we have seen before or already know
- Post-purchase rationalization, or the ability to convince oneself that
a past action or decision was a good one

Probability Biases

Fernandez identifies an additional thirty-five biases in how humans process
relations between facts or events, such as:

- Anchoring, which we have already seen entails depending on or
"anchoring" to one bit of knowledge or experience to the exclusion
of others
- Attentional bias, the "tunnel vision" effect of focusing on some
information and missing other information
- Availability, also fingered by Tversky and Kahneman as heeding
only the most common, easily accessed, or emotionally meaningful
information
- Clustering illusion, or seeing patterns in things or events that are not
really there (sometimes more technically called apophenia)
- Frequency illusion, the comico-tragic situation where people learn
or believe something and subsequently begin to see it everywhere
- Hostile media effect, a now-familiar tendency to attack the media,

or more generally anyone who disagrees with you or even questions you, as if *they* are biased

- Observer-expectancy effect, or behavior that (hopefully unintentionally) manipulates circumstances or misinterprets the facts to manufacture outcomes that the person expects or desires
- Primacy effect, or the tendency to remember and value things that happened first or ideas/beliefs that were acquired first
- Recency effect, or the opposite habit of remembering and valuing best the most recent experiences

And this list of biases does not even include perceptual illusions or outright breakdowns of thought, as in neurosis and psychosis. At the extreme are the clinical delusions, such as Capgras delusion and Cotard delusion: in the former, sufferers are sure that close friends or relatives have been substituted with look-alike impostors, while in the latter the victim literally believes him/herself to be dead. If humans can be so fooled by their brains, then we should be very wary of them.

Unfortunately, a sense of certainty does not make one's thinking any better and can actually make it worse. Marcus states that "confidence is no measure of accuracy,"[36] which is obviously true because people have been confident of all sorts of ridiculous things throughout history. Importantly but depressingly, Robert Burton goes so far as to analyze certainty as a brain-state of its own. He calls it the "feeling of knowing," quite separate from the knowledge that one claims to possess (or from having correct knowledge); it is a psychological or emotional sureness about what we think, and, once settled, this belief in the correctness of one's beliefs "is not easily undone. An idea known to be wrong continues to feel correct."[37] This sort of overconfidence in one's own thinking is especially pestilent and resistant to facts and argument because, as Justin Kruger and David Dunning forcefully argued, the people who are most certain and intransigent in their beliefs are often the least skilled in critical thinking—and the least aware of their lack of skill, which is why they labeled such people "unskilled and unaware."[38]

CHRISTIANITY AS COGNITIVE BIAS

To summarize, Thomas Kida boils the array of cognitive biases down to six lessons: humans prefer stories and anecdotes to facts and statistics; we prefer to confirm rather than question our preexisting beliefs; we do not fully comprehend the role of chance and coincidence in life; we literally misperceive the world; we tend to think in an overly simplistic manner; and our memories are too often inaccurate. His message is straightforward: don't believe everything you think![39]

Just as the cognitive science of religion has demonstrated that religion is nothing special as cognition and as culture (that is, it is based on the same evolved brain functions and mental habits as any other behavior), so it inevitably dictates that Christianity is nothing special as religion. To be sure, Christianity makes some unique religious claims vis-à-vis other religions, but it is not essentially different from other religions and merits no special treatment. It is a product of the same klugey brain as any other idea or belief.

We can begin with agency detection, the mother of all cognitive biases. The Christian god is nothing more than the Christianity's preferred non-human agent; in fact, the renowned theologian Richard Swinburne defined the Christian god as "a person without a body (i.e., a spirit) present every-where . . . knowing all things, perfectly good . . . immutable, eternal"[40]—in other words, the ideal optimally counterintuitive concept. (Swinburne neglected to recall that this god is not entirely without a body: Exodus 33:22 confirms that he has a face, a hand, and a back or butt.) This god also has all of the human emotional foibles, from anger and jealousy to a penchant for roasted meat.[41]

Further, Christianity presents a rather classic case of seeing "faces in the clouds"—sometimes literally, as random cloud formations are (mis) interpreted as the god's visage. Even more inclined to show his face is Jesus, whose likeness is as likely to appear on a wall, a tree, or a slice of toast as in the clouds. This is nothing more than the evolved human search for facial patterns in nature run amok.

Moving to the Tversky and Kahneman's heuristics, Christianity not only stands on them but depends on them: without representativeness, availability, and anchoring there would be no Christianity or any other religion. Here is an occasion to offer one critique of CSR, which often portrays humans as intrepid lone thinkers puzzling out the world with their inherited (and klugey) brain/mind. However, religions without exception institutionalize themselves so that each individual does not have to arrive at belief anew; the religion tells them what to believe and indeed goes to great lengths to instill its doctrine, from Christmas pageants to Catholic catechisms. Whether or not any god is omnipresent, religion is, or strives to be; it wants to be the most universally available of all ideas, creeping into the language (like "God bless you" when you sneeze or *adios* for "goodbye" in Spanish), the calendar, personal names, public space, and life's events big and small (including "christenings," marriages, funerals, and "grace" before meals and prayers for sleep). In Christianity-dominated societies, it is not just easy to think about Christianity; it is difficult not to think about it.

Representativeness is a special prerogative of religions, and Christianity is no exception. For the devout Christian, potentially every moment, every decision, every event is representative of or a repetition of biblical stories or messages. In a word, religion lives on repetition—it is inherently compulsive—which explains weekly church services, annual celebrations, and mealtime and bedtime prayers. At the extreme, the individual's life should be a repetition and representation of religious exemplars. Christianity, like all religions, is the quintessential model for individual life, the founder-hero (Jesus in Christianity, Muhammad in Islam, Gautama in Buddhism, etc.) providing the ultimate paradigm upon which to model oneself. Christians often search their scriptures for a passage that seems to represent the current moment and to give an answer or direction. And, by penetrating the language, shaping the calendar, naming people and places (like Corpus Christi, Texas), and defining human life in the short- and long-term, Christianity seeks to *remake* the present as a representation of the past. Christian fundamentalists, like all religious fundamentalists, are only the epitome of this impulse to construct the present in the image of the past; the "Christian Reconstructionist" movement is the most overt example.

Both its availability and its representativeness contribute to the crucial anchoring function of Christianity. (In other societies, a different religion provides the anchor.) If anchoring as a heuristic device is a shortcut to decision-making by relying excessively on one idea or bit of information, then WWJD does the trick. The worst of Christian teachers, like the early church father Tertullian, were explicit that Christians should anchor their thought to religion and admit no other sources: "After Jesus Christ," Tertullian wrote, "we have no need of speculation, after the Gospel no need of research. When we come to believe, we have no desire to believe anything else; for we begin by believing that there is nothing else which we have to believe."[42] As if religious ideas were not dangerous enough on their own, theologians from Augustine to Martin Luther taught that believers should distrust and avoid knowledge and reasoning, which they warned—quite rightly!—would undermine religious faith.

It is only too apparent that most if not all of the identified cognitive biases are at work in Christianity. Christians definitely suffer from an in-group bias, preferring members not just of their religion but of their particular sect or denomination (don't forget how Catholics and Protestants fought long bitter wars across Europe, and many Christians today continue to disdain Mormons, who are just latter-day Christians). They certainly say some nasty things—often false, always oversimplified—about other religions' believers and all nonbelievers. A sadly amusing mistake that Christians make is the false consensus effect, failing to grasp that "Christianity" is a dizzyingly diverse congeries of sects that disagree with each other to the point of incompatibility (and, as just mentioned, violence). One of the worst sins of Christians and all believers is confirmation bias, only listening to "facts" that endorse their beliefs and disregarding observations or arguments that would serve to challenge them. Christianity, of course, acts as the frame or anchor against which all other ideas and experiences are measured, and one of the worst elements of this frame is the sense of imminent and constant danger—from the end of the world to the yearly "war on Christmas"—which keeps members in a tizzy and impairs their thought processes.

In fact, as an exercise in anchoring or framing, Christians often reinterpret threatening facts as tests of faith by their god or as temptations by their devil, *and come away strengthened in their convictions when they should have been shaken from them.* "Faith" itself is the definitive anchor or frame: faith is a value in itself, and standing up to challenges to their faith actually can and frequently does affirm that faith—the precise opposite effect that such facts and arguments should have and that we intend them to have. This is why, while it is not always profitless to argue with a believer, it is certainly not as effective as it should be; arguments and disconfirming facts are filtered through all of the cognitive biases we have named, and many more, to produce "knowledge" that is openly refuted by the facts. At its very most egregious, Christianity keeps its followers unskilled and unaware, in a cage of certainty, where even doubt is more faith. Faith then is its own self-perpetuating cognitive bias, the sum of all cognitive biases and errors, preempting the very rational functions that might free thought from religion.

IN PRAISE OF SKEPTICISM

In the seventeenth century, the philosopher René Descartes made one of the most important declarations in history—that our perceptions, intuitions, and received traditions are so undependable that we should adopt a stance of radical skepticism, doubting everything that is not absolutely certain (which is essentially everything). Unfortunately, he then spuriously readmitted the thing that is the most unlikely, the existence of a god, tethering all of his formerly (and still) untrustworthy thoughts to the flimsy premise of a trustworthy supreme being.

Descartes' first instinct was correct, and it has been confirmed again and again by scientists of every discipline. Our evolved history has left us with a thinking organ of unparalleled power and of frustrating fragility. We can use it to great advantage, but we must use it carefully and always be chaste about what it thinks it knows.

Religion is only the most conspicuous and bizarre case of faulty thinking, which is then institutionalized and packaged as an intellectual and social good. But religion is not alone as a flawed product of a flawed brain/mind. Economist Dan Ariely entertainingly but convincingly illustrates how people are "predictably irrational" in the practical realm of economic decision-making.[43] In another domain where choices are a matter of life and death, the United States Army is intensely aware that "military professionals will increasingly be expected to make critical decisions under conditions of uncertainty" when "people are more susceptible to make predictable errors in judgment caused by cognitive biases" and has taken steps to address the problem.[44]

Fortunately, science and practical experience have already told us most of what we need to know to avoid the worst of the traps that our brain/mind has laid for us. Gary Marcus, among many others, has condensed the lessons down to a few suggestions, easy to understand but difficult to implement; Carl Sagan actually constructed a "baloney detection kit" to arm us against the demons that haunt our minds.[45] As much as possible, Marcus urges, consider other explanations and ways of thinking. This entails especially trying out a different frame—"thinking outside the box," as we like to say these days. Beware of your own impulses and biases. Definitely do not make your most important decisions when you are tired or distracted or otherwise emotionally agitated. Step back from the problem and from your own customary perspective (as John W. Loftus preciously insists, evaluate the claim or belief from the position of the "outsider," honestly judging how it would appear to someone who did not already believe it—what he calls the "outsider test of faith"[46]). Watch out for "the vivid, the personal, and the anecdotal," as appealing as they are. And above all else, "try to be rational."[47] This means, to a large and painful extent, using your brain against your brain. The reward is better economic decisions, fewer unwise wars, and less religion.

Chapter 3

WHAT SCIENCE TELLS US ABOUT RELIGION

Or, Challenging Humanity to "Let It Go"

Sharon Nichols

> *It is better to grasp the universe as it really is than to persist in delusion, however satisfying and reassuring.*
> —Carl Sagan, *The Demon-Haunted World: Science as a Candle in the Dark*

> *No drug, not even alcohol, causes the fundamental ills of society. If we're looking for the source of our troubles, we shouldn't test people for drugs, we should test them for stupidity, ignorance, greed and love of power.*
> —P. J. O'Rourke, *Give War a Chance*

How can religion best be countered in the world's current season of religious insanity? How can unreason be shorn of its persistent hold on humanity? Examining recent news accounts of knowledge denial will demonstrate the need for challenging the tendency for "willful ignorance." In America, Christian extremism drives denial of science and reason; therefore, to counteract it, we will examine the ultimate "cause" of religion (anthropomorphism). Examining the nature and role of culture as the locus of belief will aid our understanding of its endurance as we explore the nature of monotheistic religion and how it functions; consider the characteristics of religious founders and why they may experience religious delusions; model critical thinking using the "analyzing underlying assumptions" exercise; scrutinize the nature of science and contrast the rational versus nonrational modes of thinking; identify key trends in global culture; appraise the "power of doubt"; and make recommendations for combatting sloppy, supernatural, woo-laden, nonrational thinking. The goal is to show that humanity's future is best served through the use of reason.

Experts from many disciplines have addressed the issues related to the "different ways of knowing" (or worldviews) found in the realms of *religion* and *science*. As a social scientist with a background in sociology, anthropology, cultural geography, and geographic education, I will examine the questions of "Why religion?" and "Why can't humanity seem to just let it go?" I make no apologies for the latter question—given the greater harm than good that I believe religion poses to humanity. As a humanist, I am alarmed at attempts to turn back the clock on hard-won knowledge such as evolution, cosmology, geology, biology, and physics. Because of a minority's religious beliefs, aspects of American society are under attack, such as the notion of pluralism as a positive social good. Often these tendencies are hidden under the patina of political ideologies, but the marked division between Republican and Democrat is largely informed by religious affiliations and beliefs. Due to religious backed political/economic ideology, we are making a Faustian bargain, giving preference to short-term desire for resources and profits instead of advancing long-term policies that would address our most urgent environmental problems threatening civilization.[1] Bill McKibben says that what we are doing is in fact turning our world into a *new* planet he calls eaarth, one on which no humans have ever lived.[2]

We should be facing the future with a courageous will to act to mitigate human effects on a host of environmental problems, including climate change. Instead, the current state of affairs is that American Christian extremists are attempting to dismantle public education, particularly history and science. Education is the very foundation for solving our problems, yet it is under aggressive attack because of outmoded beliefs. Incredible!

In the name of Christian nationalism posing as patriotism, they would delete from our textbooks historical *facts* with which they're uncomfortable (or view as unsightly), and replace them with flag-waving "feel-goodism" instead. This is part and parcel of the general trend of science and fact denialism from some Far Right religious camps. I offer the following in evidence:

Item: An Oklahoma legislative committee voted to ban the Advanced Placement (AP) American History course because it's "too critical of some of America's historical actions," (or, as supporters of education might say, too historically accurate) for at least one member of the committee. This member happens to belong to an organization called "The Black Robe Regiment" (an organization open to ministers) that claims historic close ties between church and state have existed since our nation's origins.[3]

Item: Texas History books are being revised to reflect conservative Christian patriotic ideology. Evangelical Christian activist leader Cynthia Dunbar claims, "We are fighting for our children's education and our nation's future. . . . There seems to be a denial that this was a nation founded under God. We had to go back and make some corrections."[4]

Item: In 2013 Florida State Representative Ritch Workman complained that a mainstream AP World History textbook published by Pearson Prentice-Hall, contained too much "pro information" about Islam (thirty-six pages).[5] As a result, the book came under review by two school board members. What was the alleged "pro-Islam" information in the textbook? Factual descriptions of the history of Islam, its contributions to the world, the Five Pillars, descriptions of the Koran, and other sundry facts allegedly "fondly devoted" to "Muslim Civilizations."

One doesn't even know where to begin. And science denialism is even worse! Peter Gleick, CEO of the Pacific Institute, says there are two factors involved in science denialism: "political cowardice hiding behind scientific skepticism; and political pandering to special interests."[6]

Item: Senator Jim Inhofe, R-OK, is a staunch and vocal denier of anthropogenic global warming and sea level rise and has even written a book to that effect: *The Greatest Hoax: How the Global Warming Conspiracy Threatens Your Future*. Inhofe is now the Chairman of the Environment and Public Works Committee, which sets EPA regulations on the environment. He has publicly stated, "because 'God's still up there,' the arrogance of people to think that we, human beings, would be able to change what He is doing in the climate is to me outrageous."[7] This is an example of the cowardice and pandering Gleick was talking about, with silly ignorance added for effect.

Item: In Florida, Department of Environmental Protection employees were forbidden to use certain words when talking about the environment. "We were told not to use the terms 'climate change,' 'global warming' or 'sustainability,'" Christopher Byrd, a department attorney reported. Other current and former employees concurred, though Gov. Rick Scott's office denied there was an official policy to that effect. The irony is that Florida stands to be hardest hit by the effects of global warming and sea level rise.[8]

Item: Historians Naomi Oreskes and Erik Conway wrote a book called *Merchants of Doubt: How a Handful of Scientists Obscured the Truth on Issues from Tobacco Smoke to Global Warming*, which has now been released as a movie by Sony Pictures. *Merchants of Doubt* is an expose of sorts, but with a surprise twist: what it exposes is the link between science denialism and "deep-seated beliefs and ideologies—particularly those championing the free market and individual liberty (which we tend to call libertarianism)."[9] Surprised? I was expecting the writers would discuss the links between science deniers and their religion, or industry and politicians' interlinked vested interests, but instead they found what Michael Shermer called a form of "new tribalism."[10] Industry "supplies" the science naysayers with a concerted barrage of distorted facts, which they then disseminate to the public; then consumers (consumer of ideas, in this case) become the "demand" side of the equation, in essence *wanting* to hear the disinformation because it suits their own needs and ideologies.

The upshot: People don't believe in facts because they don't *want* to believe in them. Or even more sinister, they deny facts because of their own self-interests or political and religious goals. Education, environmental policy, government support for science, and our political system should not be left up to *supply and demand* of willful ignorance. No society can long hold out in that kind of economy of mind.

What is "willful ignorance"? There is a simple kind of ignorance, as in being unaware of something or unknowledgeable about it. This may spring from something as mundane as not having been exposed to the information before. You don't know because you never learned it. This kind of ignorance

is remedial; the information can be learned. There is a second kind of ignorance I call "willful ignorance": a person has been exposed to knowledge
that is readily available and should have learned it, but chooses instead to
deny the facts—or worse, even pretends to not believe the evidence.[11] S/he
does this despite experts' views that the evidence is overwhelming and that it
would be unreasonable to withhold acceptance of it (as is the case for evolution). "What normally comes along for the ride, however, is a telltale sign of
denialism: that these alleged skeptics usually have different standards of evidence for those theories that they *want* to believe (which have cherry picked
a few pieces of heavily massaged data against climate change) versus those
they are opposing."[12] As a result, the willfully ignorant response to facts they
don't want talked about is to deny, deny, deny.

> Depending on the nature and strength of an individual's pre-existing
> beliefs, willful ignorance can manifest itself in different ways. The prac-
> tice can entail completely disregarding established facts, evidence and/or
> reasonable opinions if they fail to meet one's expectations. Often excuses
> will be made, stating that the source is unreliable, that the experiment
> was flawed or the opinion is too biased. . . . In other . . . more extreme
> cases, willful ignorance can involve outright refusal to read, hear or
> study, in any way, anything that does not conform to the person's world-
> view. Michael Specter even wrote a book about denialism, entitled *Deni-
> alism: How Irrational Thinking Hinders Scientific Progress, Harms the
> Planet, and Threatens Our Lives*, in which he said that this amounted to
> a war on progress and stemmed largely from a distrust of institutions,
> particularly government and science.[13]

This kind of ignorance is inexcusable, dangerous, and reeks of
hubris. The source may be religious indoctrination, political motivations,
economic rationale, or some kind of combination thereof. It may even be
lying to oneself, but this kind of ignorance is the most destructive simply
because there is no excuse for it, and it harms society because problems
are not addressed, since even the fact that there are problems is denied.
Unfortunately, willful ignorance is proliferating and is even encouraged
and condoned in some circles, among them adherents to various religious

ideologies, government officials, political parties, school boards, social media, and news networks. Sadly, another name for "willful ignorance" is "tactical ignorance."[14] That's when it's mighty convenient to be able to say, "I'm not a scientist, but . . ." as some science denialists in Congress have done, particularly regarding anthropogenic climate change.

> Denialism is the employment of rhetorical tactics to give the appearance of argument or legitimate debate, when in actuality, there is none. These false arguments are used when one has few or no facts to support one's viewpoint against a scientific consensus or against overwhelming evidence to the contrary.[15]

Not only is this show of haplessness a sham, it is in fact "framing" talking-points for political circles and news cycles. Oh, the innocent eyes and the "whatever can I do about it?" shrugs that some politicians fake so adeptly. No wonder we think our government is broken.

There is another kind of ignorance, called "higher ignorance," which refers to "knowing you don't know but wanting to know."[16] Whenever I think of this kind of ignorance, I think of great thinkers, mathematicians, and scientists (like cosmologists Neil de Grasse Tyson and Brian Cox), whose awe for the unknown inspires humanity to reach for the stars. This third type of ignorance springs from feeling humbled by the very vastness of space; it compels us to find newer ways to find answers to its mysteries. This so-called higher ignorance impels us to fill in the gaps in our knowledge—and then to search for new mysteries and problems to solve. By embracing the unknown, we bring light to the darkness so that we can see ourselves in the universe "and know the place"—and our place in it—"for the first time."[17] There are responsibilities that come with knowledge; pretending not to know doesn't negate those responsibilities. This third type of "ignorance" isn't ignorance at all, but the clear mark of humans using their brains to reveal true things.

What happens to a society that is at war with itself over its history and its science?[18] Do facts simply lie as a smorgasbord where cultural members dine at their own whim? What becomes of knowledge in this cultural free-for-all? Since the advent of the Internet and social media, it seems all opinions are equal. But facts don't work that way. They aren't in a popularity contest—or at least they shouldn't be. What is happening is a culture clash between religion and reason. So where do we go from here? How do we challenge religious-based resistance to science and keep politico-religious ideologues from turning back the clock on human knowledge? I think we have to go all the way back to the cognitive beginnings of religion itself for an answer. Using science, we can unpeel the layers of religious belief until we arrive at its original core cognitive structural underpinnings. Science addresses the way our brains mislead us into belief in the first place by setting us up for the concept of nonexistent Ultimate Agent(s), i.e., god(s). This then evolved into the very religions that today challenge science! If we can understand that part of human nature that is so ready to believe, perhaps we can inoculate ourselves against the tendency toward knowledge denialism and combat the unreason that accompanies it. If we can do that, then we will have gone a long way toward letting go of religion and accepting science as the means of understanding the nature of reality.

THE EVOLUTIONARY ORIGINS OF RELIGION: COGNITIVE SCIENCE OF RELIGION

We want to understand the underlying *ultimate cause* of belief in religion in humans and the *proximate cause* of the persistence thereof. To achieve this goal, we will explore the relationship between human cognition and hypersensitive (or hyperactive) agency detection (HADD), anthropomorphism, and religion.

One way to understand belief in supernaturalism is to view it as an error of our brain's "rush to judgment" in attempting to explain the unknown. We will explore the role of culture in perpetuating religious belief and contrast the non-rational/religious mode of thinking to that of science. We will use the lens of cultural geography and anthropology to see culture as the operational human milieu.

Between 100,000 and 30,000 years ago (YA), modern Homo sapiens evolved an extraordinary cognitive linguistic capacity. The sudden evolution of an enlarged human brain size gave humans a decided evolutionary advantage. Chimpanzees are our nearest relatives, with whom we share 98.8 percent of the same DNA, but the difference in neural capacity due to increased human brain size has astonishing effects: "Humans have far more white matter in the temporal cortex, reflecting more connections between nerve cells and a greater ability to process information."[19] According to the Smithsonian Institution's Human Origins Initiative, a comparison of chimpanzee brain size to the modern human brain is 0.85 pounds to 2.98 pounds. We are the "top-heavy" species, with our larger brains that act as costly energy hogs: "A big brain gobbles up energy. Your brain is 2 percent of your body's weight but uses 20 percent of your oxygen supply and gets 20 percent of your blood flow."[20]

The details and the order of events are still debated, but the increase in size included the development of a neural-cognitive network of much greater complexity than that of our forebears or fellow *Homo* species. This complex neural-net is what's responsible for human language. These changes occurred in the neocortex, and they control language and emotion, which are tied to increased social complexity.[21] The degree to which this constituted a Great Leap Forward for humanity cannot be overstated, for what we are talking about is early *Homo sapiens* evolving and becoming fully modern humans.

Language is the tool used in the construction of the human inner version of the outer world, allowing humans to make both *imagery* of the world and to engage in *reflective thought* about it, for example, the Theory of Mind (ToM).[22] ToM refers specifically to "being able to infer the full range of mental states (beliefs, desires, intentions, imagination, emotions, etc.) that cause action. In brief, having a theory of mind is to be able to reflect on the contents of one's own and other's minds."[23] Through language, or symboling (in which one thing is used to represent or stand for something else),[24] humans took a great leap into another way of *being* in the world. This conceptualizing toolkit provided a means of storing and sharing knowledge accumulated over vast periods of time—"A major advantage of human language is that it is infinitely flexible."[25] This flexibility is seen in the "portability" of knowledge, among other things, which is readily seen as pivotal in the rise of humans as the dominant species. Among other things, it allows human inferences, flights of fancy, and imagination in the way humans interpret the world.

These skills and more led to the evolution of culture as the primary means of human survival.[26] Religion began to appear in the human cultural toolkit contemporaneous with language, which isn't surprising since language gave humans the ability to create cultural narrative traditions that were at first based upon anthropomorphism, as we'll show.

ANTHROPOMORPHISM, OR SEEING WHAT ISN'T THERE— OR IS IT?

According to Stewart Guthrie in *A Cognitive Theory of Religion* (1980) and *Face in the Cloud*s (1993), the accidental by-product of this increased brain capacity is the cognitive development of a specific brain "module" with which humans experience incidents of anthropomorphism. Anthropomorphism is the brain's tendency to attribute humanlike features, characteristics, or personalities to nonhuman entities in ambiguous situations, for example, the snap of a twig or a sudden movement in the forest or savanna. The search for an interpretation of what is "out there" happens instantly, since it is psychologically intolerable for the potential threat (or unknown phenomena) to remain ambiguous. The brain assumes "agency" (someone or thing acting with intention), and applies humanlike models or templates first to the phenomena, because that is the phenomena humans know best. The fact that "agency" is detected is startling: when it occurs, our minds do not ignore it, for to do so is literally at our own peril. In the prehistoric lives of humans, there were many fearful predators, not least of which were other humans, and the things that go bump in the night might very well have ill intentions. Divining those intentions constitutes placing a Pascal's wager on survival: upon seeing a dark large shape in the night, the template is applied, and the bet is made as to whether the entity is a threat. If it is, and one responds accordingly, the reward is life; if it isn't a threat, then no loss, no gain. A reflexive application of the Hyperactive Agency Detection

Device (HADD)[27] assigning "agency" to the ambiguous unknown is thus an evolutionary plus—because it leads to survival.

OUT OF THE DARKNESS

To "shed light on" something means *to bring it out of the darkness*. The *ambiguity* of the unknown causes tension whenever HADD perceives *something* is "out there." The brain's systematic anthropomorphism, more often than not, makes false positives in identification of an alleged agent. It is an all-too-easy leap for the human mind to fill-in-the-gaps with mistaken and made-up entities, such as animistic spirits, ghosts, witches, and gods. That leap, that momentous leap of the imagination meant only to reduce ambiguity, populates the world and the human mind with narratives of nonexistent, often supernatural beings whose capricious nature must be assuaged by our propitiations, magic words (prayer), and ritual acts. It sets up the preconditioning of the mind to invent religion(s). If religion initially springs from cognitive evolution and language, then these physical neurocognitive structures constitute the very underpinnings (or *ultimate* cause) of religion. Guthrie clearly thinks so: "anthropomorphism is inevitable in human thought; that it is, though by definition mistaken, not especially irrational; and that it characterizes religious thought in particular."[28] Further, he concludes that "anthropomorphism is . . . and must be a cultural universal."[29]

Because religion springs from the nature of how our minds work and after thousands of years of cultural evolution it should be no surprise that it exists in so many forms, in such diversity of divisions and sects, or that it has persisted for so long.

"The explanation for religious beliefs is to be found in the way all minds work."[30] That there are/have been so many religions (animistic, tribal, polytheistic, monotheistic, and mixed tribal/Christian) shows that humans can conceive of many different *kinds* of agents and ascribe to them different roles and abilities. Religious diversity is the norm, rather than uniformity of belief, suggesting that we are merely using our capacity to acquire, create, or invent religions as psychological, conceptual and cultural systems we use to "explain" the world.[31] We use our inference systems, but arrive at different cultural conclusions. Once religions are invented, many cultural processes act to perpetuate them and from that point on, cultural evolution takes over.

We may tentatively conclude that CSR, HADD, and ToM account for the occurrence and deep entrenchment of religion as a human universal. Early in our prehistory, we believed in the supernatural because our cognitive systems functionally prepared us to believe. Beyond the cognitive science lies the contributions of social networks, social structures, and psychological mechanisms of how we learn and exist in our cultures.[32]

These physical brain tendencies unintentionally inculcate what gets interpreted through culture as religion. Religion starts in the mind as the speculated existence and causal powers of non-observable entities and agencies acting in the world. In culture, religion is treated *as if* the entities, agencies, and their powers are real. This is reinforced culturally through the added power of religious authority figures, shamans, or priests; teaching and reinforcement through community hierarchies, the family, and the religious group, with members gathering and sharing in places of worship; the use of memorable rituals; the setting forth and teaching of canonical doctrine; music, dance, and art; teaching narratives meant to memorialize and mythologize origin stories, genealogies of leadership, hero-izing the derring-do of particular leaders; and solidifying the privileging of religion or religious groups within the culture.

The point is that where *cognitive* science leaves off in explaining the existence of religion, *culture* takes over in explaining the development and forms of *specific* religions, adding culture traits as they arise and develop. These traits change over time, accrete new aspects and beliefs, and become more complex systems of belief, evolving into religious culture trait complexes and culture trait systems. That's what we call organized religion.

CULTURE—OR *GIVE ME THAT OLD-TIME RELIGION*

First used by Germans as "Kultur" in a dictionary from 1793, the term "culture"[33] has since absorbed many meanings:

> Culture is an entirely theoretical concept developed by anthropologists to describe the distinctive adaptive system used by human beings. Culture is our primary means of adapting to our environment . . . culture is nongenetic [in the sense of hard-wired in]. . . . A culture is a complex system, a set of interacting variables—tools, burial customs, ways of getting food, religious beliefs, social organization, and so on—that function to maintain a community in a state of equilibrium with its environment. . . . No cultural system is ever static.[34]

Culture can also be defined as the "total legacy the individual acquires from his group,"[35] or even "all those historically created designs for living, explicit and implicit, rational, irrational, and non-rational, which exist at any given time as potential guides for the behavior of men."[36] Anthropologist E. B. Tylor established the modern meaning: "Culture . . . is that complex whole which includes knowledge, beliefs, art, law, morals, customs, and any other capabilities and habits acquired by man [humans] as a member of society."[37] A. L. Kroeber said, "Culture thus is at one and the same time the totality of products of social men [humans], and a tremendous force affecting all human beings, socially and individually."[38] Culture is one of those spongy concepts used by various disciplines, but here it is employed in the anthropological sense, referring to the entire way-of-life of a people. Culture includes *material* (touchable) culture traits, such as tools, clothing, and buildings; and *non-material* (non-touchable) culture traits, such as ideas, dietary preferences, taboos, and laws. A *culture trait* is the smallest discrete unit of a culture and refers to anything used by the culture;[39] *cultural diffusion* is the flow or movement of traits (as different groups come into contact); *cultural landscape* is the imprint of human activities on the earth's surface; *culture history* is the evolution through time of the culture traits; *trait complexes* are the whole of interconnected cultural systems; and *culture regions* are areas of homogeneity, such as the Midwest, the Bible Belt, or the English Midlands.

One thing that anthropologists tend to agree on is that key themes are recurrent in the numerous definitions found within the literature—that it is learned, is shared by members of a culture, is based on traditions within a culture, is dynamic (i.e., changes over time), consists of systematized patterned behavior, acts as an integrated whole, is based on modes of thought, utilizes energy systems, includes the full range of human behavior, and is based on symboling (the assigning of meaning).[40] At one time, there would have been an added feature on this list—that culture is restricted to humans—but continuing research suggests that other animals possess culture or something that approximates it.[41] The difference is that humans have *written* language—and included in our modes of thought is *religion*.

One of the most identifying features within cultures is language. That is to say, the culture "lives" through its language. Other systems that, when taken altogether, describe the full breadth of what makes up human culture, include religion, economics, technology, politics, arts and music, behavioral systems, and social organization. Culture includes the full range of human behavior, including current attitudes, dietary preferences, mode of dress, types of work, taboos, and all else that answers the questions of who we are as a people, how we do the things we do, with what tools, for what reasons, and so forth. Humans live in the environment (nature) but adapt our ways and means of survival through culture. "A culture" is always place- and time-bound. American culture in 1929 is not the same as that of 2015. Cultures are always changing—else they will die (collapse, be subsumed by some other culture, or go by the wayside).

From the dawn of human prehistory, hunting and gathering was the primary way of life, which supported small kinship groups of people acting in close cooperation. But with the gradual development of sedentary life and agriculture (10,000–8,000 years ago), riverine urban centers were established (8,000–5,000 years ago). Small group settlements grew into hamlets, villages, towns, cities, and finally to empires, each with an ever-larger population, supported by a more assured irrigation-based food supply; socially stratified society with specialization of skills; a hierarchical ruling class of religious leaders as political leaders (theocracies); polytheism; and monumental architecture, some of which was devoted to the "Sacred." These groups developed a written language (in some, but not all cases); developed mathematics and surveying; kept records for control of agriculture, irrigation, and taxation; developed extensive, far-ranging trade networks; developed a standing military; and began to accumulate a larger body of *persistent* knowledge. The names of the city-states, empires, and civilizations are familiar: Sumer, Ur, Urdu, and Babylon; the Minoan, Egyptian, and Assyrian empires. A great deal of culture contact—through migrations, wars, exploration, and trade—occurred, resulting in cultural diffusion, adaptations, new inventions, discoveries, innovations, and culture *change*. Each city-state, empire, or civilization enjoyed its own panoply of gods and traditional god-narratives, religious centers, and other landscape features.

The precise nature and sequence of cultural borrowings and adaptations can often be deduced, as with comparisons of origin and flood mythologies and narratives. Often, the names of deities or origin myths are so parallel to each other that to rule out cultural borrowing seems unreasonable.[42] Nevertheless, through the passage of time, the stories behind such contacts are lost to us. In the absence of clear explanations as to the move from polytheistic supernaturalism to monotheism, cultural narratives take their place. Those narratives, and some scattered archaeological document remnants, are virtually all humanity has left of the ancient origins (ca. 3500 BCE) and early diffusion of two of the monotheistic faiths, Judaism and Christianity. Islam is a relative latecomer. (Mohammed lived ca. 570 to June 8, 632 CE, and Islam began in 610 CE, when he was forty.) We at least can verify some of Islam's history, particularly relating to the Sunni/Shi'ia divisions, 75–90 percent and 10–20 percent of Muslims worldwide, respectively, and their impact on the modern world. Judaism and Christianity share several traits in common, aside from belief in one god. Their origins are murky, with contradictive narratives from canonical texts that are of rather late origin, after the alleged events (by decades or even hundreds of years). These texts are of dubious provenance, and almost all of them were by anonymous authors writing about people and events for which evidence is almost nonexistent and uncorroborated by independent contemporaneous sources. The Judeo/Christian faith traditions have lost a great many followers, especially in Europe, largely for the reasons just given, and Islam might have as well, except that to be an apostate is to risk facing death itself. No atheists in foxholes? If you're atheist in Islamic nations with sharia law, you must be very invisible, very quiet, trust no one, and live a lie to avoid a hideous death. It's that or emigrate to a non-Muslim nation.

Cultures (and subcultures) perpetuate religious beliefs—and it is clear that in so doing their adherents are only too willing to act in horrendous ways to protect such beliefs from outsiders. The reason is clear, for in a multitude of ways, the culture is so permeated with religion, either explicitly or implicitly, that for many people it is inconceivable to have the one without the other. The only problem is that there is very little agreement as to how to define religion, much like the sponginess we saw with defining culture.

DEFINING RELIGION

Since there is no agreed upon universal definition of religion, the most useful for our purposes is given by anthropologist Clifford Geertz in his essay "Religion as a Cultural System." Geertz writes, "Religion is a system of symbols which establish powerful, pervasive, and long-lasting moods and motivations in men [humans] by formulating conceptions of a general order of existence and clothing these conceptions with an aura of factuality that the moods and motivations seem uniquely realistic."[43] Anthropologist E. B. Tylor thought of religion as "belief in spiritual beings" as a way of understanding and explaining the natural world. For our purposes, monotheistic religion can be seen as institutionalized belief systems, opinions, and practices relating to the human relationship with, belief in, and worship of the divine (or supernatural deity).

> *Religion: (4) a cause, principle, or system of beliefs held to with ardor and faith.*
> —**Merriam-Webster's Collegiate Dictionary**

Some dictionaries confuse the issue by asserting that "any strong belief" is a religion. Rather than setting a boundary around a word so that it is clearly defined, this latter use of the term makes "religion" meaningless; *if all strong beliefs are religions, then none is special*, making the term useless.

These are just some of the characteristics of monotheistic religions and are not meant to refer to other types of religions. A monotheistic religion

- is a cultural narrative; a product of a culture that purports to tell its story.
- usually (but not always) starts out as oral traditions handed down in a preliterate society.
- usually involves belief in things not in evidence; rests on "faith."
- is either tribal, ethnic, or universal (archetypal).
- has a schedule of daily and annual events (i.e., follows a religious calendar).
- is normative; its tells a culture the details of its "ought-to-do/be."
- is mystical; without mystery, there is nothing to catch the imagination.
- is counterintuitive; if it were easy, it wouldn't be so memorable.
- builds a backstory; this provides a sense of group identity.
- is emotion-laden; if it cannot move our emotions, it is too ordinary to be remembered.
- gives positive feedback for believers who adhere to its tenets.
- has outlasted many or most other religions.
- usually has a "Founder's Tale"; overcoming great odds, or unusual incidents are related.
- lays forth proscriptions and prescriptions for behavior, diet, even thoughts.
- lays the foundations for "privileging" (e.g., Christian privilege in the United States).
- may have sprung from a need to bind people together under a unified cultural umbrella.
- uses a plethora of emotional triggers to reinforce the faith and control the faithful.
- uses narratives that pretend to be reveal historical facts.
- uses a sense of awe to incite or reinforce belief (e.g., European cathedrals with their soaring ceilings, stained glass windows that "tell the stories," statues, carvings, etc.).

- relates stories that seem to be about morality but can just as easily be seen as asserting authority over the followers.
- may fulfill social needs of individuals; give mores, norms, values; add a sense of social belonging through social reinforcement by fellow believers; fulfill a sense of purpose; followers may serve community and people through social and missionary work and entire life may revolve around the faith.
- provides the social structures to meet those needs (family and community; church, synagogue, mosque).
- may provide a sense of (false) certainty in an uncertain world, working to reduce untenable ambiguity by insistence on belief without evidence—on faith alone.
- may diminish the disconnect between the individual and society at large (alienation).
- may start with the role of charismatics, or founders, who often
  - ° have visions, dreams, nightmares, waking dreams (night terrors).
  - ° may use substances to induce or enhance visions.
  - ° may have brain conditions or disease; tumors, stroke, or epilepsy.
  - ° use deliberate cunning, trickery, and manipulation to gain followers; may be somewhat along the line of scammers, con men, and illusionists.
  - ° may suffer from mental illness.
  - ° may suffer from sleep deprivation since they frequently have their "visions" after long periods of fasting and lack of sleep.
  - ° may suffer from other medical conditions that affect thinking, moods, and behavior (e.g., thyroid problems can cause unaccountable episodic depression or rage).
  - ° have had a foundational experience that is not repeatable, which makes it special.
- develops approved texts; canonization of the narrative occurs.
- develops over time; is subject to interpretive revision, if not of actual content.
- eschews and actively discourages critical thinking and questioning; punishes those who stray.

- uses words and phrases *as if they are magic*; may involve pious lies to maintain followers' belief; uses recitations of creeds or religious texts.
- went through a period of time of state sponsorship (theocracy) or stamp of approval.
- may involve elaborate rituals, dress, iconography.
- uses its doctrines to "protect the faithful" from contamination by nonbelievers.
- affects the organization and perception of space (cosmogony); divides sacred space from profane (secular) space; affects the structure of religious landscapes.
- may depict the religion as under cultural attack (e.g., the "war on Christmas," Muslim fears of the West and new Crusades).
- may call for a "return" to a presumed "earlier Edenic past" or return to "purity" of original version of the faith (Fundamentalism).
- relies on certain *unchallenged* underlying assumptions:
  - that there is a deity.
  - that there is a hierarchy of special underling followers.
  - that there are prophets and they speak for god.
  - that the prophets have been told exactly what to say and have not deviated from their instructions.
  - that the texts, revelations, or sayings are divinely inspired (or actual words of god/Allah).
  - that the Founder is not making up, faking visions, or suffering from a mistaken case of "deity-contact."
  - that heaven, the afterlife, and hell are real places.
  - that by following certain rules, regulations, laws, and commandments, one may go to heaven in the afterlife.
  - that all the characteristics of god are not mutually exclusive: he is omnipotent, omniscient, omnipresent.

MODELLING CRITICAL THINKING

What tool can we use to discern truth from fiction, facts from error, and science from pseudoscience? How can we go about knowing if what we think we know is even true? We happen to have a model for that: It is "thinking about our thinking while we are thinking"—that is, a model for *critical thinking*. Under no misapprehensions regarding our propensity to believe what we want to believe rather than to be critical thinkers, we have to start with the requirement that to model critical thinking is therefore necessary. But this is no panacea.

Readers will undoubtedly (correctly) point out that critical thinking will not safeguard us from unreason. But for the sake of this discussion, let us admit that in a perfect world, it *should* be a requirement for human beings—if we are to ground ourselves in reality. There is no inoculation from believing in inane, silly, unreasonable, woo-laden, or even dangerous beliefs, including religion, but learning some basic skills, we hope, will help us suspend the tendency to believe without supporting evidence.

One of the key skills needed to do so is learning not only that we should examine underlying assumptions but how to do so, if we are to have *informed confidence* in what we believe to be real and true.

Challenging Underlying Assumptions

The following exercise is a way to bring home the notion that our thinking can be lazy. We are immersed in a cultural world that operates with underlying assumptions, yet most of the time we fail to realize it and therefore fail to question our thinking. Take a moment to read the following sentences, then jot down a note on anything that bothers you in the sentences. Ask yourself "What are the *underlying assumptions* and the *implications* of the statement?" And what might be the consequences if the ideas contained in the statements are followed?

1. If only the people of India would eat their cattle, they wouldn't go hungry.

2. We should always respect other people's religious beliefs.
3. I know it's true because it feels right to me.
4. Humans should have dominion over the earth.
5. Cheap oil is a necessity to the West's way of life.
6. Technology will always be able to solve our problems.
7. We should always teach both sides to an issue.
8. Anything is possible.
9. Religion and science are incompatible.
10. Knowledge requires you to have sufficient justification for what you think is true.[44]

We could choose any of these to analyze more deeply, but let's take number two: "We should always respect other people's religious beliefs." What is wrong with this statement? Events in the last few years should illustrate what happens when the values of one culture system and another clash. The weight that the West places on freedom of speech is in direct competition with Islamic theocracies and their laws against free speech (even in allegedly democratic nations). We have to ask, what do we mean by "respect"? What constitutes respect? How do we go about "respecting religious beliefs" other than our own? What happens when we do not do that to someone else's satisfaction? Are there any bad consequences? (More Charlie Hebdo-like terror attacks on newspapers that purportedly do not show respect?) Which traditions are we respecting? The ones we think are consistent with our own? What about the ones we abhor? And other conflicting cultural violations may occur if we are "ordered" to respect something we don't really respect. (Can "tolerance" be forced?) There is even a proposal in some quarters to make disparaging any religion a hate crime—which kills free speech and therefore threatens American democracy at its core. To ban religious dissent and critical commentary would give a free pass to such extreme political correctness that it would kill biblical and Qur'anic critical scholarship altogether. When something gets shrouded in the cloak of special pleading and dubious protectionism, then you know the world is in trouble. If we are to combat ignorance in all its forms, we need critical thinking, freedom of expression, and a society willing to face facts. To be critical thinkers, it helps to get in a bit of mental exercise in analyzing our everyday way of thinking, specifically looking for the hidden underlying assumptions in things we hear in the news, read in newspapers and books or online, are exposed to through social media, and get from our friends, family, and coworkers.

This exercise is a great tool for generating deeper thinking and discussion of how the world really works, who has power and who doesn't, the cultural problems inherent when conflicting values "inform" different culture groups, and a host of political, philosophical, moral and ethical issues relating to our belief systems. Nobody ever said being human would be easy. What this brief exercise does for us is that it points out the need for:

1. an informed population (educated with full and free access to information).
2. a willingness to actively seek out and understand opposing viewpoints.
3. explicitly modelling critical thinking skills.
4. practice in learning how to apply critical thinking to solve problems.
5. free speech to test ideas in the court of public opinion.
6. good communication skills, including "checking it out" and reflexive thinking.
7. a willingness to admit it and learn from others when one is wrong.
8. an acknowledgment that we may need to get it wrong in order to get it right (hopefully, we learn from our errors).

Critical thinking involves clearly stating a thesis, considering a variety of viewpoints, gathering information from credible sources, and assessing whether there are alternative theories. It requires asking whether there are good reasons for the belief, demanding the evidence for it, then remaining open-minded and unbiased in our approach. Critical thinking asks if we are being reasonable and avoiding wishful thinking; remaining objective (if it is so, it should be able to withstand many tests to verify its validity: we should be asking ourselves, "Does the evidence support the belief?" and "Can the evidence stand the acid test of further scrutiny?"); assessing the bias level of the data (the more unbiased data that backs up a claim, the better); asking whether a tentative conclusion can be reached; learning to be comfortable with ambiguity; and, lastly, distrusting absolutes.[45]

WHAT IS THE SCIENTIFIC METHOD OF INQUIRY?

Science assumes that *the universe exists*; that it consists of *natural phenomena* that are *governed by natural laws*; and that *these laws and phenomena are discernable and explainable* (can be learned and understood) through a *rational, systematic pursuit of knowledge*.[46] "Further, the laws of the universe are constant; the rules of the universe apply."[47] In other words, you do not get to repeal the laws of the universe in order to support an otherwise non-supportable belief. Science is not based upon belief, but on *evidence* from careful (repeatable) *observations* and *methodological experimentation*, *testing hypotheses*, and *making predictions* based on the knowledge gained as a result of the process.

The thing about terminology used in science is that it is difficult for the lay reader (nonscientists) to understand the specificity of the meanings used in science, as opposed to those used by the general public. The rational scientific model of thinking is an approach to problem-solving and gaining knowledge and contrasts sharply with the nonrational or religious mode of operation (which rests its case solely on belief). By contrasting the religious mode of thinking with the scientific, I think we can see the problem of why religion persists. Most people are ill-equipped to distinguish facts from opinions or science (and history) from pseudoscience (and pseudohistory). They fail to understand that a claim or an assertion is not a fact.[48] Whereas religion leads to thinking in absolutes, with no gray area, science allows for the possibility of getting it wrong and starting over. That is why it's called self-correcting; science only has a vested interest in getting it right. Here, we see the contrasting modes of thinking styles.

Non-Rational	Rational
Ex., religion, pseudoscience, paranormal	Ex., science and rational skepticism
Operates in non-rational mode	Operates in scientific mode
Involves narratives without evidence	Uses facts with supporting evidence
Fantasy proneness is exhibited	Reality-based; the natural world
Offers esoteric explanations	Natural laws; scientific explanations
Exhibits a strong "Will to believe"	Reserves judgment until facts are in
No alternative explanations considered	Considers alternative explanations
Engages in degree(s) of credulity	Engages in skepticism; rationalism
Reaches supernatural, nonscientific conclusions	*Reaches tentative, scientific conclusions*

Putting Away Childish Things

> *"When I was a child, I spoke as a child, I understood as a child, I thought as a child; but when I became a man, I put away childish things.*
>
> **—1 Corinthians 13:11**

There is a lot at stake in contrasting the two modes of thinking. Are we going to have a society that believes anything it wants, or are we going to have some standards for assessing whether a belief is warranted? We have to do a better job of "putting away childish things."

There are three trends playing out on the world stage today: anti-intellectualism, anti-science, and anti-modernity. Isaac Asimov said, "Anti-intellectualism has been a constant thread winding its way through our political and cultural life, nurtured by the false notion that democracy means that 'my ignorance is just as good as your knowledge.'"[49]

Anti-intellectualism consists of devaluing knowledge, intelligence, and education, as well as the facts that flow from them. "Intellectual" is considered a pejorative term when used by political populists and religious conservatives, in the same way as they use "elitist," "closed-minded," or "know-it-all" against those whose ideas they fear. I think anti-intellectualism really comes from a deep-seated sense of inferiority but is publicly played out with a thick mask of libertarian superiority. "I can get along fine without you telling me what to think, thank you very much!" seems to be their mantra.

The **anti-science trend** is as much about promoting political ideologies as it is of protecting religious dogma from fact-checking. When politicians deny science, it is too often the case that protecting turbo-capitalism is the driving force. The same politicians who protect their industrial donors' interests deny science for religious reasons. Science-denialism comes from the very conservative Far Right (denying anthropogenic global climate change, global warming, sea level rise; that fracking causes earthquakes and that industrialism has caused increased CO_2) and from the ultra-liberal Far Left (falsely claiming that vaccines cause autism, or that New Age naturopathy is as good as medicine). Anti-science is the foot-stomping insistence of *self-interest and uber-individualism* over what is best for the group or even for the species.[50]

Anti-modernity is the ultimate betrayal of knowledge. It is the rejection of the modern world—its institutions, its progress, its science, and even its people. It is about garnering power to control other people's lives, legitimized through religious extremism. It is particularly hateful toward women, denying them basic rights and identities. It is at once anti-individual and anti-group, since it destroys—unthinkingly, utterly. You cannot hope to change such groups. This trend seems intractable. Some say it is best left to the group's religious and political leaders to renounce and abjure such extremism. Others throw up their hands in surrender to the notion that we are in for endless religiously motivated insurgencies and wars based on hatred of the modern world. But a danger ignored is not a danger diminished; instead, the danger to the rest of us *grows* in the face of our reluctance to confront it and answer its dreadful, mindless religiously motivated killing. Make no mistake: the need to combat religious extremism is a duty of people of all faiths and no faiths, in all cultures, in all nations, *no matter whose extremism it is*.[51]

CONCLUSIONS

I live in the South. Here, the first question people ask you upon meeting you is "What church do you go to?" I have decided that my blow-away answer is, "Oh, we don't go to church much," which lets the asker off the hook without committing me to something I don't believe in (weddings and funerals are still in churches, after all!). I try not to close the door on the discussion unless someone is being rude. Don't be afraid to engage the religious in discussion. Don't debate though, because that only tends to harden already-held beliefs on both sides. Prepare a few points ahead of time that you can state in nonjudgmental terms, such as: "I am a naturalist" (and/or humanist, atheist, agnostic, etc.), reflecting your true stance: "I follow science rather than believe in religion" or "I find much greater mystery in science than in religion." Be true to yourself—without going so far as to place yourself in danger.

How do thinking people who rely on reason counter anti-intellectualism, anti-science, and anti-modernity trends? It is incredibly frustrating to attempt to deal with anti-intellectualism and anti-science; they are so contrary to common sense and reason. A strategy that is less fraught with frustration is that of "planting seeds." Exploit an opening and use communication skills to plant seeds of reason and doubt. Seeds can crack boulders; surely they can root out unreason. Have you ever seen a blade of grass growing in concrete? It is the same with doubt. Think of religion as a *differential terrain*: some of it will "wash away" in the same way that weaker rock layers erode before stronger rock layers will. This can eventually lead to canyons of doubt. Every drop of doubt that is added erodes religion further, until religious belief is no longer tenable.

The more books and articles revealing religion's weaknesses, pious lies, and evils, the more likely someone teetering on the edge of doubt will eschew religion, and step into the light of reason. It may happen gradually, much too slowly for some of us, but it will happen. It is already happening. The spate of religion-protection laws in the United States are part of the backlash caused by the religious realizing they are losing ground.[52] Let us hope that we may see the end of religious privilege in America in our lifetimes.[53]

I believe we must be in the Real World and not that of make-believe, wishful thinking, and unreason. The alternative is to turn the corner on knowledge itself, and I for one do *not intend to sit idly by while the human cultural world slides into the abyss of willful ignorance and chaos*.

The Hammer

I have seen
The old gods go
And the new gods come.

Day by day
And year by year
The idols fall
And the idols rise.

Today
I worship the hammer.

Carl Sandburg, 1914[54]

Part 2

SCIENCE AND CREATIONISM

Chapter 4

CHRISTIANITY AND COSMOLOGY

Victor J. Stenger

THE MESOPOTAMIAN COSMOS

The Christian faith emerged two thousand years ago out of the older faith of the Hebrews, a tiny tribe of desert dwellers in Canaan. The civilizations of Mesopotamia and Canaan all had similar concepts of the cosmos and humanity's place in it.[1]

The Hebrews divided the universe up into Heaven, Earth, Sea, and the Underworld, as shown in figure 1. Earth is a more-or-less flat disk floating on water, with the vault of heaven resting on foundations on the edge of the sea. Beneath Earth are the waters of the abyss. God sits at the apex of a series of heavens above the vault of the sky.

ATOMS AND THE VOID

The view of the cosmos that we hold today could not be more dissimilar to that of the ancient Hebrews. But it is almost as old, at least conceptually. In the modern view, the universe is composed of elementary material particles moving around in an otherwise empty void. Moreover, there are good reasons to conjecture that our universe is just one of a multitude of universes termed the *multiverse*, which is limitless in time and space—infinite and eternal. However, this remains unproven, and I will limit the discussion in this essay to the universe that is visible to us with our eyes, telescopes, and other instruments—and what can be solidly inferred from that data.

Figure 1. The ancient Mesopotamian model of the cosmos as adopted by the Hebrews. Image © Michael Paukner; reproduced by permission.

The notion that everything is just "atoms and the void" was proposed in the fifth century BCE by Leucippus and Democritus, two pre-Socratic Greek thinkers living in Ionia at the time and place where science and Western philosophy had their beginnings.

The most influential philosophers of ancient Greece, in particular Plato and Aristotle, rejected material atomism. It conflicted too much with their ingrained beliefs in the world of gods and myths that had been passed down through the ages.

However, the Athenian philosopher Epicurus, who lived a generation after Aristotle, built a materialistic philosophy of life based on atomism. Epicureanism was mischaracterized as debauched by those who maintained ancient superstitions and saw atomism as the threat it really was to any kind of supernaturalism.[2] In fact, it was a philosophy of moderation that would be highly beneficial to individuals and society if adopted today.

Although he wrote volumes, most of them were lost, and we would know little of Epicurus and ancient atomism today were it not for the lucky survival of a copy of a magnificent poem titled *De rerum natura* (*The Nature of Things*), written in Latin hexameter at the time of Julius Caesar by the Roman poet Titus Lucretius Carus.[3]

ANCIENT COSMOLOGY

Besides the atomists, several other ancient Greek and Roman thinkers developed highly sophisticated, scientific cosmologies. Despite the common belief that people thought Earth was flat until 1492, when Columbus sailed the ocean blue, it was long known among scholars that Earth is spherical. Aristotle said the form of the universe had to be a sphere, because it is the most perfect solid.

Aristotle's cosmology is illustrated in figure 2.[4] He made a major advance in proposing that the planets and stars were actual physical bodies (they were previous viewed as gods), although still perfect spheres.

Note the sphere of the "Prime Mover" in the figure. Aristotle's physics required that all motion have a cause and the ultimate cause, the Prime Mover, was the "First Cause Uncaused."

Ancient cosmology achieved its highest level of mathematical sophistication in the second century CE with the geocentric model of Claudius Ptolemy of Alexandria, shown in simplified form in figure 3.[5] While resembling the cosmos of Aristotle in figure 2, Ptolemy's model was completely quantitative. Knowing how to use Ptolemy's model, you could predict the positions of planets with remarkable accuracy.[6]

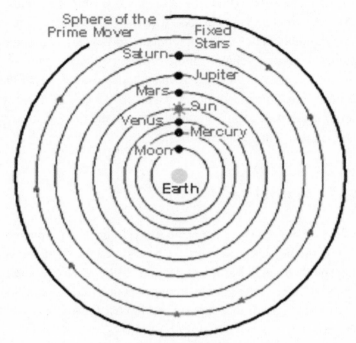

Figure 2. Aristotle's cosmology

THE DARK AGES

Unfortunately, the Ptolemaic system and scientific cosmology in general were largely forgotten in Europe after the Catholic Church gained control of the Roman Empire in the fourth century and ushered in the thousand-year decline known as the Dark Ages. At that point, Christian cosmology returned to childish images taken from the Bible.

Early Church leaders flatly rejected the Greek conception of the world. Tertullian asked, "What indeed has Athens to do with Jerusalem?" The influential bishop Lacantius, advisor to Emperor Constantine, dismissed the sphericity of Earth as a heretical belief, as well as the ridiculous notion that people on the other side "have footprints above their heads" and that "rain and snow and hail fall upwards."[7]

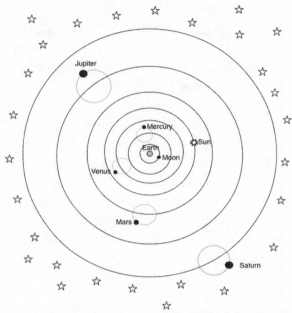

Figure 3. The Ptolemaic model of the solar system.

Not only was Earth non-spherical according to early Church fathers, so too was heaven. Rather, according to their interpretation of scriptures, it was like a tent or tabernacle. Furthermore, some argued that Earth could not be the center of the universe but must rest at the bottom because of its heaviness. Obviously, heaven is the center of the universe.

The works of Epicurus and Lucretius were also suppressed during the Dark Ages. It was only by sheer luck that a manuscript of *De rerum natura* survived to be rediscovered in a monastery in Germany in 1417. After a copy was taken to Florence, where more copies were made, it became a sensation that played an important role in the nascent Renaissance and the scientific revolution that was soon to follow.[8]

The planets move in circles on the surfaces of concentric spheres that revolve in circles around a point not precisely at the center of Earth.

While Europe was dark, Islam had gone through its remarkable Golden Age and preserved much of Greek and Roman science, as well as developing a great deal of its own.[9] But Arabic scholars, also confined by their religion, still promoted a human-centered universe. They preserved Ptolemy's work, giving it the name by which it is still known: *Almagest*, "The Majestic."

Gradually, ancient scholarship seeped back into Europe. Christian theology adopted Aristotle as the authority on most subjects, especially his idea of the Prime Mover, which was interpreted by Christian theologians, notably Thomas Aquinas, to be the creator God. In fact, Aristotle's Prime Mover was not the Creator but a spiritual something that is the source of all celestial motion and, as we saw, located at the outermost part of the universe.[10] Aristotle's cosmos had no beginning and will have no end, a teaching Christianity would ignore.

In any case, the first great European universities built by the Catholic Church become so deeply entrenched in what is called *Aristotelian Scholasticism* that the scientific revolution, which repudiated much of Aristotle's science—especially his physics—occurred outside these institutions.

THE SCIENTIFIC AGE BEGINS

Finally, in the sixteenth century, Europe awoke to the dawn of the Renaissance and, with the rediscovery by Nicolaus Copernicus, to the notion of Aristarchus that Earth is just another planet rotating about the sun. The scientific age had begun.

Copernicus's masterwork *De revolutionibus orbium coelestium* (On the Revolutions of the Celestial Spheres) was published shortly before his death in 1543. At the time, the Church did not deem Copernicus heretical, and he had dedicated his book to Pope Paul III. First, the work was highly technical, so only the most mathematically sophisticated reader could follow it. Second, it did not ask to be taken as literal reality.

In a curious turn in history, an early disciple of Martin Luther, Andreas Osiander, a theologian and amateur mathematician who had helped proofread the printing of *Revolutionibus* in Munich, added an unsigned preface. There he states that the Copernican model should only be taken as a tool for astronomical calculations, with no philosophical or theological implications. Osiander wrote:

You may be troubled by the ideas in this book, fearing that all of liberal arts are about to be thrown into confusion. But don't worry. An astronomer should make careful observations, and then frame hypotheses so that planetary positions can be established for any time. This our author has done well. But such hypotheses need not be true nor even probable. Perhaps a philosopher will seek truth, but an astronomer will just take what is simplest, and neither will find anything certain unless it has been divinely revealed to him. So if you expect to find truth here, beware, lest you leave a greater fool than when you entered.[11]

The story of the resistance of the Catholic Church to the notion that Earth and humans were not at the center of the universe is well known. Less familiar, perhaps, is that the new Protestant reformers, notably Martin Luther, were also strongly opposed. Luther is reported to have remarked, when hearing about the new sun-centered universe, "So it goes now. Whoever wants to be clever must agree with nothing that others esteem. He must do something of his own [Luther himself included?]. This is what that fellow does who wishes to turn the whole of astronomy upside down. Even in those things that are thrown into disorder, I believe the Holy Scriptures, for Joshua commanded the Sun to stand still and not the Earth."[12]

Copernicus's supporters had a hard time reconciling the Copernican model with the Bible, particularly the Joshua story (Joshua 10:12–13), as well as Psalms 93:1, 96:10, 104:5, and 1 Chronicles 16:30, which declare that the foundation of Earth remain forever unmoved.

While the Church did not immediately reject the Copernican picture as a computational model, forces were coming into play at the time that would eventually lead to its condemnation in theological circles. The Reformation produced a deep split with the Roman Church on the location of the source of authority in Christendom. Although the Church revered the Bible, it did not regard it as the final authority on theological matters. That authority rested ultimately with the pope, based on the claim of an unbroken line of succession from Peter, who was given rule over all earthly matters directly from Christ. The Reformation had to find a replacement for papal authority and the only alternative was the Bible, which meant that it was to be taken literally as the word of God.

Even today, we find Bible inerrancy to be a basic dogma in many Protestant sects, which leads them to reject scientific results that conflict with what is revealed in this sacred document, such as evolution and the age of Earth. On the other hand, the Catholic Church has little problem with these conflicts. Popes have declared most scientific discoveries acceptable, as long as they do not deny divine creation and the immaterial and immortal nature of the soul. In fact, as we will see, some have seen science as confirming Catholic teachings.

Luther, and other reformers who preached that the Bible was literal truth, objected to the Copernican picture since it disagreed with scriptures. However, it should be noted that Luther died in 1546, just three years after the publication of *Revolutionibus*, when the model was far from being established on scientific grounds. Nevertheless, under the pressure of the Reformation, over the next century the Roman Catholic Church found it necessary to take more conservative stands on many issues, and that included backing off from Copernicanism.

Another thorn in the Church's side at the time was the Italian Dominican friar Giordano Bruno, whom they could only shut up by burning him at the stake in 1600. Bruno committed many heresies to earn his toasting, but his cosmology is particularly relevant to our story. Bruno proposed that the sun was just another star in a universe containing an infinite number of worlds and having no center. Furthermore, these worlds were populated by other intelligent beings.

GALILEO AND THE TELESCOPE

The story of Galileo is familiar but still widely misunderstood. Unlike Osiander, Galileo was not satisfied to simply assent that the heliocentric model was merely a useful tool for predicting celestial events. He insisted it was a fact of nature. The refracting telescope had been invented in 1608 by the Dutch eyeglass maker Hans Lippershey. The earliest versions had a magnification of only a few times, but Galileo was able to improve that to a factor of thirty and turned his superior instrument on the heavens.

The initial publication of his observations appeared in 1610 in *Sidereus nuncius,* known popularly today as *Starry Messenger*.[13] He reported seeing spots on the sun, as well as mountains and craters on the moon. He viewed ten times as many stars than are visible to the naked eye and fuzzy nebulae that he interpreted, along with the Milky Way, as collections of stars too far away to resolve.

In 1610 Galileo also made the important discovery that, similar to the moon, Venus exhibits phases that result from the partial illumination it gets from the sun as it circles around inside Earth's orbit. This observation cannot be explained in the Ptolemaic system.

But perhaps the most important set of observations by Galileo was the four moons of Jupiter. From January 7, 1610, through March 1, he sketched the positions of four bodies near Jupiter, except when clouds obscured his view. Some sixty-four such sketches can be found in *Sidereus nuncius*.[14]

These observations left no doubt that these bodies were moons revolving around Jupiter. This meant they were not revolving around Earth, implying that Earth is perhaps not the center of the universe after all.

In 1616, the Church ordered Galileo not to claim that the sun was the center of the universe or that Earth moved, which violated several biblical references stating that Earth cannot be moved. There were only so many biblical contradictions that the Church was willing to live with.

Nevertheless, Galileo continued his telescopic observations and in 1632 published *Dialogues on the Two Chief World Systems*, in which he forcefully presented his conclusions. In *Dialogues*, the character Simplicio, who represents the Aristotelian position, makes arguments that the pope himself had previously expressed. Pope Urban VIII, who earlier was a friend and supporter of Galileo, was, to say the least, not pleased. Galileo was tried by the Inquisition for disobedience, forced to recant on his knees, and sentenced to permanent but quite comfortable house arrest.

Although technically forbidden from writing any more about physics and astronomy, Galileo continued work in physics and laid the foundation for Newtonian mechanics that appeared a generation later. Isaac Newton was born in 1642, the year Galileo died.

THE CLOCKWORK UNIVERSE

Newton's laws of motion and gravity provided the final confirmation of the validity of the sun-centered model of the solar system. Kepler had introduced the idea that the orbits of the planets are not circles but ellipses and introduced his laws of planetary motion. Newton was able to prove this mathematically. When he showed his proof to astronomer Edmund Halley, Halley convinced Newton to publish (at Halley's expense) what is inarguably the greatest scientific work in history, *Philosophiae naturalis principia mathematica*. Now referred to simply as *Principia*, it presented the three laws of motion and the law of universal gravitation, from which Kepler's laws of planetary motion were then derived. Today this is a simple exercise in freshman physics.

Using Newton's laws, Halley calculated that a comet seen in 1682 was the same as one that had been recorded by astronomers as far back as 240 BCE, circling the sun in a highly elliptical orbit every 75–76 years. Halley predicted it would return in 1758. The success of Halley's prediction, after his and Newton's deaths, may have been the single most important event in history that established the power of the new science in the minds of scholars and laypeople alike.

René Descartes had earlier proposed that God created the universe as a clockwork of perfect motion that required no further intervention, a philosophy that later became known as *deism*. During the period in the eighteenth century called the *Enlightenment*, the *clockwork universe* would become associated with Newtonian particle physics and form the basis of deism.[15] Since god is perfect, so must be his laws and thus the deist god has no need to step in to make changes. Deism clearly contradicts Christian belief, although this did not seem to get Descartes into trouble when he initially made his proposal. He certainly did not reject Christianity and, indeed, withdrew a manuscript on cosmology when he heard of the trouble Galileo got himself into.

While the clockwork universe was perhaps the primary motivating factor for deism, Newton did not profess deistic views himself and remained a Christian who saw the need for God to step in from time-to-time to keep the universe running properly.

Still, Newton was an unconventional Christian who rejected the Trinity (while holding the chair of mathematics of Trinity College, Cambridge, now held by atheist Stephen Hawking). But he still was very much a believer, not only in the Christian God but also in the occult. He spent more time and wrote more on alchemy and biblical interpretation than he did on physics. And God entered his physics in important ways. Unlike Galileo, who did not mix his religion and science (but still claimed belief), Newton turned to God to provide explanations for what he could not explain himself.

Newton's gravitational theory admitted only attractive forces, which implied that the stars, which everybody at the time assumed to be fixed, should collapse upon one another as a result of their mutual attraction. Newton conjectured that God had placed them in just the right positions where their attractions balanced.

Newton also invoked God to provide for the stability of planetary motion. He was fully aware that his derivation of Kepler's laws assumed each planet moved independently of the others, while, in fact, the planets also exert gravitational pulls on one another. Sir Isaac reasoned that the orbits of planets would not maintain their regularity by blind chance. Thus, he concluded, God had to intervene from time to time to keep things in order.

Newton's archrival, Gottfried Wilhelm Leibniz, scoffed at this:

Sir Isaac Newton and his followers have also a very odd opinion concerning the work of God. According to their doctrine, God Almighty wants to wind up his watch from time to time: otherwise it would cease to move. He had not, it seems, sufficient foresight to make it a perpetual motion.[16]

Newton and Leibniz also quarreled over the priority of the invention of calculus, which both developed independently. We still use Leibniz's superior notation today.

We have seen that a conflict has long persisted over the location of what is termed "the center of the universe." For the most part, we humans have found it easy to position ourselves at the center. All living species are self-centered—if not always the individual organism then what Richard

Dawkins termed their "selfish genes." Species would not have survived if they weren't selfish at some level.

There is also a good empirical reason for us to think we are the center of the cosmos. When we look at the sky, everything seems to revolve around us. The planets sometimes turn around and go back the other way, but they soon turn back again and resume circling Earth.

Of course, we know today that the sun is not the center of the universe either, as it was thought to be when Copernicus's picture of the cosmos was one of seven planets surrounded by a rotating shell of fixed stars called the "firmament." As telescopes improved, astronomers discovered that our sun is just another star and that no point in space can be identified as the center of all things.

A century after Newton, the French mathematicians, astronomers, and physicists Pierre-Simon Laplace and Joseph-Louis Lagrange independently proved that the solar system was highly stable despite the gravitational effects of planets on one another. The French mathematicians Siméon Poisson and Henri Poincaré later confirmed this to greater precision. This removed one of the objections Newton had to the clockwork universe: God is not needed to step in to keep the solar system in order.

Laplace was able to account for all of Ptolemy's observations within a minute of arc, including the motions of Jupiter and Saturn that were inconsistent with previous calculations. Thus, Laplace showed that Newton's laws were, in themselves, sufficient to explain the movement of the planets throughout previous history.[17] This led him to propose a radical notion that Newton had rejected: *nothing besides physics is needed to understand the physical universe.*

One often hears stories of Laplace's encounter with the emperor Napoleon in or about 1802. Here's one version, but it is still disputed, and the whole story is probably apocryphal:

Laplace had an audience with Napoleon and presented him with a copy of his book *Mécanique Céleste* (Celestial Mechanics). Someone had told Napoleon that the book contained no mention of God. Napoleon received it with the remark, "M. Laplace, they tell me you have written this large book on the system of the universe, and have never even mentioned its

Creator." Laplace, answered, "Je n'avais pas besoin de cette hypothèse-là." ("I had no need of that hypothesis.") Napoleon, greatly amused, told this reply to Lagrange, who exclaimed, "Ah! c'est une belle hypothèse; ça explique beaucoup de choses." ("Ah, that's a beautiful hypothesis; it explains many things").

Laplace never wrote anything denying the existence of God, and he may have been a deist. As discussed previously, deism—as opposed to theism—supposes a creator god who set the universe in motion and then left it alone to carry out the instructions coded into the natural laws by that god. The quotation above, if true, may simply say that Laplace did not need to hypothesize anything beyond the laws of mechanics and gravity to describe the motions of the heavenly bodies.

In his 1796 book *Exposition du système du monde*, Laplace quotes Newton as saying, "The wondrous disposition of the Sun, the planets and the comets, can only be the work of an all-powerful and intelligent Being." Laplace expresses the deist position when he comments that Newton "would be even more confirmed, if he had known what we have shown, namely that the conditions of the arrangement of the planets and their satellites are precisely those which ensure its stability."[18] Even the great Newton was not immune to God-of-the-gaps arguments.

Laplace agreed with Leibniz's criticism of Newton: "This is to have very narrow ideas about the wisdom and the power of God . . . that God has made his machine so badly that unless he affects it by some extraordinary means, the watch will very soon cease to go." We will see this mistake being made today by those who say that the universe is so imperfect that it had to be fine-tuned by God so that life could evolve. Laplace and Leibniz would counter, "God is smarter than that." I would put it another way: The universe is smarter than that.

DEISM AND THE ENLIGHTENMENT

Deism became prominent during the period called the Enlightenment in the eighteenth century when science and reason began to hold sway over theology and revelation. Deists, following the opinions of Leibniz and Laplace I have already quoted, regarded it as illogical that a perfect god needed to step in at any time after the creation to fix things that may have gone awry.

Many prominent people of the time were either avowed deists or regarded as deists based on their published views. In Europe, these included Adam Smith, Frederick the Great, James Watt, and Voltaire. In America, Benjamin Franklin, Thomas Paine, and at least the first four presidents—George Washington, John Adams, Thomas Jefferson, and James Madison—were deists.

However, the Enlightenment and its deterministic version of deism did not survive into the nineteenth century for a lot of reasons that had little to do with science. The impersonal deist god did not provide the comfort of religion sought by the average person. Christian revivalism, which appealed to rich and poor alike, with emotion in place of reason, spread throughout Europe and America. At the same time, the intellectual world of art and literature reacted against scientific rationalization and replaced it with an emphasis on intuition and emotion that was termed Romanticism. It should be mentioned that Enlightenment deism based on the Newtonian clockwork universe is pretty much ruled out by the uncertainty principle of quantum mechanics.

THE UNIVERSALITY OF PHYSICS

The nineteenth century saw many advances in astronomy as both technology and mathematical theories continued to improve. In particular, the universality of physics became firmly established. Newton had made the first great leap in that direction with his universal law of gravity. Previously, it had been assumed that one set of laws applied on Earth and another set in the heavens. Newton proposed that the force that causes an apple to fall from a tree to the ground was the same force that caused the moon to fall around Earth.

The universality of physics was corroborated in the nineteenth century with the observation that the spectral lines in the light from stars were the same as those observed from hot gases in laboratories on Earth.

When it was discovered that many of the observed stellar spectra maintained the same pattern but were Doppler-shifted in wavelength, it became possible to measure the velocities at which stars were moving toward or away from Earth. Most were shifted toward the red, indicating those stars were receding from us.

In the early twentieth century, Edwin Hubble and others established that the universe was much vaster than previously assumed and that the Milky Way, within which we reside, is just one of countless "island universes" called *galaxies* that contain hundreds of billions of stars. Furthermore, Hubble showed that the larger the redshift of the stars in a galaxy, the further that galaxy is from us. By 1920, galaxies as far away as a million light-years had been observed. But they still were just a drop in the bucket.

THE EXPANDING COSMOS

The twentieth century also saw the application of Einstein's special and general theories of relativity to cosmology. Although Einstein himself resisted the notion, it was soon established that general relativity predicted that the universe is expanding, in agreement with the observation of galactic redshifts.

The first scientist of the time who seemed to grasp the developing connection between mathematical cosmology and the remarkable telescopic observations that were coming in at the same time was a Belgian Jesuit priest and physicist, Georges Lemaître. In 1927, Lemaître published an article in French titled, "Un Univers homogène de masse constante et de rayon croissant rendant compte de la vitesse radiale des nébuleuses extragalactiques" (A homogeneous Universe of constant mass and growing radius accounting for the radial velocity of extragalactic nebulae).[19] In the article, he showed that Einstein's general relativity leads to an expanding universe that accounts for galactic redshifts. In fact, Lemaître says explicitly, "The receding galaxies are a cosmical effect of the expansion of the universe."

Lemaître's paper lay unrecognized for several years. However, it began to receive attention in 1931, when it was translated into English.[20] Einstein finally came around and, by 1933, so had the astronomical community.

THE BIG BANG

During a conference in London in 1931 on the relation between physics and spirituality, Lemaître proposed that the universe expanded from an initial glob of nuclear matter in an explosion that the British astronomer Fred Hoyle would, in 1948, derisively term the "big bang."

Much has been made of the fact that Lemaître was a Jesuit priest and that his notion that the universe began a finite time ago with a giant explosion was informed by his religious beliefs. Perhaps it was. But he always insisted that the primeval atomic nucleus was purely a scientific hypothesis not yet confirmed by the data. Indeed, Lemaître's scenario sounds more like the Chinese creation myth of a "cosmic egg" than what is described in Genesis.

Nevertheless, against his expressed wishes, Lemaître's big bang has been used by theologians and believing scientists to claim scientific evidence for a divine creation. On November 22, 1951, Pope Pius XII spoke before the Pontifical Academy of Sciences, where he announced that modern science and the Church are converging on the same fundamental truths.[21]

Introducing a theme that would be expanded upon by religious apologists in succeeding decades, the pope said, "According to the measure of its progress, and contrary to affirmations advanced in the past, true science discovers God in an ever-increasing degree—as though God were waiting behind every door opened by science."

With respect to the cosmos, the pope said, "Everything seems to indicate that the material universe had in finite times a mighty beginning, provided as it was with an indescribably vast abundance of energy reserves, in virtue of which, at first rapidly and then with increasing slowness, it evolved into its present state."

The pope concluded:

[Modern science] has followed the course and the direction of cosmic developments, and, just as it was able to get a glimpse of the term toward which these developments were inexorably leading, so also it has pointed to their beginning in time some five billion years ago. [This was a reasonable estimate at the time.] Thus, with that concreteness which is characteristic of physical proofs, it has confirmed the contingency of the universe and also the well-founded deduction as to the epoch when the cosmos came forth in the hand of the Creator.

And, in the lines that have been quoted frequently in the decades since:

Hence creation took place in time. Therefore, there is a Creator. Therefore, God exists.

Interestingly, in no place in his speech does the pope refer to specific Catholic doctrines or to the Bible. While he talks of "cosmic developments" that point to the "beginning of time some five billion years ago," he does not mention that Genesis implies a much more recent creation—on the order of 10,000 years. He also does not mention the doctrine of the immutability of species unmistakably present in Genesis.

Of course, that's the advantage of being pope. It is he who determines the official doctrines of the Catholic Church, not the Bible. Poor Protestants. They do not have available to them the same continuous line of authority going back to Christ and must rely on an obviously flawed book of myths thousands of years old.

Even so, in 1951 there still was no convincing evidence for the big bang, so it wasn't proof of anything. As historian Helge Kragh points out, "At the time the pope gave his presentation of cosmology, the field was not characterized by harmonious agreement, but on the contrary, by a fierce controversy." The pope's message was actually quite misleading, leaving the impression among laypersons that "the biblical Genesis had literally been proved by big-bang cosmology." Lemaître knew this was not the case and managed to get the pope to slightly moderate his views in later speeches.[22]

THE INFLATIONARY UNIVERSE

With the discovery of the cosmic microwave background in 1964, and the many other advances in both observational and theoretical cosmology that have continued into the twenty-first century, we can now describe the evolution of the universe in remarkably quantitative detail. During its first tiny fraction of a second, the universe underwent an exponential expansion called *inflation*, during which the universe increased in size by many orders of magnitude.

We can now plausibly explain the production of the first elementary particles and how they combined to give the lightest chemical elements as the universe cooled. Thanks to Hoyle and collaborators, we also understand how the heavier elements are formed in the death throes of stars.

We can also now describe how the first stars and galaxies formed as gravity brought atomic matter together in clumps, aided by a still unidentified but clearly present component in the universe termed *dark matter*. All of these theoretical accounts are based on the data from every type of telescope in space and on Earth, as well as on the exquisite knowledge of elementary particles we now have from almost a century of experimentation with particle accelerators.

I don't have the space to provide the details of all these recent scientific developments and will stick to the theological implications of our current picture of the cosmos. Most Christian apologists are content with inflationary big-bang cosmology and continue to argue, as did Pope Pius XII, that it conforms to the modern Christian view of a created cosmos. However, as I have already noted, the big bang bears little resemblance to the cosmology presented in the scriptures.

The best that theologians can make out of big-bang cosmology is that our universe has a beginning and so a divine creation is not ruled out—but not proved. A number of scenarios for a purely natural origin of the universe can be found in the published literature.[23] These scenarios have been worked out with mathematical rigor and are consistent with our best knowledge of physics and cosmology, showing that a divine creation is not required by either the data or theory.

CHRISTIANITY AND THE NEW COSMOS

Let's now ask how the Christian faith fits in with our current scientific understanding of the cosmos, which far exceeds that of just a few decades ago. The universe visible from Earth contains an estimated 150 billion galaxies, each galaxy containing on the order of a hundred billion stars. It is known to be 13.8 billion years old, with an uncertainty of less than 100 million years. The most distant object we can in principle see is now 46 billion light-years away, taking into account the expansion of the universe during the time the photons carrying its image traveled to our telescopes. This marks a horizon beyond which we cannot see because light has not had the time to reach us in the age of the universe.

Well-established inflationary cosmology implies that on the other side of our horizon lies a region far larger than our visible universe, which arose from the same primordial seed.

And that's just our personal universe. Besides that, there may be an eternal multiverse containing an unlimited number of other universes. However, for the purposes of this discussion I will ignore the possibility of other universes, which is still debatable, and just stick with the one that indisputably exists—our own.

So, let us consider two possibilities that are so far consistent with our best knowledge:

1. Intelligent life exists on only a single planet, Earth.
2. Life is rare, and intelligent life even rarer. But the universe is so vast that there still are countless numbers of intelligent beings in our universe.

1. We Are Alone

Christians are told that they are the special creation of a lone divinity that created everything that is. While making sure every photon and electron in the universe behaves properly without us noticing, their God is also listening to their every thought and helping them to do the right thing—as he defines it.

This involves God controlling momentous events, such as telling a president of the United States to go to war, as George W. Bush said God did,[24] or redirecting a tennis ball off a Christian lady's racquet so it wins her the point as she shouts, "Thank you, Jesus!"

Of course, a God of limitless power could do all that. By why would he have waited until just 150,000 years ago to create humans? And why would he have confined them to a tiny speck of dust in a vast ocean of space, with no chance, at least with our present physical makeup, of ever traveling much beyond the environs of Earth? If he desired the worship of humans so badly, then you would think he should have made it available to him for all times and from all places.

Apologists make an illogical argument that they think is the clincher for the existence of God. They claim that life in the universe depends very sensitively on the value of a large number of physical parameters. According to this view, since the specific values of the parameters needed for life cannot possibly have resulted from chance (they can't prove that), they must have been "fine-tuned" by God in order to produce us.

Surely any God worthy of the name would not have been so incompetent as to build a vast, out-of-tune universe and then have to delicately twiddle all these knobs so that a single planet is capable of producing human beings. It would have made a lot more sense for him to have enabled us to live anywhere in the universe, even outer space. But the fact is—he did not.

2. We Are Not Alone

New estimates based on the brightness variations of thousands of stars measured by the Kepler Space Telescope suggest that within our horizon there may be as many as 5×10^{21} planets capable of sustaining biological life of some form.[25] We have seen that beyond our horizon exists a far vaster region of space. It is estimated that this region that contains a *minimum* of twenty-three orders of magnitude as many galaxies as those inside our horizon but is likely to be much bigger.[26] This gives at least 10^{44} possible habitable planets in our universe.

This is not pure speculation. It is based on observations and the theory of cosmic inflation that is now empirically well established. So even if the probability of intelligent life for an otherwise habitable planet is miniscule, say one part in a trillion-trillion (10^{24}), that leaves 10^{20} planets with some form of intelligent life.

So, let us consider the scenario of a Christian God presiding over a universe containing a multitude of intelligent life-forms, all created in his image (being unlimited, he has an unlimited number of images) living on other planets. Of course, the Christian could simply say that this adds to the magnificence of her God. But the God of Christianity does not hold a patent on magnificence. The gods of all the major religions are equally magnificent. Christianity is more than the worship of an infinite being. It is the worship of a highly personal God who so loved the world that he sent his only begotten son there to die an excruciating death on the cross to atone for the sin of the first human beings eating from the Tree of Knowledge.

In *The Age of Reason*, Thomas Paine wrote, "Are we to suppose that every world in the boundless creation had an Eve, an apple, a serpent, and a redeemer?" Of course, Paine was unaware how vast this "boundless creation" really is. Jesus must be continually dying on the cross, every nanosecond or so, on some planet in our universe, in order to save from eternal damnation every form of life that evolved sufficient intelligence to eat from the Tree of Knowledge.

In the 1930s, astronomer Edward Milne suggested how multiple crucifixions across the universe could be avoided by Jesus dying just once, here on Earth, and the information of his act of atonement then transmitted by radio to other civilizations. However, it would take well over 46 billion years for a signal sent from Earth to reach every planet currently within our horizon, and even longer to reach those beyond. So most intelligent beings would have to wait an awful long time to learn of the Atonement.

While I know of no official policy on this question provided by the Catholic Church or any Protestant denomination, the Vatican is not unaware of the problem. Speaking unofficially, in 2008 the director of the Vatican Observatory, Fr. José Gabriel Funes, told *L'Osservatore Romano*, the Vatican newspaper, that there may be other intelligent beings created by God. However, he suggested, "They could have remained in full friendship with the creator." He likened humans to "lost sheep" for which "God became man in Jesus to save us." The other intelligent beings did not necessarily need redemption. (The fact that none have communicated with us yet could indicate that they avoided eating from the Tree of Knowledge and, as a result, never developed radio.) According to Funes, "Jesus became man once and for all. The Incarnation is a single and unique event."[27]

So, if Funes is accurately reflecting Catholic thinking, humans remain the favorite of God—although a strange kind of favorite who needs redeeming while the countless other intelligent life-forms do not.

My guess, based on what one hears from pulpits across America, is that Protestants would take the same view. Of course, fundamentalists still believe literally in the cosmology of the Bible and think scientists are a bunch of frauds, so they have no self-contradiction here. Science is just wrong. There is just one universe created six thousand years ago (and no evolution or climate change, which are just "hoaxes").

Moderate Protestants, on the other hand, have yet one more conflict between science and religion among the many that they must reconcile in order to both accept the findings of science and also still hold on to some semblance of Christian faith.

The Judaic-Christian-Islamic God is a mighty God when viewed from the perspective of the desert tribes in the Middle East that conceived him. But that God is not mighty enough from the perspective of modern science.

Religion claims to teach us humility. But it really trades in a false pride, telling people they are the children of God, that they are the center of the universe and the reason for its existence—that they'll live forever if they just follow instructions. But it's an unearned pride. So, when science shows that we, for a brief time, occupy but a tiny mote in space and time, the religious recoil from this lesson in humility.

Still, the very fact that in the short period of a few thousand years humans have been able learn so much about the universe by just looking up in the sky and at the world around them, and reasoning over what they saw, testifies that we are unique among the millions of species on this planet. A dispassionate alien observer of life on Earth could only conclude that it is meaningless, brutal, and short—with one possible exception.

We cannot yet compare ourselves with whatever intelligent life-forms might be out there. But we are special, at least on Earth and the solar system. Even though magical thinking and hubris may still destroy us, we can hope that our unique abilities will lead us to a better future.

Chapter 5

BEFORE THE BIG BANG

Phil Halper and Ali Nayeri

An ancient argument for the existence of God, the Kalam argument, has been revived in modern times by master debater William Lane Craig. The argument states:

1. whatever begins to exist has a cause.
2. the universe began to exist.
3. therefore the universe has a cause: God.

But did the universe really begin to exist, or was there a *before the Big Bang*? Only a thorough understanding of gravity can resolve this ultimate question.

In J. J. Abram's *Star Trek* reboot, a Romulan destroys the planet Vulcan. But actually it was Albert Einstein that wiped it from existence. In reality, Vulcan was hypothesized to account for anomalies in the orbit of Mercury based on Newton's theory of gravity. But Einstein's new theory of gravity, General Relativity (GR), gave corrections that could account for these anomalies without supposing the existence of the hypothetical planet.[1] It was the first of many signs of the supremacy and importance of GR. In 1922, the Russian cosmologist Alexander Friedman suggested that GR implied a dynamic universe that is either contracting or expanding.[2] But Friedman died in 1925, mostly ignored. In 1927, Georges Lemaître developed a similar idea, implying the universe originated from a "primeval atom."[3] In modern terms, the initial singularity is where space-time curvature, pressure, temperature, and density are all infinity. Continuing the evolution of the universe before the singularity is impossible, so it represents the beginning of time. Two years later, Edwin Hubble seemed to find evidence consistent with Lemaître's model.[4]

But these early singularity theorems had the implausible assumption of a perfectly symmetrical distribution of matter.[5] Lemaître correctly predicted Hubble's discoveries but incorrectly predicted that cosmic rays were from the early universe.[6] Hubble's observation showed a relationship between the speed of galaxies' recession and their distance. Using this, it was possible to work out when they were all on top of each other. Unfortunately the data implied the universe began 1.8 billion years ago,[7] a date impossible to take seriously. This became known as the "age crisis." It's no surprise then that Hubble refused to endorse any cosmological interpretation of his data, and Einstein stated in 1945, "one may not conclude that 'the beginning of the expansion' should be a singularity."[8] So an alternative to the primeval atom was developed: the steady state. Here, new matter is created in the void between receding galaxies; the universe is expanding but has no beginning.[9]

One Christian apologist stated, "Consider a number of reactions from atheists, as they encountered the evidence of the Big Bang for the first time. For instance, in 1931, Arthur Eddington wrote, '. . . the notion of a beginning is repugnant to me.'"[10] But Eddington was no atheist; he was a fervent Quaker, and his religious beliefs almost landed him in jail during World War I.[11] Bernard Lovell, the first director of the great Jodrell Bank telescope was also a Christian steady statesman.[12] Philip Quinn, a Catholic philosopher, argued that the steady state backed the notion that God creates the universe not in one instant but continuously.[13]

William Lane Craig, however, claims, "The steady state theory never secured a single piece of experimental verification, its appeal was purely metaphysical."[14] But the primeval atom appeared to predict an age of the universe that was younger than the Earth; that alone was a good scientific reason to consider the steady state. Craig says, "Typically, atheists have said that the universe is just eternal."[15] But this view was held both by believers and nonbelievers, from Newton to Einstein. It's not even clear that the Bible describes a universe created from nothing. The Hebrew words *Shamayim* (heavens) and *Eretz* (Earth) are claimed by Craig to be a Hebrew idiom for "universe," arguing that the Bible uniquely predicts creation from nothing (ex nihilo).[16] But a more literal translation is "sky" and "land."[17] Hebrew scholars could not agree that God created ex nihilo; it seems he moves over the face of the waters before light is created. According to the Midrash (a classic text of early rabbinical commentaries), there are many things created before the world, including the Torah, two thousand years prior.[18] On the other hand, the Kiowa Apache creation story, as described in 1907, states, "There was a time when nothing existed to form the universe, no earth, no sky, and no sun or moon to break the monotony of the illimitable darkness,"[19]casting doubt on Craig's claim on uniqueness.

In the 1960s, Roger Penrose and Stephen Hawking developed new singularity theorems that did away with the many unrealistic assumptions of their predecessors.[20]They did assume energy conditions that imply gravity is always attractive and that GR is the right theory of gravity. New evidence developed by George Gamov and Ralph Alpher showed the Big Bang model correctly predicted the abundance of light chemical elements.[21] These scientists were all nonbelievers, and they were key authors of the modern Big Bang theory. Meanwhile, William Fowler and Fred Hoyle were able to model the origin of heavier elements in stars. More evidence, such as the CMB (cosmic microwave background) and the distribution of galaxies, mounted, and the "age crisis" was getting closer to resolution as Hubble's data was revised.[22] So it's fair to say that by 1979, when Craig published his Kalam argument, the Big Bang had defeated the steady state.

In the same year, Alan Guth and Henry Tye were looking at the production of exotic particles known as monopoles in the early universe. Theorists believed that because of early phase transitions the Big Bang should generate them in abundance, but none were to be found. The two suggested that a particular type of phase transition known as super cooling could prevent their formation. Guth found this led to a brief stupendous growth spurt, an inflationary era where the observable universe doubled in size every 10^{-37} seconds.

Inflation solves a number of other troubling problems for the standard Big Bang cosmology in a single stroke.[23] The first is known as the (particle) horizon problem: distant regions of the universe are at the same temperature despite the fact they could not have been in causal contact with each other in the past. Inflation solves this by its rapid stretching of smaller regions that were in causal contact. Another difficulty is the origin of structure problem: galaxies and stars form from gravitational instabilities so, if there was a uniform expansion of space, where did the instabilities come from? Inflation amplifies quantum fluctuations and turns them into the seeds of galaxies.

The last problem is known as the flatness problem. Flatness in this context refers to whether hypothetical triangles in space have angles that sum to 180 degrees. In curved space they do not. The universe was known to be approximately flat, but this was unstable; as time goes by, the universe should move away from flatness, so it would have to start with incredibly fine-tuned flatness from the beginning. Theists had quoted this fine-tuning as evidence of God.[24]Guth found that the equations behind this instability depend on gravity being attractive, but in inflation the vacuum energy provides a repulsive force that turns these equations upside down. Inflation drives the universe toward flatness regardless of its prior state. He wrote, "Spectacular realization: This kind of super cooling can explain why the universe today is so incredibly flat—and therefore resolve the fine-tuning paradox."[25] But the repulsive field associated with inflation, the inflaton field, violates the energy condition of the Penrose-Hawking theorem.[26] Perhaps inflation happens before the Big Bang and removes the singularity?

To understand when inflation happened, it is important to define the concept of the Big Bang. One definition is to say the Big Bang is the theory that the universe evolved from a hot dense state. Another is to say that the universe evolved from an initial singularity. If we use the singularity definition, then inflation happens after the Big Bang. But if we use the hot dense definition, inflation must have happened before the Big Bang because the universe is very cold during inflation. At the end of inflation, the inflaton field dumps its energy, creating a hot soup of matter and radiation. Hence Guth describes inflation as a prequel to the Big Bang.[27]

The decay of the inflaton field, like any other decay process, is characterized by a half-life. But what is happening to the remaining field that has not decayed? It's exponentially expanding, so the total amount of inflaton field need never decline. Inflation then seems to make not one Big Bang but an infinite number of them. This is known as eternal inflation.[28]

Inflation radically changes our conventional picture of cosmology. It violates the singularity theorems and, according to Alex Vilenkin, the Big Bang "is no longer a one-time event in our past: multiple bangs went off before it in remote parts of the universe, and countless others will erupt elsewhere in the future."[29] Perhaps eternal inflation is only eternal into the future and not the past. That is the conclusion of the famous BGV (Borde, Guth, and Vilenkin) paper.[30] The paper doesn't say the universe (they mean the multiverse) must have a beginning, but that inflation alone cannot avoid the initial singularity. Vilenkin has gone further, claiming it as a proof of the cosmic beginning.[31] Craig and the Big Bang theists seized on this to show that even if there is a multiverse it must have a beginning.

But many professional cosmologists do not agree. Most prominent of these is Leonard Susskind, one of the pioneers of string theory. Susskind argues, "an inflating universe which is future-eternal must also be past-eternal."[32] Even if there was a first Big Bang, it will be infinitely more likely that any observer will be in its infinite future, so any hypothetical first Big Bang must be infinitely far into the past. Anthony Aguirre also criticized the conclusion, constructing a model of inflation that he cheekily called "steady state eternal inflation."[33] Aguirre was later quoted by Craig to support his case for an absolute beginning,[34] but in fact Aguirre was refuting the past eternal status of another model known as "the emergent universe."[35] That doesn't imply he supports a past finite universe, though; Aguirre argues the universe may have no beginning[36]. According to Chris Wetterich, in the earliest epochs of our universe particles have no mass and so time has to be redefined, and the result of that process is an eternal past.[37] Yasunori Nomura, a collaborator with Alan Guth, states, "By requiring that the principles of quantum mechanics are universally valid and that physical predictions do not depend on the reference frame one chooses to describe the multiverse, we find that the multiverse state must be static—in particular, the multiverse does not have a beginning or end."[38] Sean Carroll, another Guth collaborator, proposes the quantum eternity theorem, which argues that the rules of quantum mechanics require an eternal universe as long its net energy is not zero.[39]

Alan Guth stated in 2015, "I don't know whether the universe had a beginning. I suspect the universe didn't have a beginning. It's very likely eternal but nobody knows."[40] Vilenkin himself seems to have backed off the idea that the BGV proves a beginning, conceding that theorem can be violated. But he thinks such violations are unrealistic.[41] Science resolves disagreements amongst theorists by putting them to experimental test. But Guth has said such a hypothetical beginning would "be completely washed out by the eternal evolution of the universe. Thus, there would be no way of relating the properties of the ultimate origin to anything that we might observe in today's universe."[42] So, unlike the conventional Big Bang, our cosmic origins, according to inflation, is unlikely to ever be confirmed by experiment.

Craig originally used the Penrose-Hawking theorem to show the universe had a beginning, but when inflation showed the theorems were vulnerable he latched onto another singularity theorem published in 1994.[43] But that too was vulnerable[44] so, since 2003, he has used the BGV theorem. But, as we have seen, theorists don't agree this theorem proves an absolute beginning, and experiments cannot confirm this.

It's ironic that Craig uses the works of inflationary theorists; responding to the claims of Guth that fine-tuning is alleviated by inflation, Craig stated, "Yet inflationary models are extremely speculative. They rest on so-called Grand Unification Theories (GUTs), themselves speculative. No positive observational evidence establishes that the universe underwent an inflationary phase. In fact, inflationary models predict a universe possessing critical density, whereas observation supports a value ten times lower than that. . . . And no inflationary model has yet succeeded in starting and stopping inflation so as to allow for galaxy formation. Most importantly, inflationary models require the same fine-tuning which some theorists had hoped to eliminate via such models."[45] Craig made this statement in 1996, and in 2011 he repeated the allegation that there was no evidence for inflation.[46]

Craig was incorrect. Revisions to the critical density of the universe necessitated by the discovery of dark energy implied that inflation does match observations.[47] Theorists Lawrence Krauss and Michael Turner had actually predicted this.[48] Secondly, NASA's CMB satellite, WMAP, provided "compelling evidence that the large-scale fluctuations are slightly more intense than the small-scale ones, a subtle prediction of many inflation models."[49] This is only one of many inflationary predictions and, in 2009, *Nature* stated, "Inflation has passed every observational test to date."[50] Ironically though, when an experiment called BICEP 2 mistakenly implied they had found definitive evidence for inflation, Craig argued it confirmed his views of creation all along.[51]

Inflation may require fine-tuning in GR, but in 2011 quantum gravity theorists claimed that fine-tuning is removed in that framework.[52] Craig's claim that inflation cannot lead to galaxy formation is one we find bizarre, as even some of inflation's most vocal critics, such as Roger Penrose, accept that is the theory's greatest strength.[53] Craig claims inflation rests on speculative GUTs, but Guth disagrees, saying that inflation does not require them.[54] However, it is true to say the underlying physics of the inflaton field is unknown. Without this understanding, it's possible that inflation has been misunderstood; perhaps it's not really eternal, or maybe it didn't even happen. But so far the consensus from observational cosmology is that inflation did happen and from theoretical work that it is eternal. However, the final test for inflation will be the search for the correct spectrum of gravitational waves, and here some argue the theory will fail. There are alternatives to inflation; could they restore the classical Big Bang picture?

Andy Albrecht and Paul Steinhardt are two of the early pioneers of inflation. But in recent years they have sought alternatives, with Steinhardt being particularly critical of the theory he helped to found. Albrecht teamed up with the self-styled "bad boy of cosmology," João Magueijo. The two examined a proposal that Einstein had toyed with but rejected, that the speed of light (c) is not always constant. If it were much higher in the early universe, this would mimic a period of inflation, providing an alternative solution to the horizon problem.[55] They also claimed a varying speed of light (VSL) could answer the other problems of the standard Big Bang and have a profound additional consequence. If c dramatically changed, then the law of conservation of energy could be violated and the Big Bang would be the result. In VSL, dark energy is converted into matter, and in the far future, when dark energy dominates the universe, the same process will begin again, creating a cyclic universe. As Magueijo says, "this process goes on forever in an eternal sequence of Big Bangs."[56] Small variations in something called the fine structure constant have been observed.[57] This is dependent on the speed of light, so this could be a sign of VSL. The results remain on the edge of detectability, and we shall have to wait for more data to see if the claims are robust. If they are, there are profound consequences; not only might a cyclic universe be confirmed, but the laws of physics might be shown to vary through space and time.

While Albrecht worked on VSL, Steinhardt has developed multiple inflation alternatives. A recent proposal known as the Higgs Bang is inspired by the discovery of the Higgs boson.[58] One might expect the Higgs field to exist in its lowest possible energy state. But the value of the Higgs discovered at CERN implies that the vacuum might be in what's known as a metastable state. According to Fermilab physicist Joseph Lykken, "It may be that the universe we live in is inherently unstable."[59] When the vacuum eventually decays, it will release a huge amount of energy. Our Big Bang may be a bubble formed from this type of violent event. A brief antigravity phase is also envisioned, which mimics inflation and creates a cyclic scenario. The authors claim the model is geodesically complete to the past, despite the BGV theorem. A geodesic is an analogue to a straight line in curved space-time, traced out by a particle; if the model is geodesically complete to the past, then the universe has no beginning.

Roger Penrose, one of the authors of the classic singularity theorem, no longer believes it applies to our universe. His new model is based upon the concept of conformal invariance.[60] The angles of a triangle sum to the same number irrespective of size, so they are conformally invariant. In relativity, the length of an object and the passage of time depend on its speed: traveling at c, length goes to zero and the conventional passage of time is lost. Penrose notes that if the universe is filled with particles that have no mass, then they all travel at c, which means space-time itself becomes conformally invariant. This enables the singularity to be smoothed out, implying a pre–Big Bang existence. If the universe, driven by dark energy, expands to infinity, then, Penrose argues, the same condition will exist in our far future. So the universe undergoes a "rescaling." Our Big Bang era is just one "aeon" in a cyclic universe that he claims need not have any beginning.[61] The dark energy we see today plays the role of inflation in the next aeon. Penrose controversially claimed to see signs of the prior aeon in the CMB.[62] This was widely dismissed[63] so CERN's Krzysztof Meissner conducted a more robust study and found a similar signal.[64] But with Penrose having cried wolf once, few were prepared to listen again.

One might expect Craig to try and point out some of Penrose's difficult assumptions, but instead Craig described the proposal as "a multiverse model in which you have, so to speak, twin expanding universes coming out of a common origin point. So you do not have one universe chronologically prior to the other; rather they both share a common beginning before which there is not anything, and then you have a sort of branching or multiverse structure."[65] Penrose himself said that Craig's description is "very inaccurate." It is not a multiverse model, it doesn't have a branching structure, nor two universes coming from nothing, and the signature predicted by the model depends on the aeons being chronologically prior. Penrose thinks it has no beginning and is cyclic.[66] The clue, of course, is in the model's name: Conformal Cyclic Cosmology.

Other alternatives to inflation include the matter bounce scenario,[67] where there is a period of contraction driven by exotic matter that does not obey the singularity energy conditions. Another is suggested by Nikodem Popławski,[68] who claims black holes give birth to new universes. This harks back to Lee Smolin's proposal of "cosmological natural selection."[69] In this scenario, each universe born of a black hole has a different value for the constants of nature. Those universes that make more black holes will be "naturally selected" as they make more new universes in turn.

Alternatives to inflation, then, do not seem to restore the classical Big Bang picture. But what of the other assumption in the singularity theorem—that gravity is described by GR? If the Big Bang is true, then the entire observable universe was smaller than an atom. In this regime, the strange effects of QM (quantum mechanics) will become impossible to ignore. In QM, particles can be in a "superposition" of states, and quantities are described by probabilities rather than having definite numbers as they do in GR. So a quantum theory of gravity is required if we are to understand the Big Bang. When inflation was proposed in the early 1980s, the leading candidates for quantum gravity didn't exist. But some cosmologists attempted to apply quantum principles to the beginning of the universe anyway. Such approaches are known as semi-classical.

In 1982, Alex Vilenkin[70] adapted a proposal from Edward Tryon.[71] In QM, particles can pop in and out of existence from the vacuum, the lifetime of the fluctuation being related to its energy. But gravity is negative energy, and so perhaps in our universe this negative energy balances the positive energy of matter and the entire universe could be a fluctuation from the vacuum. But where did the vacuum come from? Vilenkin suggests that perhaps if space-time is treated quantum mechanically it too can fluctuate into existence from "literally nothing," then inflation can take over and generate the multiverse.

A year after Vilenkin's proposal, Stephen Hawking and James Hartle[72] argued the initial singularity could be removed and turned into a smooth surface by introducing a concept in mathematics known as imaginary numbers (e.g., the square root of -1). In the Hawking-Hartle model, time becomes imaginary at the Big Bang. There is a beginning, but only in the imaginary time coordinate. In the real-time direction, the universe may be eternal and might even feature a bounce.[73]

Another semi-classical approach comes from creating a quantum version of the equation at the heart of the singularity theorem, the Raychaudhuri equation. Ahmed Farag Ali and Saurya Das did this in 2015[74] and argued that a repulsive force is created, preventing the formation of singularities and ensuring the universe is eternal into the past and future. This is a conclusion not shared by Aron Wall, who applied something known as the Generalized Second Law (applying thermodynamics to GR) to conclude that a singularity is still present at the Big Bang.[75] Both papers argue their results may hold in full quantum gravity. Craig used Wall's paper in a debate with Sean Carroll to imply there is still a beginning in quantum gravity.[76] But neither scientist can demonstrate their results hold on full quantum gravity. Wall admitted, "This is quantum gravity, so none of us really know what we're talking about!"[77]

But none of the above scenarios were developed from a genuine quantum gravity theory. The two leading candidates for this are loop quantum gravity (LQG) and string theory. LQG was developed in the 1980s based on a mathematical breakthrough made by Abhay Ashtekar, who would later be elected president of the Society for General Relativity. Ashtekar found a way to reformulate GR in quantum mechanical form. The new theory implied that space was made of loops that are the gravitational equivalent of magnetic force lines. It took twenty years of development before loop quantum cosmology (LQC) could be formed. In LQC, space cannot become infinitely dense, so when density reaches its limit, gravity becomes repulsive.[78] This acts much like a sponge, which will absorb water poured on it, and then, once full, it switches from becoming absorbent to repulsive. This repulsive force implies the Big Bang is really a big bounce from a prior contraction. This differs from the matter bounce we mentioned earlier because being driven entirely by quantum geometry it does not require any exotic matter.

Thousands of papers have been published on LQC.[79] One of the most intriguing results shows that the probability of inflation starting is almost 1:1, whereas in GR it may be $1:10^{85}$.[80] Another shows that fluctuations in the CMB are compatible with the bounce picture.[81] Even if it's true that inflation does not avoid the singularity, the LQC bounce does. Vilenkin has admitted the BGV theorem can be violated by prior contraction but thought it would lead to "messy singularities."[82] In LQC, singularities are removed. Previous bouncing models had to avoid a type of instability near singularities known as BKL chaos; even Craig admitted, "This has been shown to be calmed by a loop quantum approach." However, he also stated that LQC "seems to be ruled out by accumulation of dark energy, which would in time bring an end to the cycling behaviour."[83] Just as Craig was wrong to state CCC was not a cyclic model, he's wrong to associate LQC with cyclic behavior.

Much of Craig's critique of LQC focuses on a cyclic version of LQC. But mainstream LQC does not predict cyclic behavior. (That would have to rely on dark energy being a variable, which has nothing to do with LQC.) Instead, the universe is seen as an hourglass with an expanding and contracting phase mirroring each other, joined by a quantum bridge at the big bounce. In one debate, Craig claimed that even if LQC was true it still implies a beginning.[84] But, since singularities are removed, it's hard to see a justification for this statement.

One possibility to detect the bounce is in distortions to the polarization of the CMB.[85] Light polarization happens when wave vibrations occur in a single plane. One type of polarization is known as the B mode, and finding it could give us access to information about incredibly early states and reveal a bounce signature. Recently, astronomers have been seeing incredibly short but powerful Fast Radio Bursts (FRB) in intergalactic space. The source objects are thought to be only a few hundred kilometers in size. The origin of these FRBs is a mystery. Loop theorists have recently analyzed black hole singularities and found these also bounce. If there were primordial black holes in the early universe, we would see these now bouncing, giving off short intense radio bursts.[86] Are FRBs bouncing black holes? The exciting explanation in physics is often not the right one, so we remain cautious. The important point is not that this is the signature of quantum gravity (it's far more likely they are from astrophysical sources) but that bounce signatures are in principle possible and that scientists are working hard to look for them.

String theory is the most dominant approach to quantum gravity. It works by replacing point particles by strings that can be open or closed. Closed strings were shown to have the properties of the long sought after graviton. Other vibrational modes of strings represent other matter and force carrying particles. So string theory can claim to be a TOE (theory of everything) unlike LQG, which is concerned only with quantizing gravity. However string theory equations imply extra compact dimensions and super symmetric particles to work.

Just as the inventor of LQG has pioneered attempts to use the theory for cosmology, so the first string theorist, Gabriele Veneziano, has initiated string cosmology in a model he calls the Pre–Big Bang (PBB).[87] String theory has a rich mathematical structure, and a key feature for cosmology is called T-duality. T-duality is similar to conformal invariance such that in the presence of extra dimensions, large and small are dual solutions to the same equations. One way to describe the expansion of space is to say that a function called the scale factor is increasing over time. A higher scale factor represents more distance between distant galaxies. At the classical Big Bang, the scale factor is zero. But in the PBB, the scale factor cannot become zero due to the fact that strings have a minimum size. The PBB posits a duality for the scale factor so that on the other side of the bang is a dual universe: ours is expanding to infinity; the dual is contracting from infinity. The PBB can make similar predictions to inflation but without producing B mode polarization, so if these are detected the PBB will have to be modified or abandoned.[88] Craig agrees that the model evades the BGV theorem but thinks that the infinite past is a mathematical artefact.[89] Needless to say, Veneziano does not agree and wrote an article in the *Scientific American* describing "the myth of the beginning of time."[90]

Another model inspired by string theory is the Ekpyrotic model.[91] Here, the Big Bang is the result of a collision between postulated higher dimensional objects called branes. We agree with Craig that the model is vulnerable to the BGV theorem, and it's not clear that it can resolve the singularity. However, Craig is wrong to say that the PBB and Ekpyrotic are the only two models of string cosmology to aspire to an infinite past.[92] One of the authors of this chapter has worked on a promising approach known as String Gas Cosmology(SGC).

SGC was initially proposed in 1989 by Brandenberger and Vafa.[93] It is a scenario of the evolution of the very early universe based on basic principles of superstring theory. Specifically, SGC is a scenario that makes use of the new degrees of freedom and new symmetries that distinguish particle-based theories from string theory. String theory is much richer than that of elementary point particle theories. Strings have a tower of oscillatory modes whose number increases exponentially with energy. In addition, there are new quantum "winding modes," which can be viewed as characterizing the number of times that a string can wind a compact spatial section. The existence of the many new string degrees of freedom, will lead to a very different thermodynamic behavior of the very early universe. If we imagine a gas of closed strings in thermal equilibrium and start this box off with a large radius R, then most of the energy will be of a type known as "momentum modes." As the box slowly shrinks in size, the energy of the momentum modes increases (since it scales as $1:R$), so the temperature increases. However, as the temperature approaches a critical value called the "*Hagedorn* temperature," then the increasing energy density will lead to the production of oscillatory modes, rather than to a further increase in the energy of the momentum modes (and hence to an increase in temperature). The implication of this is that the temperature of a gas of closed strings cannot become infinite as in the classical singularity. In fact, if the size of space decreases below the string scale, then the winding modes become light, the energy of the system will flow into those modes, and the temperature will decrease. There is in fact an important stringy symmetry: physics at large values of R is equivalent to physics at small R. This is one aspect of the T-duality symmetry of string theory. Later, it was shown that the presence of this extra degree of freedom, a field known as the *dilaton*, makes the dynamical equations that govern the evolution of the universe to be T-dual as well.[94] That is, there is a dual universe to (our) expanding universe, which is contracting. The two universes meet at the so-called self-dual point. After 2005, SGC was further developed to accommodate the origin of structure formation and gravitational waves. In the Nayeri-Brandenberger-Vafa setup,[95] the mechanism that leads to the generation of matter fluctuations and gravitational waves in string gas cosmology is completely different than one can find in other models of early universe. First, during the Hagedorn phase, the state of matter is a dense thermal gas and not a matter vacuum state as in the case of inflation. Hence, in string gas cosmology, the fluctuations (the seeds for galaxy formation) are thermal in origin. Second, matter is given by a gas of closed strings and not by a gas of point particles. Hence it is string thermodynamics and not point particle thermodynamics that determines the spectrum of the fluctuations that are generated.

SGC and the PBB model both imply a contracting dual universe on the other side of the Big Bang phase and no singularity. But they are not identical. In cosmology today we observe there is a nearly scale invariant fluctuation in temperature of the CMB. In order for the PBB to generate this, one needs to invoke another field known as the curvaton.[96] But in SGC they are determined by the specific heat capacity of a gas of closed strings. The overall amplitude of the spectrum is given by the ratio of the Planck length to the string length, and it is thus also a prediction from string theory. SGC predicts a feature called red tilt in these temperature fluctuations. This means the energy decreases with scale as happens in inflation. However, in SGC there is induced a blue tilt in the gravity wave spectrum (i.e., the strength of the signal increases at smaller scales). Inflation predicts a red tilt in the gravity-wave spectrum, which could be inferred from B mode polarisation.[97] It is also distinct from the Ekpyrotic and PBB models. In the former, gravity waves are suppressed, so we do not expect to see primordial gravity waves or B modes. In the latter, there is some suppression such that we might see high frequency gravity waves, but there are not B modes.[98] So measuring not only the presence (or lack of presence) of gravity waves but also their spectrum could be the key to early universe cosmology. Another attractive feature of SGC is that it only works if three of the compact spatial dimensions in string theory become large, hence explaining why we don't see the other hidden dimensions.[99]

There are many approaches to quantum gravity other than the dominant loops and string. But to our knowledge, the only other one that has been applied to cosmology is Horava-Lifshitz gravity. In this model, the link between space and time is cut at the high energies of the Big Bang.[100] Yet again, this implies a bounce.[101] What we find remarkable is that in multiple independent approaches to quantum gravity the bouncing solution is found. This doesn't mean we know this happened, and these scenarios could still be wrong. But it does make the bounce a very strong candidate for our cosmic origins. In science, theorists can show ideas are promising, but it takes robust observations to turn them into facts. All statements about the very early universe are speculative, including both bounces and singularities. We (the authors) see no prospect of an observational signature of a singularity at the Big Bang. The same is not true for the bounce.

A further argument used by Craig centers around entropy. This can be roughly thought of as a measure of disorder, although a more technical definition should be related to the number of micro states that are associated with a given macro state. The most entropic objects in the universe are black holes; if you could scramble up their constituents they would always look the same. Entropy is overwhelmingly more likely to increase into the future, which means it must have been lower in the past. In the far future, the universe will be in a cold diffuse state, with no usable energy, a maximal entropy state. So if the universe has an eternal past, why aren't we already in this state? Why was the entropy so low at the Big Bang?

Alan Guth has a simple solution: "We don't know if the maximum possible entropy is finite or infinite, so let's assume it's infinite. Then no matter what entropy the universe started with the entropy would be low compared to its maximum. . . . [A]n interesting feature of this picture is that the universe need not have a beginning."[102]

Guth is likely to be thinking in terms of eternal inflation; new pocket universes are always being born, so the multiverse as a whole will never come into thermal equilibrium, even though each pocket will. In the Carroll/Chen model the entropy of the universe is always increasing but the entropy density isn't.[103] Small pockets of low entropy can emerge, and these can give rise to baby universes that bud off from the parent universe. The low entropy baby universes do not decrease the entropy of the whole multiverse, which always increases. Ashtekar argues that before the bounce a horizon develops that engulfs the collapsing universe. Unlike some models, the entropy is not decreasing close to the bounce point, it's increasing. But this horizon disappears after the bounce—"this is a mathematical fact about LQC"—and so "one is led to reset the entropy clock."[104]

As entropy is about counting macro and microstates, it can be argued to be an observer-dependent effect.[105] As no observer can ever go through the bounce, one only ever observes rising entropy. Penrose points out that the vast majority of the entropy of the universe is tied up in black holes. When black holes evaporate, they take their entropy with them, and so a cyclic universe is possible for eternity.[106] The issue of whether information (and entropy) is destroyed in black holes or not, is known as the "information paradox." At the moment, there is no definitive way to solve this paradox, but Penrose's view is a minority among physicists. In the Higgs

Bang model, the bubble that forms the new universe is empty of matter and radiation and so does not inherit the entropy of the prior state.[107]

A deeper question is where does the arrow of time come from? The laws of physics are time reversible (processes described by physical laws such as billiard balls colliding can be described in both directions of time). Yet we see a world with a definite arrow of time, we remember the past and not the future. One solution is that the arrow of time can emerge in an eternal universe from dynamical processes, but the arrow may have two directions pointing away from a middle point. A team led by Julian Barbour[108] has simulated particles interacting under gravity, and during these simulations there would always come a point when the particles clump together. The point of maximal clumping can be thought of as analogues to the Big Bang and its minimum entropy. But Barbour's simulations run in both directions of time. So an observer on the other side of our Big Bang also sees entropy increasing. Guth's model seems simpler, and without gravitational interactions, but the results are similar[109]. Craig has argued that models that have this bi-directional time still have a beginning.[110] They may have a thermodynamic beginning, but they don't have a geometrical beginning. The universe exists for all points in time. There is no creation ex nihilo. As Sean Carroll said: "It is hard to express the extent to which I think this is grasping at straws. The axis for time goes from the top to the bottom and it goes forever. The only sense in which this universe is not eternal is that there is a moment in the middle where the entropy is lowest. . . . [T]hat has nothing to do with the kind of beginning you would need to give God room to work."[111]

There is no shortage of ideas from some of the worlds' leading physicists about how to construct an eternal universe consistent with the second law of thermodynamics. In fact, some of them may even be able to explain why we have an arrow of time or why we appear to have a low entropy condition in the early universe, rather than just assuming them.

Craig argues that it's impossible to have an actual infinite, hence the past must be finite.[112] He points out many paradoxes of the infinite, but these are due to using the wrong mathematics, treating infinity as a number—which it is not. Craig, however, uses singularity theorems to provide the physical basis for a beginning.[113] Giving up the infinite past by embracing the singularity involves swapping one infinity (the past) with four more (pressure, density, curvature, temperature).[114] If one cannot transverse the infinite, how is it possible to go from infinite density at the singularity to the finite density we observe today? One escape clause might be to state that the singularities are only potential infinities (an unbounded sequence) rather than actual infinites, but in cosmology the universe arrives at the singularity in a finite amount of time, hence John Barrow states, "These infinities, if they do exist, would be actual infinities."[115]

Craig often quotes the great mathematician David Hilbert as saying that the infinite is nowhere to be found in nature.[116] Hilbert's justification was that "*Euclidean* geometry necessarily leads to the postulate that space is infinite. Although Euclidean geometry is indeed a consistent conceptual system, it does not thereby follow that Euclidean geometry actually holds in reality. . . . Einstein has shown that Euclidean geometry must be abandoned. . . . [A]ll the results of astronomy are perfectly compatible with the postulate that the universe is elliptical."[117]

But Hilbert was writing in 1925, before the geometry of space (not to be confused with space-time) had been properly measured. In the twenty-first century, these measurements have finally been performed, and we can now say that Hilbert was wrong. The results of astronomy show no convincing deviation from a flat Euclidean space.[118] This doesn't prove the universe is infinite, but it proves that Hilbert was wrong to exclude the possibility based on astronomical observations. Cosmologists do not usually rule out infinities as being unreal. Alan Guth states ,"In an eternally inflating universe, anything that can happen will happen; in fact, it will happen an infinite number of times."[119] What they do object to is infinities for observable quantities such as the energy of a collision. Penrose points out that only massless particles survive for eternity. They do not feel the passage of time, so there is no problem for them to sail straight through infinity.[120]

Another problem that springs from relativity is that there is no unique sense of "now" that the universe has to get to. Craig admits this and so has to rely on a nonstandard approach to relativity known as the Lorentzian interpretation. This implies an "A" theory of time, which restores a unique sense of "now," unlike the "B" theory, which denies it.[121] But this model of time is the least popular among professional philosophers[122] and has been severely criticized by historians of relativity as being incompatible with Einstein.[123] Craig's interpretation of relativity implies FTL (faster than light) travel is possible even without wormholes or extra dimensions. When FTL neutrinos were thought to be detected, Craig claimed "the triumph of Lorentz."[124] But hopes of vindication were dashed, as the result was found to be due to faulty wiring.[125]

It is not clear how Craig can make his views on FTL compatible with the BGV theorem he relies upon. Here is Vilenkin's description: "A space traveler has just zoomed by the earth at the speed of 100,000 kilometers per second and is now headed toward a distant galaxy, about a billion light years away. That galaxy is moving away from us at a speed of 20,000 kilometers per second, so when the space traveler catches up with it, the observers there will see him moving at 80,000 kilometers per second. If the velocity of the space traveler relative to the spectators gets smaller and smaller into the future, then it follows that his velocity should get larger and larger as we follow his history into the past. In the limit, his velocity should get arbitrarily close to the speed of light."[126] But if the speed of light can be exceeded, then how can the theory be applied?

Craig is a professor at Biola University, which has a doctrine statement saying those who reject Christ "shall be raised from the dead and throughout eternity exist in the state of conscious, unutterable, endless torment of anguish."[127] This might be considered only a potential infinite except that God knows the future, in which case he can count an infinite amount of moments that nonbelievers will be in anguish from. So it seems Craig's own doctrine leads to an actual infinity.

Cosmology does not tell us the universe had a beginning, a fact pointed out even by many Christian academics, such as Princeton philosopher of science Hans Halvorson[128] and Oxford cosmologist John Barrow.[129] Don Page, an evangelical Christian who earned his PhD in cosmology under Stephen Hawking, wrote an open letter to Craig: "We simply do not know whether or not our universe had a beginning. . . . I myself have also favored a bounce model."[130] Craig's response is that science doesn't require us to "know" in the sense of certainty, only that it be more plausible than not.[131] But the combined probability of two statements with 51 percent probability is only 26 percent. The standard in a criminal trial, for example, is not certainty, nor is it more than 50 percent probability; it is evidence beyond reasonable doubt. Otherwise, jails might have 49 percent of their inmates guilty. The same is true in science; statements must be shown to be true beyond a reasonable doubt. A finite universe is far from being established beyond reasonable doubt.

But if the universe did begin, does it need a cause? Causes happen before effects. But if the universe has a beginning, then there was no before. If there is no before, no arrow of time, nor laws of physics, then how can we demand that causality still exists? Craig claims that causes can be simultaneous with their effects and gives Kant's example of "a metal ball continually causing a depression in a soft material."[132] But in contemporary physics, all forces are carried by particles that move at a finite speed, which includes the force that creates the depression, hence there is no simultaneous causality here. A variation considers that the ball has been there from eternity past. But if that is allowed, then the Kalam is defeated. If causes are not prior to their effects, the very structure of causality can be questioned. Some scientists have even suggested that closed timelike curves allowed in relativity enabled the universe to create itself.[133] So denying the prior nature of causality threatens claims of theistic creation. How would one distinguish God creating the Big Bang versus the Big Bang creating God?

Some cosmologists have suggested QM allows the universe to be created from nothing. Craig has little time for such proponents: "In these models of the universe, the universe comes into being out of the vacuum; it doesn't come into being from nothing. . . . [T]o tell lay people that in this case something comes from nothing is simply a distortion of these theories and, as I say, an abuse of science."[134] Ironically, it is the cosmologist that Craig seems to quote the most often, Vilenkin, who argues for this; his 1982 paper opens with, "A cosmological model is proposed in which the universe is created by quantum tunneling from literally nothing."[135] Vilenkin explains, "If there was nothing before the universe popped out, then what could have caused the tunneling? Remarkably, the answer is that no cause is required . . . the behavior of physical objects is inherently unpredictable and some quantum processes have no cause at all."[136]

Vilenkin's model is not a fluctuation from the vacuum that requires the prior existence of space-time. It's a fluctuation of space-time itself. Craig complains that some interpretations of QM are not indeterministic but some are. Since we can't currently experimentally probe the right interpretation, it's not clear that causality is an essential property of the universe. Craig claims the vacuum isn't nothing, but, in that case, what observation can ever confirm something can't come from nothing? If creation ex nihilo is impossible, why does it suddenly become possible for a disembodied mind? Do we even need to assume a past finite universe came from nothing? This implies a prior state the universe came from. But if there is simply a first state of time, perhaps there is no need to postulate a universe popping into existence at all.

But what if there is a beginning and a cause? Why must it be God? In Vilenkin's model, the universe owes its existence to physical laws. Craig assumes that such abstract objects as physical laws have no causal powers. Here, he's smuggling in a particular philosophical assumption that physical laws are simply descriptions or properties of objects rather than commands objects obey. Alan Guth seems to takes the opposite view, "If laws are just properties of objects . . . how can those laws continue to operate when the object is not really there?"[137]

The alternative cause Craig proposes is a disembodied mind. But neuroscience implies minds come from the activity of physical brains.[138] No mind ever created from nothing; no mind ever acted without a physical body. We have no good reason to believe disembodied minds exist. Not so for the laws of physics. Craig has said a prefrontal cortex is essential for self-awareness.[139] So how could this exist without material reality? If there is a mind behind the universe, why should it be that of a perfect being that answers prayers?

In conclusion, we cannot say the universe had a beginning; the singularity may be only as real as the planet Vulcan. But even if it did begin, we cannot say it must have a cause. And if it did have a beginning and a cause, we see no reason to think that the cause is a perfect being that loves us and answers our prayers.

Chapter 6

INTELLIGENT DESIGN ISN'T SCIENCE, AND IT DOESN'T EVEN TRY TO BE SCIENCE

Abby Hafer

THE *WEDGE STRATEGY* AND THE DISCOVERY INSTITUTE'S PLAN TO DESTROY SCIENCE

The *Wedge Strategy*[1] is a document that its writers wish you didn't know about. In fact, it was marked "Top Secret" and "Not for Distribution." The only reason we know about it is because it was leaked to the Internet in 1999. Written in 1998, the *Wedge Strategy* is probably the first document in history to propose defeating the very idea of science by using a public relations campaign.

The people who wrote this document are from the Discovery Institute (DI), a Seattle-based organization that has received a great deal of funding from multimillionaire Howard F. Ahmanson Jr. Other wealthy conservative and religious entities also contribute to the Discovery Institute.

Why did they write the *Wedge Strategy*? Here's a quote from Mr. Ahmanson himself:

My goal is the total integration of biblical law into our lives.[2]

The *Wedge Strategy* (or Wedge document) outlines how the Discovery Institute plans to split science and rationality away from American culture

141

and use intelligent design (ID) as the "thin end of the wedge" to start the process. The reason for this is that the people at the Discovery Institute hate science. In the introduction to the *Wedge Strategy*, they say:

> The proposition that human beings are created in the image of God is one of the bedrock principles on which Western civilization was built.

and

> Yet a little over a century ago, this cardinal idea came under wholesale attack by intellectuals drawing on the discoveries of modern science.

In other words, they say that western civilization itself is being attacked by modern science. Here's a quick reminder from me: western civilization *invented* modern science.

They further believe that this has caused endless harm. The introduction to the *Wedge Strategy* also says,

> The cultural consequences of this triumph of materialism were devastating.

Keep in mind that when they say "materialism," they mean finding out about the material world using material means. In other words, science. Their basic message is that the world has gone to moral hell, and it's science's fault.

So their goal is simple: they want to destroy science. Not only do they not like science themselves, they don't want anybody to use it as a means of finding out facts about the world. Here are the goals set out in the Wedge document:

Governing Goals

- To defeat scientific materialism and its destructive moral, cultural and political legacies.
- To replace materialistic explanations with the theistic understanding that nature and human beings are created by God.

The Discovery Institute invented intelligent design in order to use it to destroy science. Here is a quote from the "Five Year Strategic Plan Summary" section of the *Wedge Strategy*:

> Design theory promises to reverse the stifling dominance of the materialist worldview, and to replace it with a science consonant with Christian and theistic convictions.

In short, the Discovery Institute wants children to be taught intelligent design in science classes in order to help destroy the idea of science itself.

How Does This Fit in with Creationism?

The Discovery Institute refers to intelligent design in public as though it were different from creationism. Don't be fooled. Intelligent design was a strategic rebranding of creationism, done when creationism was declared to be religious doctrine by the Supreme Court in 1987.[3]

This is made clear in a famous textbook example, which was literally found in a textbook. The textbook was *Of Pandas and People*, a creationist book that the Discovery Institute was developing and hoping to market as a biology text to public schools. When the Supreme Court declared creationism to be religion, its promoters realized that their so-called biology textbook wasn't going to sell very well. So they invented intelligent design. Then, they changed their textbook in just one way: they changed the word "creationism" to "intelligent design," and the word "creationists" to "design proponents."

But they didn't do a very good job. In one case, they put in the words "design proponents" but didn't completely remove the word "creationists." The result was the tell-tale phrase "cdesign proponentsists," which was found in one of the drafts of the supposedly nonreligious "intelligent design" textbook.[4] This is one proof among many that intelligent design is nothing more than a strategic rebranding of creationism and, ironically, an excellent example of a transitional species.

Intelligent Design's Success

It's clear from the Wedge document that the Discovery Institute is using intelligent design as a means of starting its campaign to overthrow science. It wants to control what you think and what your children learn. It also wants to be the one providing the unscientific school materials to your children. It intends to get laws passed in order to allow it to do that.

It has been too successful. Intelligent design is now an accepted "alternative" to evolution in public-school science classes in Louisiana and Tennessee. The Discovery Institute helped the legislatures in both those states write the legislation that made this possible.[5] In 2014, Missouri,[6] Ohio,[7] Oklahoma,[8] South Dakota,[9] and Virginia[10] all had bills promoting Intelligent Design and Creationism introduced in their state legislatures.

What's worse, this drivel is being exported to the rest of the world.[11]

Intelligent Design's Perverted Little Problem

The Discovery Institute is antipathetic to science. However, its promoters have a perverted little problem. Even while they have contempt for legitimate science, they must *pretend* that intelligent design is itself legitimate science. Why? Because it has to be treated that way to be taught in science classes in American public schools. So the Discovery Institute must persuade people that ID is science, even while they privately hold science in contempt and blame it for many of the ills of the world.

If intelligent design is rightly identified as a religious explanation rather than a scientific one, then it can't be taught in American public schools, because the Constitution, as consistently interpreted by the Supreme Court, forbids the promotion of any particular religion by the government. Since public schools are ultimately run by the government, this means that if intelligent design is exposed as an unscientific religious idea, teaching it would be limited to private schools outside the government's purview.

Since promoters of ID desperately want to indoctrinate all young people at the earliest age possible, they want ID to be taught in all schools. If ID can only be taught in private, church-affiliated schools or Sunday schools, then the churches and their followers will have to pay for it. But if ID can be *wedged* into the public-school science curriculum, then it will be taught at taxpayer expense to everyone's kids. This is why the "wedge" metaphor is so chillingly apt.

ID's Claims Are Unsuccessful Scientifically, but Successful Politically

ID's claims of scientific legitimacy have been widely refuted by biologists. These claims were also rejected by Judge John E. Jones in the famous *Kitzmiller* case.[12] However, proponents of ID have continued to insist that ID is science. And they continue to have success in persuading legislators, school board members, teachers, and others involved in decisions regarding science education, both in the United States and elsewhere in the world. Therefore, ID cannot be considered a spent force.

Articles Published by the Discovery Institute Itself
Form the Basis for Its Claim That ID Is Science

To bolster their claim of being legitimate science, ID's proponents point to the many ostensibly scientific articles and books that have been published on the subject. Based on these articles, they insist that the question of how biological organisms and species came into existence remains open.

Although articles promoting ID rarely appear in peer-reviewed scientific journals,[13] proponents of ID still claim that many of their articles are "peer-reviewed." Articles promoting ID often claim to present original research. However, they rarely contain descriptions of the methods by which any research was done, or present any experimental and/or quantitative results. The explanation of scientific methods that I will give in this chapter will make it clear why this is important.

If ID is actually scientifically valid, then it should not be rejected simply because it is ideologically odd. However, for it to even approach scientific validity, a central question must be answered in the affirmative: has ID produced any quantified results as evidence to reinforce its hypothesis? In other words, do they have any data, or do they just spend their time arguing?

This is the question that I addressed in my research.

But first, some explanation of how science is done is in order.

HOW SCIENCE IS DONE—SOME BASICS

Science is simply a method of investigation. We use it to obtain a better understanding of reality. It involves a set of rules and methods that are needed to move forward in this process. In this section, I will describe a few of these rules and methods.

The Roles of Measurement and Quantification in Science

Measurement and quantification—putting numbers to things—is crucial to the scientific process.[14] Although scientific investigations in a given field may start with careful *observational* studies, measurement and quantification are expected to follow not long thereafter if the field is to be productive.

The Example of Research on AIDS

How AIDS came to be understood is an excellent example of how observations lead to quantification, which leads to fruitful research. The study of Acquired Immune Deficiency Syndrome (AIDS) began in 1981 with the observation that five gay men in the Los Angeles area had rare lung infections and generally weakened immune systems. By the end of that year, 270 gay men were reported with severe immune deficiencies. Careful quantitative studies were what revealed how AIDS is transmitted, identified the Human Immunodeficiency Virus (HIV) as its cause, led to commercial diagnostic tests, and, by 2000, produced drugs that keep people with the disease alive for many years.[15] At this point, we can estimate how many people have died when the drugs have been refused.[16]

Thus, in twenty years, AIDS went from being a deadly mystery to being preventable and largely treatable. From a scientific standpoint, it went from being an *observation* that some homosexual men had unusual lung and immune problems, to being a thoroughly documented, understood, and largely treatable disease, all because quantitative procedures were used to test hypotheses about the observations.

How Does This Compare to Intelligent Design Research?

The idea of the intelligent design of biological organisms was formally presented in 1802 by William Paley in his book *Natural Theology*,[17], though he did not use the exact phrase "intelligent design." More recently, it was proposed in 1984 in *The Mystery of Life's Origin: Reassessing Current Theories* by Charles B. Thaxton, Walter L. Bradley, and Roger L. Olsen.[18] The idea was further promoted in the school-level biology textbook *Of Pandas and People*,[19] which was edited by Dr. Thaxton and first published in 1989. If something is being presented in a science textbook meant for school children, it is reasonable to assume that the field is well developed.

ID has been around for more than two centuries and actively pursued for at least twenty-five years. This is enough time for a field of scientific enquiry to pass beyond the initial observational stages, and progress into hypothesis testing and quantification.

However, it has been the experience of this author that ID articles do not read the way that articles in biological journals do. In a biological journal article, a subject is introduced, a hypothesis is often described, the methods by which an investigation was done are explained, the results are given, and conclusions are expressed. ID articles frequently seem to this author to be all introduction. Particular viewpoints are described and then argued for or against. This goes on for many pages, with new information generally not being introduced. The articles seem to be largely argument, and they generally do not even have a results section.

Other Requirements in Scientific Research

In addition to careful observation and quantification, science makes further demands on its investigators. Here are some of them:[20]

Hypotheses, Predictions, and Falsifiability

At some point in the development of a given field, hypotheses must be produced. Further, predictions based on these hypotheses, that are testable at least in principle, must be produced. An idea must be testable, at least in principle, in order to qualify as science. If an idea cannot be tested under any circumstances, then it falls outside of the realm of science.

Further, it must be possible to test the hypothesis in a way that would prove the hypothesis wrong. This is called falsifiability.

Reproducibility

It must also be possible for other researchers to get the same results if they carefully use the same methods of investigation. This is called reproducibility. For results to be accepted, they must be reproducible.

No Supernatural Explanations

Supernatural explanations are not allowed in science. This is the rule that has allowed science to advance in such astounding ways.

It may seem reasonable to invoke a deity at times, but this is a nonproductive idea. Our knowledge of the natural world has progressed specifically because science has rejected supernatural explanations for phenomena in favor of natural ones. Verifiable explanations based on reality have then been found.

If supernatural explanations of natural phenomena had been considered sufficient, further research leading to productive, natural answers would never have been done. For instance, referring back to the history of AIDS, if scientists had accepted that the wrath of a deity was the cause of AIDS, we would never have discovered the HIV virus, the real cause of AIDS. Finding the real cause was *productive*: it led to effective prevention and treatment.

Intelligent Design, Supernatural Explanations, and Predictability

ID breaks a number of the rules outlined above.

First, and obviously, it resorts to supernatural explanations. That, in fact, is entirely what ID is—a supernatural explanation for how species came into being, in contrast to the nonsupernatural explanation of evolution by natural selection.

Second, by resorting to supernatural explanations, it would appear that ID makes prediction impossible. Dr. William A. Dembski, a Senior Fellow at the Discovery Institute's Center for Science and Culture, has written as much:

Yes, Intelligent Design concedes predictability.[21]

Intelligent Design, Quantification, and Experimentation

Despite admissions by its proponents that ID is a supernatural explanation and concedes predictability, these proponents still insist that ID is science and should be taught in public schools as such. They base this extraordinary claim on written work by ID proponents, which they say is scientific research. Much of the work sounds scientific to the untrained reader, since it addresses scientific and technical subjects and uses scientific terminology.

So, supposing that supernatural explanations are possible, how would a legitimate scientist find this out? By doing careful observations and controlled experiments, in order to show that no explanation other than the supernatural one is possible. It is reasonable to expect that a field that insists its work be taught to school children in science classes and placed in science textbooks would have completed many controlled, quantified experiments to justify its extraordinary claims.

One would expect to see a plethora of experimental data, and a corresponding plethora of references to it.

MY RESEARCH

That expectation is what animated my research.

I should explain that scientists find it very easy to explain why intelligent design articles are not scientific, but they find it very difficult to do so in a quick and concise way. It is possible to do point-by-point analyses of ID articles, but this can become very tedious for most audiences. ID articles rarely have just one mistake in them. More often, even one article is so full of mistaken assumptions, misrepresentations of facts, inappropriate comparisons, the mistaking of similes for facts, erroneous logic, and inept conclusions that pinpointing all of them and explaining how they are wrong can be very slow reading. When this is multiplied by the sheer number of ID articles that have been written, this is far more than any general reader wishes to take on. This includes the many general readers who wish to support the causes of teaching evolution and good science education.

With this in mind, I decided to try a more concise approach. I analyzed all Discovery Institute papers that claim to be scientific and peer-reviewed to see if they contained that necessary element of scientific research articles: data.

My question was: Do intelligent design authors have any data, or do they just spend their time arguing? Given my descriptions of how science works, you can see why this question is so important. This analysis had to be done in a careful and dispassionate way. To this end, I downloaded ID articles and simply counted how often the word *data* appears in them.

By contrast, argumentation in scientific papers, though present, is a fairly minor element. Different schools of thought may be mentioned in the introduction to a research paper, but only as a prelude, as a means of explaining why a particular investigation took place. The rest of the paper will describe methods, results, and conclusions. So I also looked at ID articles to see how much argumentation they contained.

Because I was going to investigate a large body of work, it seemed wise to set limits on the investigation. For this reason, I chose to search ID articles for two words only: *data* and the word-root *argu* (covering argument, argues, argumentation, and so on). I understand that it is not necessary to use the word *data* in order to have data. For instance, the word *results* may be used instead. Nonetheless, from many years of experience, I know that scientific articles often use the word *data* when referring to the results obtained. Likewise, words formed from the word-root *argu* generally describe differences of opinion. Limiting the scope of this investigation to the words *data* and *argu* seemed like the most fruitful approach.

The Need for a Control Group

The Discovery Institute claims that its articles promoting ID are scientific articles. In order to properly assess this, it is necessary to compare them to articles already known to be scientific articles. For this reason, I needed a control group of articles from an established source of original scientific research articles.

The Control Group

The fairest comparison, as I saw it, would be to compare the articles from the Discovery Institute to articles from another institute, one already known for its high-quality scientific research.

For this, I selected a large set of articles from the Smithsonian Tropical Research Institute (STRI). The STRI is known for doing high-quality biological research, and its articles were all available online. An entire section of the STRI website is devoted to research in evolutionary biology. These were the articles that I used as the control group in my investigation. The articles that I obtained from STRI were also searched for *data* and *argu*.

My Hypothesis

To recap the problem so far: the articles and books that ID writers write can sound scientific. They use scientific-sounding words. ID writers often claim to do scientific research. So the central question is: are they really doing science?

I addressed this question by looking at the supposedly peer-reviewed or peer-edited scientific articles by ID writers and asked, "Do they rely primarily on data, or on argumentation?"

Scientific investigations rely on data. Reliance on argumentation may be acceptable in other fields of scholarship, such as philosophy, but this is insufficient in science. If you cannot eventually produce data, then science is not what you are doing.

My hypothesis was that the ID writers at the Discovery Institute do not work with data.

My Prediction

This meant that there should be very few references to data in their writing.

I therefore predicted that they would use the word *data* only rarely and the word-root *argu* more often.

By contrast, I predicted that the articles by the legitimate scientific researchers at STRI would do the reverse.

Research Methods: How Were the Articles Obtained?

To do such a study impartially, it is necessary to set rules for how the articles will be selected prior to looking at them. This way, the decision to use an article was based on prior rules, not on whether I liked the article.

I laid out specific methods for accessing the articles, based on the structures of the websites for STRI and the Discovery Institute as they existed in August 2010. This impartial means of selecting articles meant that I could make an unbiased comparison between articles from the two sources. All the articles for this study were downloaded between August 9 and 11, 2010. Only articles in English were used.

Obtaining Intelligent Design Articles

The Discovery Institute is the leading institutional proponent of ID. It has provided the language for successful antievolution legislation. During my study, an entire section of the Discovery Institute's website was devoted to allegedly scientific articles about ID produced by authors with whom the Discovery Institute was associated. For these reasons, ID articles were all obtained from the website of the Discovery Institute.

My exact procedure:

1. I used Google to search for the phrase "Discovery Institute" and clicked on the Discovery Institute's website (Discovery Institute, 2010).
2. There, I clicked on the section labeled "Science and Culture," which brought me to the Center for Science and Culture website. (The Center for Science and Culture is a Discovery Institute organization.)
3. There, I found the heading "Scientific Research and Scholarship," under which was a subheading that read,
4. "Peer-Reviewed & Peer-Edited Scientific Publications Supporting the Theory of Intelligent Design (Annotated)."

This was followed on the site by a defense of its somewhat unusual designation:

> Editors' Note: Critics of intelligent design often claim that design advocates don't publish their work in appropriate scientific literature. For example, Barbara Forrest, a philosophy professor at Southeastern Louisiana University, was quoted in *USA Today* (March 25, 2005) that design theorists "aren't published because they don't have scientific data."
>
> Other critics have made the more specific claim that design advocates do not publish their works in peer-reviewed scientific journals—as if such journals represented the only avenue of legitimate scientific publication. In fact, scientists routinely publish their work in peer-reviewed scientific journals, in peer-

reviewed scientific books, in scientific anthologies and conference proceedings (edited by their scientific peers), and in trade presses. Some of the most important and groundbreaking work in the history of science was first published not in scientific journal articles but in scientific books—including Copernicus' *De Revolutionibus,* Newton's *Principia,* and Darwin's *Origin of Species* (the latter of which was published in a prominent British trade press and was not peer-reviewed in the modern sense of the term). In any case, the scientists who advocate the theory of intelligent design have published their work in a variety of appropriate technical venues, including peer-reviewed scientific journals, peer-reviewed scientific books (some in mainstream university presses), trade presses, peer-edited scientific anthologies, peer-edited scientific conference proceedings and peer-reviewed philosophy of science journals and books. We provide below an annotated bibliography of technical publications of various kinds that support, develop or apply the theory of intelligent design. The articles are grouped according to the type of publication. The first section lists featured articles of various types which are of higher interest to readers, which is then followed by a complete list of the articles. The featured articles are therefore listed twice on this page (once in the featured articles section and again below in the complete list).

This paragraph was followed by the heading:

5. "Featured Articles"—the papers that the Discovery Institute itself claims are scientific research.

Since these were the articles that the DI claimed were its scientific articles, I took the DI at its word and used these for my research. This study was interested only in original scientific research, so I rejected review articles and books. However, chapters of books that the Discovery Institute claimed were original research were included.

All articles from this section of the Discovery Institute's website that met these criteria were used. In each case, the entire article was analyzed, including the abstract (if there was one) and the captions of tables and figures (if any). I did not include reference lists in the analysis.

Obtaining Evolutionary Biology Articles

The articles that formed the control group were all obtained from the website of the STRI. As I explained earlier, STRI is known to produce high-quality scientific research.

My exact procedure:

1. I used Google to search for the Smithsonian Institution (Smithsonian Institution, 2010), then
2. clicked on the Smithsonian's Home Page, then
3. "Research" at the top of the Home Page, then
4. "Tropical Research Institute (STRI)," then
5. "Programs" at the top of that page, and
6. "Evolution" on the left-hand side of that page.
    This led to a list, in alphabetical order, of STRI scientists who had published papers on evolutionary biology. From there, one could
7. click on each scientist's name, and a list of articles by that scientist became available.

The articles were listed in chronological order of publication, starting with the most recent. All were available for downloading. Only papers from STRI that the institute listed under the category of Evolution and that had been published in peer-reviewed scientific journals were considered. Of these, only original research articles were used. Reviews, review articles, and books were not included. Of the original research articles, four per author were used: the first four by each author that met these criteria (so that more prolific authors would not be overrepresented in the sample). The entire article was analyzed, including the abstract (if there was one) and the captions of tables and figures (if any). I did not include reference lists in the analysis.

Research Methods: Basic Analysis of All the Articles

I analyzed a total of 63,024 words in eleven articles from the Discovery Institute and 143,172 words in twenty-eight articles from STRI. In both cases, the articles were copied entirely and pasted into a separate document. After all the articles had been copied, with the Discovery Institute articles placed in one document and the STRI articles in a different one, the content analysis began. I used the "Find" tool in Microsoft Word to search each compilation of articles, first for *data* and then for *argu*. I went through each article personally, using the "Find" tool.

When counting the instances of the word *data*, all words with the root *data* were chosen when they referred to quantitative results, such as *data* and *database*; words containing *data* in reference to organisms, such as *chordata*, were excluded.

When counting instances of the word-root *argu*, all words containing the root *argu* were used, when they referred to persuasion and disputation. These included the words *argue*, *argument*, and *arguing*, for example.

Results

Among the 63,024 words from the Discovery Institute and 143,172 words from the Smithsonian Tropical Research Institute that were analyzed, the word-root *argu* was used 88 times in the articles from the Discovery Institute, but only 11 times in articles from STRI. By contrast, *data* was used 270 times in the STRI articles, but only 24 times in articles from the Discovery Institute. These results are shown in Table 1 and Figure 1.

	Argu	Data
Discovery Institute	88	24
STRI	11	270

Table 1. Numbers of times that the word-root *argu* and the word *data* occur in eleven articles from the Discovery Institute and twenty-eight articles from the Smithsonian Tropical Research Institute.

Note: The numbers in the upper left quadrant and the lower right quadrant are both very high, while the numbers in the upper right and lower left quadrants are very low; these strong diagonals are an indication of statistical significance.

Statistical Analysis

I performed a Pearson's chi-square analysis on these results. The chi-square test is appropriate for categorical data such as these. The sizes of the control and experimental populations do not have to be the same when the chi-square test is used.[22] The result of this analysis was extreme:

$$p < 1.9 \times 10^{-53}. \; (p < .00000000000000000000000000000000000000000000000000019)$$

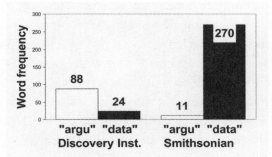

Figure 1. Numbers of times that the word-root *argu* and the word *data* occur in eleven articles from the Discovery Institute and twenty-eight articles from the Smithsonian Tropical Research Institute (STRI). Scientists at STRI made heavy use of data but rarely used or cited argumentation in their articles. By contrast, Discovery Institute writers rarely referred to data and never to testing a hypothesis, but they referred heavily to argumentation. The difference is quite significant: $p < 1.9 \times 10^{-53}$.

The p-value is a statistical measurement of how likely it is that an experiment's results are the result of simple random chance. The lower the p-value, the more convincing the results are. So a p-value as tiny as the one gotten through this analysis means that the results in this experiment are not a fluke.

Further Analysis

The above results alone seriously damage ID's claim to be a well-developed branch of science. The rarity of references to quantified data in these articles indicates a field that is either not well developed as a science or not a science at all. However, it seemed appropriate to take this investigation a step further and examine every instance in which Discovery Institute authors used the word data at all, to see whether this usage ever referred to testing a hypothesis. After all, a field of science well developed enough to merit inclusion in science textbooks and school instruction should be well developed enough to have done some hypothesis testing.

In essence, I was giving ID another chance. If any of the Discovery Institute's references to data led me to a report of active hypothesis testing, it would at least indicate that some writers at the Discovery Institute were attempting to do legitimate scientific research. One or two tested hypotheses would not alone make ID worthy of inclusion in biology textbooks, but it would at least indicate integrity on the part of the Discovery Institute writers who claimed that they were doing valid research.

Results of Further Analysis

I found that out of the twenty-four instances in Discovery Institute–published papers in which the word *data* was used, nineteen referred to data generated by other people, usually data on Cambrian fossils. This was not active scientific research. These articles simply talked about other people's work, without any predictions or testing being done by the authors of the ID articles.

Of the remaining five articles, four referred to data as a concept. That is, they talked about what data might look like if they had any. Finally, there was one paper that had original data, but there was no hypothesis testing. In short, in all of the purportedly peer-reviewed ID literature that the Discovery Institute had published, there was not a single instance of hypothesis testing.

Research Conclusion

Proponents of intelligent design claim that it is science. The results here, gleaned from their own writings, strongly contradict this claim. Yet the claim that ID is valid science is having a profound effect on how science is taught in this country, as well as overseas—for instance, in Brazil[23] and Turkey.[24] Legislatures, writers of state science standards, school boards, teachers, and parents are too often persuaded by the scientific-sounding language used by ID proponents.

The results presented here are strong evidence that ID cannot be considered a scientific discipline because it does not follow the basic requirements for scientific research. First, it is a supernatural explanation. Second, it cannot be used to make predictions. Third, it relies on argumentation rather than on data and the testing of hypotheses. Given these drawbacks, it is clear that ID writers at the Discovery Institute were not doing scientific research. It may perhaps be appropriate to consider their work scholarship, perhaps of a philosophical nature, but it is not science. Likewise, instruction regarding the idea of ID might be appropriate in a class on the history of ideas, but not in a science class.

It should be reiterated that the ID papers analyzed here are the ones the Discovery Institute specifically listed *as their scientific research papers*, not opinion papers. This makes it clear that the Discovery Institute does not readily distinguish between fact and argument.

These results should inform the debate as to whether or not ID should be included in a science curriculum.

Peer Review of the Results in This Chapter

The results listed in this chapter have themselves passed peer review. The original research on which this chapter is based was published in *The American Biology Teacher*—a peer-reviewed journal.[25]

DEFENSE OF TEACHING EVOLUTION

This study's results provide a straightforward defense for teaching evolution and not intelligent design.

In order to show that ID is not scientific, it is no longer necessary to refer to point-by-point refutations of ID articles, which can be wearisome for most general readers and listeners. Instead, a teacher, parent, school board member, or other concerned citizen can point to this study, which examines many of ID's said-to-be-scientific articles, and point out that ID isn't science.

There is very little data in articles by ID writers and lots of arguing, whereas true scientists do the reverse. A complete lack of hypothesis testing points to a field that is at best undeveloped, since ID proponents have had more than adequate time to progress to quantified studies and hypothesis testing. The fact that they haven't done this despite twenty-five years of constant arguing, publicity, and lobbying for their views indicates a far greater interest in arguing, publicity, and lobbying than in doing authentic scientific research.

The fact that ID proponents want their ideas to be a part of biology textbooks before they have produced a large body of quantitative research indicates that they either do not understand what science is or do not care.

This will be a useful study for teachers to know about. It will also be useful to school board members and those who deal with school boards; and to legislators and those who deal with them. All these individuals need useful quick references for dealing with those inclined to accept ID.

If schools and legislators want to insert ID into school curricula on the grounds that ID is science, then it is very important to establish that ID is *not* science. This is important both from the standpoint of teaching legitimate science in science classes and from the standpoint of constitutional separation of church and state.

Let's Mess with Texas

For many decades until recently, it was not even necessary for ID promoters to lobby school boards across the country in order to undermine the teaching of evolution. They only had to lobby the school board for the state of Texas. The state of Texas is the second-largest buyer of textbooks in the country, and it required local school boards to choose books from a specially-sanctioned list provided by the state school board. National textbook publishers usually gave in to the demands of the Texas school board because it was the most influential buyer in the country. For many years, the Texas state school board was more than happy to pretend that creationism and intelligent design were just as valid scientifically as evolution by natural selection.

The result was that school children throughout the United States had to read the drivel that the Texas board insisted be included, as well as being deprived of a good deal of valuable and correct information that the Texas board wouldn't tolerate. The teaching of evolution across the nation therefore suffered for generations.

Delightfully, local school boards in Texas are now allowed to choose their own books. What's more, new national science education standards have been created. As of the end of 2014, these had been adopted by twelve states and the District of Columbia, with more likely to join. This means that even if Texas never adopts the new science standards, the banded-together states that have will create a market for textbooks that is as big as, or even bigger than, the one provided by Texas.

Nor do textbook publishers need any longer to publish one edition for the entire nation. Digital media and book-manufacturing technology now allow textbook companies to tailor their books for different markets, so decisions made in Texas can finally stay in Texas.[26]

What this means, however, is that actions of non-Texas state and local school boards throughout the country are now more important than ever. If state legislators can be persuaded to adopt the national science standards, then children in that state are more likely to be given a fact-based, high-quality education in the roots of biology. The research presented in this chapter may help.

The "Strengths and Weaknesses" Loophole

One of the tactics that ID promoters at the Discovery Institute love to use is to insist that teachers and science curricula include discussions of the "strengths and weaknesses" of evolution. This has the advantage, from a legal standpoint, of not sounding overtly religious. But it can allow a teacher or textbook writer to introduce ID and creationism as "alternatives" to evolution, since it encourages teachers to discuss the "strengths and weaknesses" of evolution.

It has the additional tactical advantage that it is pretty easy for a school teacher to confuse his/her students and claim that there are weaknesses in evolution that don't exist. The Discovery Institute, as I have shown in this chapter, is very good at writing things that sound scientific but are factual rubbish. However, a school child cannot be expected to know facts from rubbish, especially if the rubbish is coming out of a teacher's mouth.

In fact, the reason why we go to school in the first place is to learn the knowledge that our ancestors have figured out and be able to proceed from there. If a school child is expected to figure out everything starting from first principles and be told no facts at all, then there is little point in going to school in the first place, which may well be what the Discovery Institute wants.

The Weakness of "Strengths and Weaknesses"

The Discovery Institute's argument about "strengths and weaknesses" has a major flaw, and it's this: If one is going to teach the "strengths and weaknesses" of evolution, then one should teach the strengths and weaknesses of intelligent design as well—and ID has nothing but weaknesses.

The research in this chapter shows that, from a scientific standpoint, ID has weaknesses and no strengths at all. This finding should be trumpeted any time that people feel the need to talk about the "strengths and weaknesses" of evolution.

Additionally, what about the strengths and weaknesses of Newton's laws of motion? The roundness of the earth? The periodic table of the elements on the wall of the chemistry lab? Entire semesters could just as justifiably be wasted confusing students with false claims of scientific weaknesses in any of these.

Since ID's only power rests in politics, it is important to know how to make arguments against it that are short, easy to remember, and easy to repeat. Just reference this chapter and say:

Scientific research has shown that none of intelligent design's papers have any science in them.

This is an insurmountable barrier for anyone claiming that ID is science, as an excuse for teaching it in science classes.

Intelligent Design Promoters' Fundamental Contempt for Science

At this point, it's important to circle back to my earlier discussion of the *Wedge Strategy*. In addition to breaking many of the basic rules that have made science the incredibly successful engine for acquiring real-world knowledge that it has been for the last four hundred years, the intelligent design promoters at the Discovery Institute have announced, via the *Wedge Strategy*, their contempt and even hatred for science itself.

Do not trust the "scientific research" that is done by people who demonstrate a fundamental contempt for the whole idea of scientific research. Would you take your car to a mechanic who doesn't believe in the internal combustion engine?

The Discovery Institute's promoters of intelligent design have said in their own internal document, the *Wedge Strategy*, that they think that science is evil. They disapprove of the scientific method as a method of investigation. In fact, they have sworn to defeat it. Here's a quote from the *Wedge Strategy*'s "Five-Year Strategic Plan Summary":

> We are convinced that in order to defeat materialism, we must cut it off at its source. That source is scientific materialism.

Remember the language here—materialism means relying on evidence from the material world. In other words, the real world. It means not settling for supernatural explanations for things in the real world, and instead finding material explanations that will be productive and can be built upon.

When the Discovery Institute says that it wants to cut off "scientific materialism," what it means is that it wants to prevent the use of the scientific method as a way of finding out about the material world—even though, or perhaps especially because, the scientific method has proven to be unparalleled at finding new information.

The Discovery Institute Willfully Forgets the Benefits of Science

So this is the Discovery Institute's true stance on science: They really think that individuals and society would be better off if we had never started down the path of doing scientific research.

Never mind that science told us how diseases can be transmitted through contaminated drinking water. Never mind that science gave us safe, modern surgical techniques, which mean that most people survive surgeries. Never mind that science has allowed childbirth to progress from something with a high chance of killing the mother and the baby to something that both almost always survive. The Discovery Institute thinks that we should never have figured out such things, because they were all figured out through materialistic scientific research.

I have not heard members of the Discovery Institute volunteering to drink water contaminated with untreated sewage or volunteering to have dental work done without novocaine, but this is apparently what they wish for the rest of us.

Over the past four hundred years, the methods of modern science have improved our lives so dramatically that today they are unrecognizably long, comfortable, healthy, and safe compared to life in the Middle Ages. The modern world has its problems, but life in the prescientific world was more miserable by far.

It's All about Power

In fact, it is probably science's breathtaking success that causes the people at the Discovery Institute to dislike it so much. Science has been so successful at explaining the world and at improving people's lives that they fear that their dogma-based worldview will be extinguished entirely, and their authority over other people will be extinguished along with it. As a result, they are fighting to defeat science itself and replace it with a world-view dominated by a God whose "messages" they themselves will control and dominate.

It's their way of making people obey them. This is because they probably believe, perhaps correctly, that if people learn to depend on science for their facts about the world, they will be unwilling to swallow the non-fact-based dogma over which the Discovery Institute has control. These religious zealots are not trying to benefit people, they are trying to have power over them. They want to start in science classrooms and end with all Americans being told by the Discovery Institute, and its spiritual compatriots, what they should think and do.

The Discovery Institute Doesn't Do Scientific Research, It Never Has, and It Doesn't Even Try

My research shows not a single instance of hypothesis testing in all the "scientific" ID literature put out by the Discovery Institute.

The *Wedge Strategy* announces that intelligent design proponents at the Discovery Institute disapprove of the scientific method. Therefore, it's not surprising that they don't use it.

All I have shown here is that these intelligent design promoters are true to their original words—they don't like science, don't approve of using the scientific method, don't want anyone to use it, *and* don't use it themselves.

This is very important to remember. They may pretend to do science, but their own words (which I counted) prove that they do not.

Part 3

SCIENCE AND SALVATION

Chapter 7

SAYING SAYONARA TO SIN

Robert M. Price and Edwin A. Suominen

"If the book of Genesis is an allegory," complains the creationist carnival barker Ken Ham, "then sin is an allegory, the Fall is an allegory, the need for a Savior is an allegory, and Adam is an allegory." He observes that those "who wish to mesh evolution with the Bible must accept that there was death, suffering, diseases like cancer, and even thorns before the Fall." But, he says, with his traditional reading of that Bible, "all those things are a result of sin! The Scripture is explicit that thorns came after the Curse." And thus he vocally rejects the reality of evolution, almost certainly not due to any careful weighing of evidence but because of his fear that it "destroys the foundation of all Christian doctrine."[1] As profoundly and disturbingly wrong as Ken Ham is about pretty much everything else pertaining to evolution, we find his summaries of its theological difficulties dead on.

Unfortunately for him, Adam was indeed an allegory—more precisely, a figure in an *ethnological myth* who symbolized all humanity, the root stock from which all others descended. The closest real equivalent was an African man living many tens of thousands of years ago, who was the ancestor to every male human now alive.[2] This individual is often referred to as "Y-chromosome Adam," but he lived in Africa, not Mesopotamia, and was nothing like the loner described in Genesis. He had companions, including male rivals who carried Y chromosomes of their own,[3] and hundreds of generations of ancestors who had been members of *Homo sapiens*, too.

The woman (or women) who carried this Adam's lucky Y chromosome to an endless succession of sons was no equivalent to Eve, either. Genetic Eve actually goes back much further, living somewhere around

170,000 years ago. Today, all humans, male and female, have mutated versions of this ancient African woman's mitochondrial DNA.[4]

She, of course, had plenty of ancestors, too. The fact is that there never was any first human, male or female, who was a member of *Homo sapiens* but had parents who were not. There was just a gradual transition to the cluster of attributes we now associate with our own species: walking upright, a large brain with an ample prefrontal cortex, the presence of a hyoid bone for vocalization, to name a few.

Some of our supposedly "human" traits actually go back a lot further than us, like the use of tools and fire. *Homo heidelbergensis* was making sophisticated stone hand axes and wooden spears hundreds of thousands of years before the first humans did.[5] And there were probably human-controlled fires burning in hearths by 500,000 years ago.[6]

Even if we could isolate a single, final mutation as the emergence of humanity, it is just not possible for the entire human race to have descended from a single father and mother. Genetic evidence now makes clear that there have never been fewer than about a thousand members of *Homo sapiens* throughout the more than 100,000 years of its existence.[7] Actually, that was well understood before scientists started looking into DNA: The "paleontological record thoroughly establishes that one population is always preceded by another, making the idea of a single pair of humans procreating an entire species unthinkable."[8]

Now, with our origins tracing back to a population of gradually evolving ancestors and the Tigris and Euphrates nowhere in sight, it's abundantly clear that there never was any Adam or Eve from whom we could have inherited our supposed taint of sin-corruption. Nor was there any Eden in which a Fall might have occurred. So what are we to make of this business about "sin," about a doctrine of some original sin that goes all the way back to Paul as *the* reason God had to send a savior on our behalf?

And what is really so sinful and worthy of eternal condemnation about our sexual, selfish, and aggressive urges, anyhow? Sure, we have collectively agreed on certain moral ideas now that we have come to live in large, fixed societies rather than just roaming bands of kin. But it is *those very traits* that allowed our ancestors to pass them on, with all the rest of our DNA heritage, in a harsh and competitive world. Jesus said the meek inherit the earth, but we haven't inherited much from them, genetically. Rather, it is the "disproportionate replicators" who left their mark on us, our forebears whose drive and passion got their DNA immortalized into children who would, with enough luck as well as drive and passion of their own, continue down the line. You won't find many celibate shrinking violets in your ancestry.

PAUL PROBLEMS

As the story goes, the disobedience of Adam—eating the fruit of the Tree of the Knowledge of Good and Evil—destroyed the virtue with which he was created. It introduced a taint of both guilt and depravity that Adam and Eve would pass on to their offspring and ultimately to the entire human race. It's a grotesque extension of the inherited guilt that Yahweh threatened against those who might dare break his law given on Sinai, "visiting the iniquity of the fathers upon the children, and upon the children's children, unto the third and to the fourth generation" (Exod. 34:7). At least that version seemed to have a statute of limitations for great-great-great grandchildren.

But there is nothing for us to be saved *from* without an Adam who fell and took the human race down with him. If it were simply a matter of a generous God forgiving our trespasses, as we forgive one another, there would be no need for the cross. Unless Adam introduced a fatal and virulent toxin into the human race, no such radical surgery as the atonement provides would be needful. It is not so much a case of making the punishment fit the crime as of making the crime fit the punishment, for otherwise "Christ died in vain" (Gal. 2:21).

Evolution-accepting theologians like Peter Enns, Ansfridus Hulsbosch, Jerry Korsmeyer, C. John Collins, Denis Lamoureux, Robin Collins, and Tatha Wiley avert their eyes and pretend to find some redeeming qualities to Christian teachings about sin—especially *Original* Sin—because they must. A fundamental aspect of their religion—the act of a universal sinner requiring a universal redeemer—is at stake, though some of them try to deny or minimize that. And, regardless of the modern efforts at explaining away an ancient and ill-informed doctrine, they still need to figure out what to do with Paul.

As the major Christian apostle, what Paul says about Adam and the origin of sin is even more important to Christian thinking than anything found in Genesis. And Paul has plenty to say (perhaps too much, given the trouble it causes!) about Adam's sin.

As is well known, Romans 5:12–19 draws a typological connection between the First and the Last Adam. Adam damned all, while Christ saves all, each through a pivotal deed with universal ramifications:

> Therefore, just as through one man sin entered into the world, and death through sin, and so death spread to all men, because all sinned—for until the Law sin was in the world, but sin is not imputed when there is no law. Nevertheless death reigned from Adam until Moses, even over those who had not sinned in the likeness of the offense of Adam, who is a type of Him who was to come.
>
> But the free gift is not like the transgression. For if by the transgression of the one the many died, much more did the grace of God and the gift by the grace of the one Man, Jesus Christ, abound to the many. The gift is not like that which came through the one who sinned; for on the one hand the judgment arose from one transgression resulting in condemnation, but on the other hand the free gift arose from many transgressions resulting in justification. For if by the transgression of the one, death reigned through the one, much more those who receive the abundance of grace and of the gift of righteousness will reign in life through the One, Jesus Christ.
>
> So then as through one transgression there resulted condemnation to all men, even so through one act of righteousness there resulted justification of life to all men. For as through the one man's disobedience the many were made sinners, even so through the obedience of the One the many will be made righteous. [Rom. 5:12–19, NASB]

The passage is repetitive, so much so in fact, that it is tempting to suspect it has suffered from scribal glosses as copyists added explanatory notes in the margins, notes that were subsequently mistakenly copied into the text the next time a new copy was needed. But, assuming this is all the work of a single author, we see that he did envision Adam's sinful deed introducing death, as each human being thereafter actualized his inherited sinfulness by his own acts of sin.

The phrase "because all sinned" by itself might imply a *Pelagian* understanding that each man is his own Adam, each woman her own Eve, starting out fresh but soon repeating the Fall.[9] But this seems to be ruled out by the later qualification that all were "made sinners" by the primordial disobedience of Adam. So it does seem fair to say that the Romans passage teaches at least the basic notion of Original Sin as a will-warping taint inherited from Adam.

Paul makes the same basic point in 1 Corinthians 15:21: "For since by man came death, by man came also the resurrection of the dead. For as in Adam all die, even so in Christ shall all be made alive." Augustine recognized this; the "statement which the apostle addresses to the Romans . . . tallies in sense with his words to the Corinthians." Paul intended no difference in meaning by referring to death instead of sin in the latter passage, says Augustine, since his discourse there was focused on resurrection rather than righteousness.[10]

Naturally all this creates a major problem once one realizes there was not only no Adam, but no First Man—no matter what his name was. It seems that *the very purpose* of Pauline Christianity is to solve a supposed problem that is rooted in myth rather than fact.

Protected from theological nuance by their ignorance and denial of the scientific proof for evolution, the fundamentalists have unwittingly become some of the best expositors of the problem, as we saw with Ken Ham's protest about allegory. Similarly, Albert Mohler of the Southern Baptist Convention has this to say:

> The denial of an historical Adam and Eve as the first parents of all humanity and the solitary first human pair severs the link between Adam and Christ which is so crucial to the Gospel. If we do not know how the story of the Gospel begins, then we do not know what that story means. Make no mistake: a false start to the story produces a false grasp of the Gospel.[11]

Luther took the story's theological significance for granted, steeped as he was in Paulinism and innocent of any plausible alternative to the idea of a first human pair. The Fall created the need for the savior:

> The Son of God had to become a sacrifice to achieve these things for us, to take away sin, to swallow up death, and to restore the lost obedience. These treasures we possess in Christ, but in hope. In this way Adam, Eve, and all who believe until the Last Day live and conquer by that hope.[12]

If some among the first myriad of humans sinned and others did not, would that mean that the Sethian Gnostics were right in believing in two human races, the "kingless race" of Seth and the sinful race of Cain? On the other hand, one might admit that a few rotten apples would provoke sin in the whole barrel, but that would be Pelagianism again, and it is not clear an atonement would be either necessary or effective in repairing the damage. So the Pauline problem is twofold. First, he was wrong about there having been a First Adam. Second, his ascribing to Adam a universal taint of sin running so deep as to require not merely "spot forgiveness," sin by sin, but radical metaphysical surgery through the cruciform atonement goes up in smoke. It follows directly as a conclusion: "Then Christ died in vain" (Gal. 2:21).

Faced with this dilemma, some of our Christian evolutionists perform a bit of radical surgery of their own. They are quite willing to admit Paul was wrong about Adam and to resort to "limited inerrancy," saying that Paul's assertion is correct (Christ *does* eradicate Original Sin), but the terms in which that assertion is cast are negligible. We are to see in the Romans 5 argument a parallel to Jesus' insignificant inaccuracy in the mustard seed parable: The mustard seed turns out not to be "the smallest seed on the earth" (Mark 4:31), but so what? How does that affect the point of the parable? It's small enough! But do we really have a true parallel here? Paul's appeal to Adam and his imagined sin is not some illustrative window dressing but rather the premise of an argument. How can we jettison the logical and scriptural arguments that led Paul to his theological conclusions yet retain his conclusions as if they transcended their supports?

Peter Enns claims that "Paul's handling of Adam is *hermeneutically* no different from what others were doing at the time: appropriating an ancient story to address pressing concerns of the moment. That has no bearing whatsoever on the truth of the gospel."[13] But how can it not? If we kick away the whole Adam-Christ typology as mere first-century trappings, why did Christ have to die? The atonement of Christ is not only left hanging in midair, but it is even evacuated of all meaning: What is it he supposedly did, then, besides get executed?

And what a telling choice of words when Enns refers to "what remains of Paul's theology." Why draw a magic circle around what's left? Why not acknowledge that all the terms of his message, and therefore his message itself, are at home only in a way of thinking that we have long since discarded? Isn't the whole notion of a dying and rising savior appeasing a judgmental deity just as obsolete as the idea of a first human progenitor suddenly popping into existence—complete with endogenous retroviruses, fossil genes, vestigial traits, and a belly button?

NATURAL ELECTION

A different understanding of Adam's fall and Christ's redemption, advocated by the Church Fathers Irenaeus and Athanasius, was advocated in the 1960s by Ansfridus Hulsbosch—the *recapitulation* theory.[14] According to it, Adam did not fall from a position of righteousness but rather swerved off the course of maturation and perfection that God intended for him. Adam sidetracked the human race. Jesus Christ comes to put us back on course, pressing on toward perfect maturity. Hulsbosch contended that this model fits in marvelously well with evolution. We need not worry about a taint and how it came to infect humanity. It is rather a lack of maturity, and Christ helps us to mature.

Hulsbosch reminds us of the difference anthropologists draw between natural selection and cultural selection. Natural selection would eventually eliminate near-sightedness after enough near-sighted people got run over by cars! No more descendants to spread their defective ocular genes! It is the game of chance between mutation and environment.

But cultural selection means that we can turn the course of the river of evolution by conscious decision. We decide that individual lives mean as much as species survival, so we invent eyeglasses. They save us from getting killed, but now the species will never be rid of near-sightedness. Okay, we say (especially those of us wearing glasses): It's a fair trade.

You are what you are biologically. On this, at least, Freud was right: Biology is destiny. As Jesus said, "You cannot by trying add a single cubit to your height" (Matt. 6:27). Natural selection is over for you. But cultural selection is not. You can evolve further that way. You *can* cultivate the genes natural selection gave you for altruistic, or at least cooperative, behavior. Genes whisper suggestions rather than shouting commands, as biologist Paul Ehrlich puts it. This is what Christian theology calls *sanctification*. This is what Colossians 3:9–11 and Ephesians 2:11–15 mean when they urge you to put on the new humanity because Christ has made obsolete the old human species and produced a new humankind in his image. That's cultural selection.

How, pray tell, did this upgrade get accomplished? One might infer that a Pelagian understanding of Adam and Christ would be the most natural next step. In opposition to Augustine, with his doctrine of inherited guilt and taint and evil inclination, Pelagius held that Adam and Eve had "merely" built a society on the basis of, and according to the rules of, selfishness and sin. Born into such a world, and socialized into its ways, no human being can escape becoming a sinner, too. What Jesus Christ did was essentially to start a new, alternative community in which people might come up in a society founded upon righteousness as taught by its founder, Jesus. God's grace, then, is not some saving and sanctifying power mysteriously at work in those who partake of the Church's sacraments. It is simply God's generosity in providing this wholesome matrix in which one may thrive apart from the sinful world outside. Nothing mysterious about that.

Take Alcoholics Anonymous as an analogy, or as an example, since it is a kind of scaled-down version of the Christian Moral Rearmament movement (Buchmanism).[15] There is nothing particularly spooky about it. AA delivers alcoholics from their besotted hell by means of strong group accountability and peer support. So with the Church according to Pelagius, himself a monk who lived in a monastic community.

As Augustine argued, though, there is no real role for Christ's death in that scenario. He becomes more of a teacher than a redeemer. This is why the Liberal Protestants of the nineteenth and early twentieth centuries revived a form of Pelagianism that placed its hopes not on evangelism but on education and social reform. Having rejected the atonement as "butcher shop religion," they preferred a social, collective approach creating a new, environmental conception of sin and its remedy via the Social Gospel.

Jerry Korsmeyer, despite being a Catholic, seems to advocate Pelagianism as the natural outcome of his abandonment of Augustinian Original Sin. He applies some nuance to the word "generation" that the Council of Trent failed to convey:

> Insights from the social sciences have been helpful in consideration of the transmission of original sin, said by Trent to occur by generation, not imitation. But Trent did not discuss the meaning of "generation." For modern theologians it is considered to apply to the whole process of socialization that occurs when humans enter the world and are assimilated into the family, the local community, and the wider social sphere. The principal idea is that humans find themselves thrust into a world where sinful people, activities and structures are already in place, and influencing and training the mind long before it is capable of making fully free moral decisions. We are contaminated by evil just by being born, by our generation into the human race, and not just by imitating our ancestors.[16]

That's not Pelagianism? Doesn't Korsmeyer really mean that the modern social sciences have proven Augustine wrong and Pelagius right? That heresy has evolved into orthodoxy? He'd be closer to the truth in saying that, of course, but if he does, his book can kiss the *Nihil Obstat* and *Imprimatur* goodbye.

GARDENER'S GUILT

Despite Jewish indifference and a slow evolution of the doctrine by early Christians, Original Sin—the Fall of Man—is supposedly a clear teaching of Genesis 2–3. How could anyone not see it there?, C. John Collins wonders. Where, pray tell, does he profess to behold it? Well, it is not quite seeing either the forest or the individual trees. It is more like seeing Bigfoot sneaking between the trunks out of the corner of your eye:

> When we ask what a Biblical author is "saying" in his text we are not limited to the actual words he uses. For example, we will note that Genesis 3 never uses any words for sin or disobedience; but it would be foolish indeed to conclude that what Eve and Adam did was not "sin." The author wants us to see that it was indeed, and to be horrified.[17]

He finds it absurd to suggest that "the text does not 'teach' that Adam and Eve 'sinned,'" just "because there are no words for 'sin' or 'rebellion' in Genesis 3." God's question in Genesis 3:11 ("Hast thou eaten of the tree, whereof I commanded thee that thou shouldest not eat?") is "as good a paraphrase of disobedience as we can ask for." What about the fact that "the text of Genesis does not say that humans 'fell' by this disobedience"? Both "objections stem from a failure to appreciate that Biblical narrators tend to prefer the laconic 'showing' to the more explicit 'telling,' leaving the readers to draw the right inferences from the words and actions recorded."[18]

But what are these "right" inferences? For Collins, they are the ones that align the text with Christian theology. The far more adventurous Denis Lamoureux dares to say, "Significantly, the category 'original sin' is not in the Bible. This concept was formulated by St. Augustine." The punch line, for him, is merely that "all humans are sinners."[19] Nonetheless, Original Sin may yet be necessary to the orthodox theology that he, like C. John Collins, wants to retain. This, obviously, is why he insists on interpolating it into the Eden story.

Robin Collins considers several possible interpretations for Original Sin, which seem to us like a number of emergency fire exits arrayed around a room such that one will be nearby no matter where you're sitting. The one that aligns with science is the "biological interpretation," where Original Sin is "nothing more than biologically inherited propensities, such as aggressiveness and selfishness, that help the individual or one's kinship group survive but typically do not promote the flourishing of the larger community."[20]

This stands slightly apart from the main point of Original Sin. We are not just damned at birth, but also depraved, having inherited a tendency to heap our own sins on top of the first one. John Calvin, that cheerful theological thinker, called it "total depravity." It doesn't mean everyone is as bad as they could be all the time, that we are a species made up of Mansons and Dahmers. It just means, a la Freud, that our every action, no matter how apparently noble, has a selfish motivation, which may be hidden from oneself and others. Jonathan Edwards (another ray of sunshine; he gave the infamously sadistic sermon "Sinners in the Hands of an Angry God") explained it like this: We do have a kind of free will, but the range of choices open to us are only sinful ones. Robert Ingersoll did not spare this aspect of the doctrine his scathing wit when he spoke about the Eden story in his *Lecture on Orthodoxy*: "A god that cannot make a soul that is not totally depraved, I respectfully suggest, should retire from the business. And if a god has made us, knowing that we would be totally depraved, why should we go to the same being for repairs?"

GIMME THAT OLD-TIME SIN-DEFINITION!

This goes to the question of what "sins" are, really, when looking at ourselves as evolved primates. We'll discuss that, too, but first let's consider what the concept of sin would have meant to the Bible writers. Then we may be better able to decide whether such sin is even the sort of thing theologians claim to find in the Eden story.

The concept seems to have originated as a function of ritual taboos, having nothing, at first, to do with moral misdeeds. No one ever supposed you were doing anything "immoral" by eating a ham sandwich. The offense was rather a ceremonial transgression, violating the food taboos implied in the cultural taxonomy (just as we don't eat roaches or dogs, though some cultures do). Thus "sins" were specifically incurred against God, not other mortals. The latter is where the categories "immoral" and "criminal" applied.[21]

As of the Prophets (especially Isaiah), wrongs done against neighbor and employee began to be considered to offend God, too. Isaiah excoriates those who sashayed into the temple ready to impress God with their piety but who had engaged in unjust business dealings, etc., out in the everyday world (Isa. 1:11–18). God didn't want their stinking worship as long as they were exploiting others. This seemingly had never occurred to most people. They figured God was only concerned with sacrifice formulas, ritual propriety, etc., as when he kills Nadab and Abihu for ritual carelessness—offering an improperly mixed incense ("strange fire," Lev. 10:1–3)—or commands the execution of some poor jerk for gathering kindling on the sabbath (Num. 15:32–36), or atomizes poor Uzzah for touching the Ark of the Covenant, trying to steady it when it teetered on the brink of the ditch (2 Sam. 6:6–7). Isaiah and the others were part of the phenomenon of interiorizing and moralizing religion worldwide over a period of a thousand years, which psychiatrist Karl Jaspers dubbed "the Axial Age."[22] Henceforward, it was believed that moral wrongs committed against one another were also to be deemed as offenses to God, thus moralizing the concept of "sin."

Hand in hand with this development went another: The concept of divine "holiness" came to be moralized, rationalized. Originally, "holiness" denoted "separation" between God and the world, as well as the "Wholly Other" character of the divine, its dreadful, frightening uncanniness. God was beyond good and evil, strictures that pertained only to puny human affairs, just as our pets are bound by rules irrelevant to us—"I create peace, and I create evil" (Isa. 45:7); "Is there evil in the city and Yahweh has not done it?" (Amos 3:6). Once God became a moral being, the evil had to be sloughed off onto the back of a different supernatural being, and this is why Jewish thinkers borrowed the Zoroastrian Ahriman, the evil antigod, with whom Satan (hitherto God's deputy) was merged to give us a devil.[23]

THE ONLY EXPIATION I CAN SEE

Sacrifices "expiated" the pollution caused by ritual "sins" by "washing" it away in the shed blood of the animal's slit throat. It looks as if the sacrificial interpretation of the death of Jesus began as a way for Jewish Christians to explain how God could accept believing Gentiles who, though they now believed in Jesus, had never either offered Jewish temple sacrifices nor even observed the dietary laws of the Torah. God hadn't condemned them unless they were immoral, not even for nonobservance of Jewish law, since the Torah applied only to Jews. But if God was now to accept Jesus-believing Gentiles on a par with Jews, they had to be "cleaned up" from their long-standing ritual "uncleanness." This Jesus's sacrifice supposedly did.[24]

The rather different notion of the "substitutionary atonement" or "penal substitution" extended Jesus' sacrifice to wipe away the guilt of *immorality* insofar as it offended God. Repentance was no longer enough to save you, as it had been in the teaching ascribed to Jesus in the Synoptic Gospels. But why? Why isn't God's willingness to forgive the repentant (as illustrated in the Prodigal Son parable) enough? Because early Christians needed some reason for Jesus to have died. Individual sins, even a great number of them, might be simply forgiven, but to justify Jesus' death, something much worse had to be at stake, something requiring radical surgery.

Some of our Christian evolutionists are quite open about the seeming fact that the Original Sin doctrine was reverse engineered from belief in the salvation wrought by Jesus Christ on the cross. The solution begat the problem. For example, Jerry Korsmeyer says, "Original sin *explains* the need for Christ's coming, his death and resurrection. No Adam, no fall; no fall, no atonement; no atonement, no Savior. One can understand why this version of Christianity requires original sin."[25] Tatha Wiley concurs: "In the early Christian era, the story of expulsion from the garden became the primary revelatory text for why the forgiveness of Christ is universally necessary."[26]

The important thing is Jesus' atonement, they seem to be saying, not the various ancient rationalizations for it. Can we not retain the saving death on the cross without the ancient guesses as to why it was necessary? What they do not seem to see is that the death of Jesus cannot be understood as a solution unless there is a prior problem to solve. And evolution replaces the whole Adam "problem" with a scientific explanation.

ADAM MADE ME DO IT

Traditional theology has told us that the Fall put the taint of sin and corruption into man from Adam and Eve's screw-up. An understanding of
evolutionary human origins dispenses with such nonsense on two grounds.
First, as we saw earlier, there was no such event; it is an ancient myth with
evident pagan parallels. Second, those actions that have been labeled as
"sin" do not arise from any subsequent corruption of our nature, but *from
our very nature itself*, as the descendants of reproductive survivors in a
harsh and brutal world. All of those "sinful" traits that Christian clergy
have condemned from their pulpits ever since Clement of Alexandria's
joyless asceticism have evolutionary explanations that make a whole lot
more sense than the story that Christian theologians dreamed up.

The connection between human sinfulness and the Eden story was
very clear in Luther's mind. Adam was in a state of innocence; there was
no sin to be found because, Luther thought, logically enough, "God did
not create sin." If only Adam had obeyed God and not eaten the forbidden
fruit, "he would never have died; for death came through sin."[27] Following
this statement is a fanciful description of what the Fall- and death-free
world would have been like:

> For us today it is amazing that there could be a physical life without death
> and without all the incidentals of death, such as diseases, smallpox, stinking
> accumulations of fluids in the body, etc. In the state of innocence no part
> of the body was filthy. There was no stench in excrement, nor were there
> other execrable things. Everything was most beautiful, without any offense
> to the organs of sense; and yet there was physical life.[28]

For Luther, our supposed sinfulness is just part of Original Sin, which
"really means that human nature has completely fallen." Since "we all lie
under the same sin and damnation of the one man Adam," he asked in *The
Bondage of the Will*, his landmark rant against human goodness and free
will, "how can we attempt any thing which is not sin and damnable?"[29]

It's understandable that Christians up to Luther's time would hold this
view. "Medievals could talk without self-consciousness about an historical

Adam and Eve because an alternative view had not yet arisen in biblical interpretation or from a scientific understanding of human origins," notes Wiley.[30] But now there is no such excuse for our bad behavior, nor any inherent sinfulness to feel guilty about.

THE MAN OF SIN

All those "works of the flesh" that Christianity condemns—our lust for power and procreation, favoritism toward self and kin, propensity to hog energy in gluttony and conserve it in sloth, quick anger responses to threats and slights—are the results of millions of years of evolutionary experiment in what allows genes to replicate best in these human vehicles they've wound up in. It is all driven by the genes, Dawkins reminds us in his magnum opus *The Selfish Gene*, and the genes *don't care* about the morality of that behavior—they don't care about anything. They have no lofty morals, no sense of higher purpose; "the true 'purpose' of DNA is to survive, no more and no less."[31]

Dawkins emphasizes "that we must not think of genes as conscious, purposeful agents," even though natural selection "makes them behave rather as if they were purposeful." Both genes and natural selection itself are often described with the language of agency, but he and other scientists know what they're talking about when they do so: "When we say 'genes are trying to increase their numbers in future gene pools', what we really mean is 'those genes *that behave in such a way as to increase their numbers* in future gene pools *tend to be the genes* whose effects we see in the world.'"[32]

Daniel Fairbanks (a Mormon whose scientific writing seems utterly unclouded by dogma) cites one genetic success story that is a moral failure of the worst degree: the "tyrannical rampage" of Genghis Khan "throughout central Asia to establish the largest land empire in history." The Great Khan's "sexual exploits throughout the region he conquered are well documented. Furthermore, his male relatives, who inherited the same Y chromosome as he, ruled much of the region even after his empire fell, and they also had power for sexual exploitation." Now, one particular Y chromosome is found in about 8 percent of all men living in that region: "The pattern of distribution and the time of its origin extrapolated from the data both point to the time and the area of his reign, strong evidence that thirty million men today inherited his Y chromosome."[33]

The Bible doesn't provide a sterling example in the rape and conquest department, as must be acknowledged by anyone who has objectively read Joshua or Numbers 31:17–18 ("kill every male among the little ones, and kill every woman who has known man intimately. But all the girls who have not known man intimately, spare for yourselves," NASB). Thankfully, though, few professed Bible believers (or anyone else) sink to that level anymore. Cultural evolution has erected societies in which we largely rise above the basest impulses that biological evolution has left us with.

According to Russell Kolts, a professor of psychology and advocate of an evolution-informed approach to anger management, we share these impulses "with organisms—like reptiles—that appeared long before we did in the evolutionary chain." The human brain, he says, is

> an evolutionary patchwork quilt—stitching together a complex and varied system of structures and functions, some of which date back almost to the beginnings of life itself. Our brains have evolved with a host of extremely powerful emotions, capable of harnessing our thoughts and attention, often without our awareness.[34]

The way we act on these emotions "often seems suited to earlier times in the human story—they prepare us to fight, flee or lie down in submission when we'd be better served by pausing and taking a moment to become more mindful, and perhaps to analyze, consider or negotiate."[35] Reflection on the evolutionary heritage of our brains gives us a sense of understanding, Kolts says, both for ourselves and for others. It's certainly a more productive approach for one's mental health than the one taught by Jesus, who equated the natural, common impulses of anger and forbidden sexual attraction with murder and adultery, and said that calling someone a fool warrants a descent into the fiery hell (Matt. 5:21–28).

Hellbound or not, we all lapse into angry outbursts from time to time. This rash tendency is entirely to be expected, says Kolts. Our threat "system has evolved so that it is activated *rapidly*, because defences that come on too slowly may be too late." We have been prey more than predators, even for most of human evolutionary prehistory, and there isn't much time to react when the tiger is about to pounce. Unlike Jesus and the clergy, who have frowned and forgiven ever since, Kolts wants us to understand that having a rapid-response amygdala for threat response "is *not our fault*; it is simply the way our brains work."[36]

How do the theologians account for all this? Well, we've already seen Robin Collins's attempt to address the issue of "biologically inherited propensities," as he describes the "biological interpretation" of Original Sin. These traits (e.g., aggression and selfishness) help the individual or one's kinship group survive but typically do not promote the flourishing of the larger community. Essentially, under this view, the doctrine of original sin, the Genesis story, and the various statements in the epistles tell us nothing more than what science tells us. Advocates of this view often assume that we are purely biological and physical beings. Hence science, not theology, becomes the primary place to look to understand the nature and origin of human beings.

He admits this view to be common even among theologians but thinks it is reductionistic, making "the voices of theology, Scripture, and Church tradition" into "a sort of fifth wheel." Instead, "it is the purported findings of science that are claimed to provide the correct understanding of human nature and the human condition."[37] But isn't this just the Achilles' heel of all these theo-biological approaches?

Collins objects that humans seem to be more than merely physical creatures. Well, how could he *not* think that and still be religious? The biological interpretation also "tends to reduce evil merely to our acting on biological impulses, ignoring the particularly serious forms of evil that are made possible by our own self-awareness and transcendence—evils such as idolatry of self, viewing other people as mere objects, and the like."[38] But these traits and deeds, too, are directly rooted in our survival instincts. As the anatomist and Christian Daryl Domning points out, our "sinful" human behaviors

exist because they promote the survival and reproduction of those individuals that perform them. Having once originated (ultimately through mutation), they persist because they are favored by natural selection for survival in the organisms' natural environments. Since these behaviors are directed to self-perpetuation and succeed in a world of finite resources only at the expense of others, it is accurate to call them, in an entirely objective, non-psychological and non-pejorative sense, selfish.[39]

Vices are not unique to us as humans, either. If our supposed sinfulness is the result of forbidden fruit in Eden, then innocent nonhuman animals must have gotten the go-ahead to start screwing up in similar ways, since "there is virtually no known human behavior that we call 'sin' that is not also found among nonhuman animals." Even pride, says Domning, "proverbially the deadliest sin of all, is not absent."[40] Either that, or there was a whole world full of animal Fall dramas being played out alongside Adam and Eve; perhaps among them, the first ancestral chimpanzee guiltily munched on a piece of the Banned Bark, and then looked down in horror to see that she *wasn't* a naked ape, after all!

Of course, that is ridiculous. But no more so than it would be to deny the "unambiguous conclusion" that Domning draws from his observation that animals are "doing things that would be sinful if done by morally reflective human beings."[41]

This is a realization that is both humbling and freeing: *We are the genetic success stories of our ancestors' behavior as well as their bodies.* Forget all the ancient nonsense about being saddled with some inherent "sin-corrupt" nature, about blaming everything on some mythical Adam. You are here because you had ancestors who did what it took to survive and reproduce in a very harsh world.

It's as simple as that! What you inherited from your furrier forebears is not some taint of sin but *the very traits* that allowed them to produce you. The whole idea of sin, original or otherwise, is a shriveled vestige left over from cultural evolution, one that we can happily excise from our thinking in a world of secular morality and self-awareness.

Chapter 8

THE SOUL FALLACY

Julien Musolino

Few ideas have been as widespread and enduring as the belief in the existence and immortality of the human soul. As Mark Baker and Steward Goetz observe in their book *The Soul Hypothesis*, "Most people, at most times, in most places, at most ages have believed that human beings have some kind of soul."[1] At the dawn of the twenty-first century, soul beliefs still permeate every aspect of our lives. Results from national polls and empirical studies indicate that most Americans, and many more people around the world, believe in the existence and immortality of the soul. Most religious doctrines would be unrecognizable if stripped from the claims they make about death and the afterlife. Moreover, these beliefs are constantly reinforced by a wealth of books, articles, TV shows, and pronouncements from writers of all stripes who claim to have found credible scientific evidence for the existence of the soul.

In sharp contrast to popular opinion, most mainstream scientists have abandoned the idea of the soul. This is what Nobel laureate Francis Crick famously called "The Astonishing Hypothesis": "You, your joys and your sorrows, your memories and your ambitions, your sense of personal identity and free will, are in fact no more than the behaviour of a vast assembly of nerve cells and their associated molecules."[2] Echoing Crick's words, Harvard cognitive neuroscientist Joshua Greene explains that while most people are dualists and think of themselves as immaterial minds or souls housed in physical bodies, most psychologists and neuroscientists disagree. The sciences of the mind proceed on the assumption that the mind is simply what the brain does.[3] As philosopher Owen Flanagan explains, the primary operation of the scientific image of persons is to "desoul" human beings.[4]

At the end of his book *Descartes' Baby*, psychologist Paul Bloom observes that the magnitude of the clash between the scientific view of persons and the popular one should not be underestimated. People may eventually learn to live with the fact that we do not have souls, but it won't be easy. One way to avoid the paradigm-shifting confrontation between these two incommensurable worldviews is to find refuge in the idea that science and the soul occupy nonoverlapping magisteria. The soul is a philosophical, metaphysical, or religious claim, we are often told, and so it is safely insulated from the corrosive reach of science and cannot be affected by discoveries in biology, neuroscience, or physics.

The goal of this essay is to show that precisely the opposite is true. Soul beliefs, I will argue, are genuine scientific propositions. The importance of this conclusion cannot be overstated. If the soul hypothesis is indeed a scientific claim, and not merely a religious or metaphysical idea, then the protective layers surrounding Christianity's core melt away, opening the door for science to demystify the world's largest religion. My first conclusion then is that determining whether people have souls is as objective an endeavor as deciding the age of the earth or the origins of species. My second conclusion is that in spite of well-publicized claims to the contrary, there is in fact no credible evidence for the existence of the soul. Worse for Christianity, and religion more generally, modern science gives us every reason to believe that people do not have souls. My final, and perhaps most important conclusion, is that we have nothing to lose, morally, spiritually, or aesthetically by letting go of our belief in the soul. In fact, we even have something important to gain. This, I hope, is a conclusion that we can all embrace—Christians and non-Christians alike.

HISTORY AND THE SHRINKING SOUL

The late historian Howard Zinn liked to remind his readers of the importance of understanding history. In order to get a firmer grasp on the nature of soul beliefs, our first task will therefore be to explore their historical origins. The study of history reveals three important trends that are relevant for our purposes. The first is that the soul that we commonly speak of today started its life as a plurality of entities. Before there was one soul, the soul that we know today, there were multiple souls. There were *free souls* and *body souls*. Free souls were vagabond souls that floated away from our body while we dreamt. These souls may also have been at the origin of *surviving souls*, the kind of souls that eventually make their way to the afterlife. Body souls, by contrast, were more sedentary and liked to stick with us. Those were *life souls*, which were believed to animate us with the spark of life, and *ego souls*, to which we owe our quintessentially human psychology.[5]

The second observation is that these different souls were conjured by our ancestors to make sense of biological and psychological phenomena. As science fiction writer Arthur C. Clarke put it, any sufficiently advanced technology is indistinguishable from magic.[6] To our prescientific ancestors, the entire world must have looked like one big magic show. How come cats can do all kinds of interesting things but rocks just sit there and do nothing? What special property is associated with being alive? And where do our thoughts, dreams, and emotions come from? Can the matter in our body really feel, think, and make rational choices? To answer these mind-bending questions, our ancestors proposed that different souls are responsible for biological and psychological phenomena. This suggests that ignorance of nature's ways, coupled with a penchant for certain kinds of explanations, gave birth to our cherished soul beliefs.

Third, as scientific understanding progressed, the different souls started to melt away like snow in the sun. As physicist Jerome Elbert observes in his book *Are Souls Real?*, "the ideas for the different [souls] arise from ignorance of how life, emotions, mental characteristics, consciousness, dreams, and trances arise in nature. In a technically advanced modern society, many of the items in the list have scientific explanations. In the long run, only those things that are not understood by science are apt to retain their supernatural explanations."[7] Nowhere is this shrinking trend better described than in René Descartes's writings.[8] To many, Descartes is remembered as the father of the modern soul. In reality, he was one of the first thinkers to realize that the life souls of his predecessors were superfluous. Inspired by the new mechanical philosophy, Descartes explains in his writings that viewing the human body as a machine allows him to dispense with Aristotle's life souls; his vegetative and nutritive souls. Today, the physical basis of life is firmly established, modern biology has buried the doctrine of vitalism and, in the process, it explained away the life souls of our ancestors. The last frontier is now the human mind, and so, naturally, this is where people are looking for the soul.

THE SOUL TODAY

A lot has changed since the seventeenth century. If we were to show Descartes the GPS device in my car, a piece of equipment that knows where I am, can tell me which routes I should follow, warn me of various dangers on the road, and redirect me if I make a wrong turn, the great philosopher would have had to conclude that the device is powered by a rational soul. While Descartes's descendants would be amused by this conclusion, most of us today still believe that people have souls. In this regard, not much has changed in the last three hundred years.

The conclusion that soul beliefs continue to be widespread today is well documented. According to a 2009 Harris poll, 71 percent of Americans believe in the existence of the soul and its survival after death. In 2013, Harris ran a follow-up study and found that most Americans, 64 percent, continue to believe in the immortality of the soul.[9] A 2008 Pew report indicates that 74 percent of Americans believe in the afterlife, and that a majority of respondents are absolutely certain of the truth of this belief.[10] The Gallup organization ran polls between 1997 and 2004 and found that belief in heaven among Americans ranges from 72 percent to 83 percent.[11] Similar trends have been reported by the Barna Research Group of Ventura, California, which found that 81 percent of Americans believe in the afterlife and that 76 percent believe in heaven.[12]

Results from Gallup polls and World Value Survey indicate that soul beliefs have been remarkably steady in the United States since the end of World War II. In 1947, 68 percent of the US population reportedly believed in the afterlife. At the turn of the twenty-first century, the number reached 76 percent, with only minor fluctuations in the intervening decades. The situation in the United States contrasts with what has been reported in other developed nations, such as the Scandinavian countries, France, Belgium, Germany, and the Netherlands. The numbers documented in these countries around the middle of the twentieth century were comparable to those found in the United States, but they have since sharply declined, following a wave of secularization in most first-world countries during the second half of the twentieth century. Today, between 30 and 50 percent of respondents in those countries report believing in the soul and the afterlife.[13] In other parts of the world, such as Africa and India, the numbers are even higher than in the United States. In his book *Immortality*, philosopher Steven Cave concludes that the overwhelming majority of the world's population subscribes to the soul narrative.

The second line of evidence supporting the conclusion that soul beliefs are still very much alive today comes from more detailed and fine-grained empirical studies falling under the umbrella of a new subfield of the sciences of the mind called the cognitive science of religion. These studies have documented that soul and afterlife beliefs emerge early in childhood, persist into adulthood, and are related to other variables, such as level of religiosity and belief in the paranormal. In my own studies of the nature and prevalence of soul beliefs, I have found that most of the college students I teach at Rutgers University report believing in the existence of the soul and its survival after death. For example, in a laboratory study I ran in the fall of 2013, my colleagues and I found that out of more than two hundred undergraduate students, 88 percent reported varying degrees of belief in the existence of the soul (their responses on a six-point scale to the question "Do you believe that you have a soul?" ranged from probably yes, to most likely yes, and definitely yes).

When asked follow-up questions about the nature of these beliefs, most participants indicated that their soul was an immaterial entity distinct from the body, that it was immortal, and that consciousness was an aspect of the soul that would continue to exist after death. In response to questions about the role played by their soul, 74 percent of students indicated that it gave them a moral compass, 68 percent consciousness, 67 percent feelings, 67 percent the ability to fall in love, 64 percent free will, 61 percent their personality, and 46 percent the ability to make decisions. Most participants regarded the soul as distinct from the mind and believed that it was given to them directly by God.

Results from national surveys and the more detailed studies described above indicate that most people today believe in a soul that has three main properties.

1. The soul is believed to be *immaterial* or *nonphysical*, as well as distinct and separable from the body.
2. The soul is believed to be *psychologically potent* in the sense that it contributes to our mental lives.
3. The soul is believed to be *immortal* and to carry our consciousness into the afterlife.

This is the soul that this essay is about, the soul that captures the imagination of most people around the world. Needless to say, the word *soul* can also be used metaphorically as in *James Brown has soul, These poor souls didn't see it coming*, and a host of other phrases and idioms. However, talking about soul food or soul mates does not commit you to a particular metaphysical worldview. For that reason, I will have nothing to say about metaphorical, idiomatic or poetic uses of the word *soul*.

I also have little doubt that religious scholars can always find ways to take issue with the definition of the soul proposed above and claim that it is not the only one or even that is not theologically accurate. To that I would reply that I do not really care about theological accuracy. After all, most people aren't theologians. As a cognitive psychologist, I am interested in what most lay people think, and not in the pronouncements of a handful of theologians. The definition given above has the virtue of being empirically based and of corresponding to what most people believe. Besides, as we will discover shortly, the soul we defined above also happens to be the one that Christian apologists so passionately defend in their popular writings.

THE SOUL IS A SCIENTIFIC HYPOTHESIS

In his classic essay on science and religion, written in 1935, the great Bertrand Russell already recognized that soul claims are scientific propositions. As he explains, "There is, indeed, one line of argument in favour of survival after death which is, at least in intention, completely scientific. . . . It is clear that there could be evidence which would convince reasonable men."[14] The evidence that Russell had in mind stems from the contention that the soul is separate from the body and can continue to function after death. This claim leads to an obvious prediction, humorously captured by psychologist Steven Pinker when he wrote that "My late aunt Hilda could beam a message telling us under which floorboard she hid her jewelry."[15] Jokes aside, evidence of meaningful communication with the deceased, if it were available, would indeed give us good reasons to believe that the soul can continue to function once the body has expired.

In a similar vein, individuals who report out-of-body experiences, during which they see themselves and their surroundings from a third-person perspective, often interpret these baffling states of mind as resulting from their soul temporarily leaving their body to float about the room. If frolicking souls could provide us with an immaterial pair of eyes, they would be in a position to collect visual information not available to the body's more restricted vantage point. In the literature on near-death experiences, which are often accompanied by out-of-body experiences, this possibility is known as *apparently nonphysical veridical perception* (AVP). If evidence for AVP were available, it would lend strong credence to the idea that the psychologically potent soul can operate outside the confines of the body.

The claim that the mind can function independently from the body is in part what makes soul claims scientific propositions. But there are other reasons as well. For believers, the soul isn't just an impotent blob of spiritual stuff. Crucially, it is believed to contribute to our psychological functioning. This means that in order to perform its putative functions, the immaterial soul should be able to causally interact with the physical matter in our brains. If the elementary particles in our brains faithfully behave in accordance with the laws of physics, then their behavior is determined by the laws of physics and not by the soul. On the other hand, if the soul could somehow nudge the elementary particles in our brain to push our buttons, then these particles would not behave as predicted by the laws of physics, with detectable consequences. Claims about a psychologically potent soul are therefore claims about physics.

With these considerations in mind, we can now formulate the two competing hypotheses regarding the nature and functioning of the human mind, the dualistic and materialistic hypotheses.

Dualistic Hypothesis: Human beings are composed of a physical body and an immaterial, psychologically potent, and immortal soul. Body and soul are distinct entities and the soul can continue to exist and function after we die.

Materialistic Hypothesis: The mind, the domain of the soul, cannot function separately from the body for the simple reason that our mental experiences are caused by physical activity in the brain. What we call mind is nothing but a description of the functioning of the brain at a certain level of abstraction.

EVALUATING THE SOUL HYPOTHESIS

The scientific case against the soul, our dualistic hypothesis, is overwhelming. The key to understanding why the soul cannot be reconciled with modern science is the notion of *consilience*, the idea that evidence from multiple, unrelated sources converges to support strong conclusions. As we discussed earlier, the soul has shrunk as scientific understanding progressed. There is no formalism, apart from trivial analogies, that describes the immaterial soul substance. There is no objective empirical evidence supporting the conclusion that consciousness can operate separately from the brain. Souls fly in the face of what we know about modern science. Finally, there is overwhelming evidence supporting the materialistic hypothesis, the alternative to dualistic soul claims. In sum, the soul, like the emperor's new clothes in Andersen's famous tale, has exactly the set of properties that it should have if it didn't exist.

WHY THE PROPOSED EVIDENCE IS NO EVIDENCE AT ALL

A good recipe for writing a bestseller is to use the word *heaven* in the title of the book and offer evidence for the existence of the soul. Books such as *90 Minutes in Heaven, Heaven Is for Real,* and *Proof of Heaven* are but a few examples of a larger trove of publications aimed at convincing the masses that human beings have much in common with angels. These enthusiastic soul advocates are an eclectic group of religious apologists, new age gurus, and a handful of scientists who are unhappy about the mainstream, materialistic consensus. Among the evidence touted by these authors, one finds claims about reincarnation and the past lives of children, being able to communicate with the dead, recording the voices of the deceased, phenomena such as free will and consciousness, near-death experiences, and myriad other claims.

In spite of its apparent diversity, the evidence presented by soul advocates can be grouped into a small number of families. Here I will consider only claims based on phenomena that mainstream science recognizes as real. The idea is to consider the strongest possible evidence for the existence of the soul and show that even this is utterly unconvincing. The first family of soul claims centers on the self-examination of one's conscious thoughts. Plumbing the depths of one's conscious inner world delivers the powerful impression that our mind and body are not cut from the same cloth. As philosophers and soul advocates Stewart Goetz and Charles Taliaferro explain in their book *A Brief History of the Soul,* throughout the ages, mind-body dualism has been affirmed mostly on the basis of first-person experience.

Using a concrete example, Goetz asks us to consider the movements of his fingers on his keyboard as he typed his essay on the soul. To Goetz, it is obvious that these movements are caused by his choices to use certain words for a particular purpose. Since choices and purpose are not physical things, Goetz argues that our voluntary actions are not caused by anything physical but by our immaterial soul. He concludes that, "If our common-sense view of human beings is correct . . . I, as a soul, cause events to occur in the physical world by making a choice to write this essay for a purpose."[16] The problem with Goetz's view, or any approach that tries to understand how the mind works based on introspection, is that it reflects an extraordinarily naïve view of psychology that was abandoned by professionals in the field more than a hundred years ago.

If we could understand how the mind works simply by consulting our inner thoughts, feelings, and impressions, then the doors of all the psychology, neuroscience, and cognitive science departments around the world would immediately close. We need scientific psychology, with its reliance on the third-person perspective, *precisely* because we cannot intuit the relevant truths. As philosopher Paul Churchland reminds us, "The red surface of an apple does not look like a matrix of molecules reflecting photons at certain critical wavelengths, but that is what it is. The sound of a flute does not sound like a sinusoidal compression wave train in the atmosphere, but that is what it is. The warmth of the summer air does not feel like the mean kinetic energy of millions of tiny molecules, but that it what it is." Churchland concludes that, "If one's pains and hopes and beliefs do not introspectively seem like electrochemical states in a neural network, that may be only because our faculty of introspection, like our other senses, is not sufficiently penetrating to reveal such hidden details."[17]

The second family of soul claims is closely related to the first one and comes from a fascinating form of subjective experience called the near-death experience (NDE). Originally reported by Plato in *The Republic* and made popular by Raymond Moody's sensational 1975 bestseller *Life after Life*, NDE has captured the imagination of the general public. The subjective manifestations of NDE are eerily similar to what we imagine the afterlife to be, including an awareness of being dead, encountering deceased individuals, experiencing a life review, and moving through a dark tunnel with bright lights at the end. The phenomenon is undoubtedly real and has been experienced by millions of people in the United States and around the world. A common interpretation of NDE is that these unusual experiences offer us a tantalizing glimpse of the afterlife and prove that the soul can survive bodily death. Not surprisingly, most authors who claim that the soul is real invoke NDE as one of their sources of evidence.

In spite of the otherworldly qualities surrounding NDE, the phenomenon has failed to convince the scientific community that there is anything paranormal about these experiences. There are four good reasons to reject soul claims based on NDE reports. The first is that the phenomenon is inherently subjective and that there is no credible body of evidence supporting *apparently nonphysical veridical NDE perception*.[18] The second reason is that there is nothing extraordinary about the experiential manifestations of NDE. Neuroscientists have shown that the relevant experiences can be linked to well-established neuropsychological processes in the brain.[19] The third reason to doubt the paranormal interpretation of NDE is that the phenomenon isn't restricted to imminent death and has been reported in people whose life was in no danger whatsoever.[20] The fourth reason, as will discover shortly, is that there is a mountain of evidence supporting the conclusion that consciousness and its various manifestations are brain-based phenomena.

The third family of soul claims rests on a logical fallacy known as the God-of-the-gaps—or in this case, the soul-of-the-gaps—argument. The approach consists in pointing to aspects of our mental lives not yet understood by science, claiming that such phenomena will never be understood within a narrowly materialistic framework, and triumphantly concluding that the soul must be doing the work. The usual suspects involve consciousness, free will, our moral sense, and the creative aspect of language use. While it is, of course, true that we are far from a complete understanding of how the mind works, this line of reasoning is nevertheless entirely fallacious, for reasons that should be well-known to anyone with a minimally functioning faculty of critical thinking. As physicist Jean Bricmont explains, dogs do not understand celestial mechanics, but this does not mean that anything supernatural is involved.[21]

The last family of soul claims exploits the difficulty and complexity of serious, professional science by blurring the line between science and pseudoscience and giving fantastical interpretations to well-established scientific theories. The trick is usually accomplished by bypassing the peer-review process and peddling extraordinary claims directly to the general public. In his book *Proof of Heaven*, former neurosurgeon Eben Alexander contends that mainstream scientists are "willfully ignorant" of the overwhelming evidence supporting his dualistic conclusions. According to Alexander, all that evidence can be found in a 2007 book called *Irreducible Mind*, which touts a plethora of phenomena that mainstream science either does not recognize as real or as involving anything paranormal. These phenomena include near-death experiences, reincarnation, maternal impressions, apparitions, precognition, and sudden graying of the hair.

THE SOUL FLIES IN THE FACE OF SCIENCE

Imagine that someone were to claim that she knows a well-trained athlete who can run a marathon in under ten minutes. If the runner wasn't available to be tested, we wouldn't need to suspend judgment regarding the validity of the claim until such a test could be performed. Given what we know about human locomotion and the length of a marathon, the proposition that any human being, however well-trained, could run twenty-six miles in ten minutes simply cannot be taken seriously. A similar line of reasoning can be applied to the existence of the soul. In a *Scientific American* piece titled *Physics and the Immortality of the Soul*, theoretical physicist Sean Carroll explains that given what we know about the laws of physics, the existence of the soul, like the claim that someone can run a marathon under ten minutes, is supremely implausible.

Carroll begins by pointing out that the laws of physics underlying everyday life are completely understood, that they have been supported by every experiment ever performed, and that within these laws, the information contained in our brains simply cannot continue to exist after we die. Within this context, belief in the soul would amount to rejecting everything we know about modern physics. As Carroll explains, "The choice you are faced with becomes clear: either overthrow everything we think we have learned about modern physics, or distrust the stew of religious accounts/unreliable testimony/wishful thinking that makes people believe in the possibility of life after death. It's not a difficult decision, as scientific theory-choice goes."[22]

A similar line of reasoning comes from what we have learned about modern biology. We know with a very high level of certainty that human beings weren't planted on earth in present form within the last ten thousand years but, instead, that we evolved from more modest life-forms alongside millions of other species. This means that we are the physical outcome of a physical process operating over a set of physical raw ingredients and that there is an unbroken chain linking us to the very beginnings of life on Earth several billion years ago. The implications of these conclusions are as straightforward as they are damning for claims about the existence of the soul. As philosopher Paul Churchland explains, "There seems neither need, nor room, to fit any nonphysical substances or properties into our theoretical account of ourselves. We are creatures of matter."[23]

THERE IS OVERWHELMING EVIDENCE FOR MATERIALISM

If our materialistic hypothesis is on the right track, we can ask ourselves what we would expect the world to look like. If the mind is simply the brain described at some level of abstraction, then we should expect a number of things to follow. First, since the brain is physical, and therefore divisible into parts, we would expect the mind to also be divisible, at least in principle. Let's call this the *divisibility* of the mind. Second, if our conscious experiences arise from the operation of the brain, damaging the brain should also impact consciousness. Let's call this the *fragility* of the mind. Third, if the activity in our brain causes our conscious experiences, including our voluntary actions, then we should be able to push the mind around by directly stimulating the brain. Let's call this *mind control*. Fourth, if our thoughts ultimately boil down to patterns of electrochemical activity in our brains, then it should be possible to read people's minds by analyzing the physical activity in their brain. Let's call this *mind reading*. Finally, if our abilities to reason, think, and make decisions are properties of the matter in our brains, then it should be possible to build other physical devices that display similar, mind-like properties. Let's call these *intelligent machines.*

The divisibility and fragility of the mind, mind control, mind reading, and intelligent machines are all a reality today; precisely what we would expect to find if our materialist hypothesis is right. In 1981, neuropsychologist Roger Sperry won the Nobel Prize in physiology and medicine for his work with split-brain patients, demonstrating that individuals whose brain hemispheres had been surgically disconnected behave as though two separate selves, or minds, inhabit the same body. Cut the brain in half and you'll also split the mind. Psychological phenomena such as our feeling of authorship and free will, our inherently subjective experience of pain, and our very sense of self can be obliterated by brain damage. In a condition known as *Alien Hand Syndrome*, patients lose voluntary control of one of their hands, often using their other, normally functioning, hand to rein in their rogue limb. People with a condition called *pain asymbolia* may realize that they have been hurt, but they no longer experience the sensation of pain. Patients suffering from *Cotard's Syndrome* are convinced that they are dead, that they have no self, and that they are bodies without content. The mind is very fragile indeed.

In 2009, a group of French neuroscientists published a result in the journal *Science* that would have made Descartes rethink his entire approach to psychology. By electrically stimulating the posterior parietal cortex of their patients, the team was able to evoke "pure intentions." Patients reported a desire or a will to move and said things like, "I felt a desire to lick my lips." Amazingly, when the team increased the intensity of the electrical stimulation, patients became convinced that they had performed a movement that actually never took place.[24] This is mind control in action. Using a brain-imaging technique called functional magnetic resonance imaging, or fMRI, cognitive neuroscientists can literally read the minds of their subjects by scanning their brain and analyzing the data. In one study, for example, participants were shown pictures of human faces, cats, and artifacts such as chairs, shoes, and bottles. By analyzing the patterns of brain activity generated by the experience of viewing these different pictures, the team was able to infer with 96 percent accuracy what pictures their participants had been viewing.[25] Mind reading is still in its infancy, but it is now a reality.

LIVING IN A SOUL-FREE WORLD

Many people find the idea of a soulless person unnerving, if not utterly horrifying. In my own research, I have asked Rutgers undergraduates to tell me what they couldn't do or be if they didn't have a soul. Students responded that they would lack feelings, compassion, the ability to love, the inclination to behave morally, and the capacity to make decisions. Some have even asked me what the point of living would be if we didn't have a soul. Similar sentiments are expressed by popular authors who try to convince the general public of the reality and importance of the soul. In his preface to Dinesh D'Souza's book *Life after Death: The* Evidence,[26] evangelical pastor Rick Warren warns us that without the soul and the promise of an afterlife, "there is no basis of any meaning, hope, purpose, or significance to life. . . . The logical end of such a life is despair."[27] Are these fears legitimate? Is the materialistic outlook as bleak as its detractors portray it to be? I hope to convince you in the pages ahead that the answer to these gloomy questions is an emphatic "no."

WE LOSE NOTHING

If you worry about the demise of the soul and what that would entail, let me offer some relief. The key observation to avoid fretting over the doom and gloom scenario ominously promulgated by soul advocates is to remind ourselves of the distinction between facts and their associated explanations. Let me offer an analogy: Suppose that you were holding an apple in your hand and decided to release it. It would fall to the ground. If the word *fact* means anything at all, it is simply a fact that released apples, unless some trickery is involved, fall to the ground. How can we explain this fact? A look at history reveals that explanations have changed over time. For Aristotle, apples fell to the ground because they were compelled to return to what was believed to be their natural place. In the seventeenth century, Isaac Newton dispensed with Aristotle's teleological explanation and declared that apples fall to the ground because of a goalless gravitational force that he quantified mathematically. A few centuries later, Albert Einstein developed his theory of relativity according to which gravity follows as a consequence of the curvature of space-time.

While apples have always fallen to the ground when released and will continue to do so in the foreseeable future (the invariant fact), the reason we believe they do has changed over time (the variable explanation). Now imagine someone panicking over the replacement of Newton's theory of gravity by Einstein's, worried that all his precious material possessions will start flying off into space. This would of course be silly. Facts don't change, and apples will continue to fall to the ground when released, regardless of who is right about gravity. Let us now transpose the analogy to human beings. It is a fact that people, barring unfortunate circumstances, have complex mental lives and flexible behavior. We can experience a wide range of mental states, including a nuanced palette of feelings and emotions, behave rationally and morally, be moved by beautiful music, and develop scientific theories.

For most of human history, the explanation for such amazing capacities eluded us entirely, and so we imagined that different kinds of souls were responsible for our biological and psychological powers. Here too, the explanation has changed, but the facts about human biology and psychology are here to stay. People will continue to fall in love, be moved by beautiful music, and be jealous of their neighbors, regardless of whether we believe that their soul or the neurochemistry of their brain is the relevant causal factor. In letting go of the soul then, we lose only a potential explanation, and none of the facts that we hold dear. This is the good news. The bad news, soul advocates would be quick to point out, is that our souls are also supposed to give us free will and immortality. Let go of the soul, then, and we become biological puppets with an expiration date.

The soul is believed to give us some of our psychological powers, and cause us to do the things that we do, but, being immaterial, it is not subject to the physical laws of cause and effect. The soul, then, is the ultimate uncaused cause and allows us to escape physical determinism. On this scenario, called libertarian or contra-causal free will, our will is free in that it is not caused by anything, and this freedom from causation is what is believed to provide the basis for moral responsibility. Get rid of the soul, the story goes, and you must also abandon any meaningful notion of moral responsibility. Fortunately, this conclusion is yet another fallacy. If I were to invite you to my house to discuss free will over a drink and, completely out of the blue, I threw my martini in your face, the first thing you would look for is a reason for my baffling behavior. If I told you that my showering you with vodka and olives had nothing to do with what you said or I how felt—that it was not caused by *anything*—you would conclude that I am insane, not free.

Earthquakes and tsunamis are fully caused, and we do not hold them morally responsible for their devastating fury, but people can escape the laws of physical causation because of their nonphysical soul, and that's why, dualists tell us, we hold each other morally accountable. If being or not being fully physically caused were the only imaginable difference between earthquakes and people, the only basis upon which we can secure our notion of moral responsibility, then letting go of the soul would indeed be problematic. But there is more to people and earthquakes than meets the eye. Human beings have rational powers that earthquakes and tsunamis lack. People can consciously contemplate the consequences of their actions, engage in moral reasoning, treat other human beings as autonomous moral agents, and respond to threats, blame, and praise. As is recognized within the law, this is enough to securely anchor our notion of moral responsibility.

What about immortality? Do we necessarily need to subscribe to the soul narrative in order to find peace and comfort in the face of death? Two separate lines of evidence suggest that the stakes are not nearly as high as soul advocates would have us believe. The first is that the beneficial aspects of religion do not seem to come from religious belief per se—belief in a supreme being, a world of cosmic justice, survival of the soul—but rather from the kind of social structure and support that religious communities provide, as well as the kind of societies in which people live.[28] If so, there is a clear alternative to praising the virtues of dualistic and afterlife beliefs, namely improving societal conditions and giving people a sense of belongingness. Second, nations like the Scandinavian countries, where religiosity levels are some of the lowest in the world, also happen to have some of the highest levels of human development and happiness. If the Swedes and the Danes, most of whom do not believe in the existence of the soul, are not paralyzed by mortality-related anxiety, there is hope for the rest of us.[29]

WE HAVE MUCH TO GAIN

In medieval times, most people believed in what Michael Shermer calls the witch theory of causality, the idea that "women cavorting with demons caused assorted maladies and calamities."[30] Such beliefs, no doubt sincerely held by our pre-scientific ancestors, led to practices and attitudes toward women now only seen in horror movies. We have made much moral progress since the Middle Ages, and one of the important lessons we have learned is that the wrong theory of causality leads to the wrong policies and practices. This conclusion remains as valid today as it was several hundred years ago. In order to devise the wisest, most efficient, and most humane policies, we must do so on the basis of our best understanding of the world and not rely on old myths that have lost all scientific credibility. One such myth is the soul narrative. And so, like the witch theory of causality, it too must be abandoned because it gets in the way of progress and a more just and humane society. Not only do we not lose anything by letting go of the soul, we in fact have much to gain.

During the 2012 presidential campaign, one of the billboards I drove by was an ad that read "Obama supports gay marriage and abortion. Do you?—Vote Republican." At the heart of the perennial debate over women's reproductive rights, passionately rehashed every election cycle, lies the question of when life begins. Religion teaches people that the moment of conception is the relevant benchmark because it is when the embryo is allegedly ensouled. An in-depth analysis of public opinion on the question of abortion carried out by the Gallup organization reveals that the strength of one's religious beliefs is the major force driving attitudes on abortion.[31] That, in the twenty-first century, an argument ultimately based on the question of when the soul enters the body can be used as a political wedge to divide people over the question of abortion is one of the scandals of American life.

At the other end of the continuum, when life approaches its deadline, questions about the right to die with dignity arise. At the time of this writing, only a few countries, and a handful of states in the United States, have legalized physician-assisted suicide for the terminally ill. While public opinion surveys reveal that Americans are divided over the question of whether such laws should be passed, they also show that those opposed to these laws are the more religiously inclined respondents. Here again, the link to soul beliefs is all too clear. Christian educational material asserts that "God alone has the right to initiate and terminate life,"[32] and reminds their readers that the process of dying is important because it brings the soul closer to God and interrupting it would therefore be a grave mistake. Christianity is not unique in this regard. Most religions are against assisted suicide for similar reasons.

During the run-up to the 2012 presidential election, Texas governor Rick Perry was asked by NBC news anchor Brian Williams how he felt about having overseen the execution of 234 death-row inmates in the Lone Star State. Surprisingly, the crowd erupted into cheers at the mention of that number. When asked why he thought the audience reacted that way, Perry answered, "I think Americans understand justice." This may be so, but Americans should also know that we have the most dysfunctional criminal justice system in the developed world. We incarcerate more people than any other nation on Earth, and unlike all other first world countries, we still have the death penalty. In his book *Thinking about Crime*, criminal law and policy expert Michael Tonry describes the US criminal justice system as "a punishment system that no one would knowingly have built from the ground up. It is often unjust, unduly severe, it is wasteful, and it does enormous damage to the lives of black Americans."[33]

In an influential article titled "For the Law, Neuroscience Changes Nothing and Everything," cognitive neuroscientists Joshua Greene and Jonathan Cohen argue that our dualistic intuitions lie at the heart of our retributive impulses and therefore that dualism has a direct influence on our criminal justice system. As Greene and Cohen put it, "New neuroscience will change the law, not by undermining its current assumptions, but by transforming people's moral intuitions about free will and responsibility."[34] As we saw earlier, soul beliefs lead to a libertarian or contra-causal view of free will and moral responsibility. On this account, retribution is warranted, because the contra-causal soul can, at will, decide to override the effect of any causal factor. Remove the soul from the equation, accept the conclusion that people are fully caused, and retribution loses its *raison d'être*. As Greene and Cohen conclude, "We foresee, and recommend, a shift away from punishment aimed at retribution in favour of a more progressive, consequentialist approach to the criminal law."[35]

The late science fiction writer Barrington Bayley opened his fascinating book *The Seed of Evil* with a short story called "The God Gun," in which two Englishmen manage to kill God and in the process destroy every living creature's soul. As a consequence, life loses all its beauty and meaning. For Bayley, at the beginning of his journey, the soul is a blessing. By the end of the book, Bayley recounts another story, called "Life Trap," in which the main character discovers what lies beyond death. The vision truly horrifies him as he learns that he is bound to relive exactly the same life because his soul is trapped in a never-ending loop, inexorably bouncing between the poles of birth and death. For Bayley, this is the dark secret of the soul. It wasn't a blessing after all but a terrible curse. For us today, much the same can be said. Far from being the blessing that so many take it to be, belief in the soul is actually what stands in the way of progress and a more just and humane society.

Chapter 9

FREE WILL

Jonathan Pearce

Whether or not free will exists is something which has concerned thinkers of all varieties for rather a long time. And the debate still rages. For some. For others, like myself, it is a case of accepting an illusion of sorts and wondering more about what we do with that knowledge, pragmatically, for ourselves and wider society.

So what is free will? Well, defining the term, as is often the case with matters philosophical, is the problem. In fact, it is such a problem that many debates about free will get mired in equivocation. What I will seek to do in this chapter is spell out a version of free will that not many philosophers agree with or adhere to, but, importantly, one that is favored by a good many theologians and apologists. I will explain why this concept is incoherent and go on to mount a robust case that it is also debunked empirically by the sciences. Yes, I have to do the philosophy before I do the science; hard luck to those readers who are averse to such a privilege! Having set out the rational evidence for its negation, I will show how this is terminal for any belief in a judgmental god, as is entailed by Christianity. In short, that free will is an illusion fundamentally debunks most notions of the Christian god.

DEFINING OUR TERMS

For your average layperson,[1] and this particularly includes your religious laity, free will is simply defined by the ability to do otherwise. Being a little tighter on the definition, I would state that it is:

> The real, rational, and conscious ability to do otherwise in any given sce-
> nario, all things remaining equal.

In other words, this ability is derived both consciously and rationally, and an agent could really decide otherwise, as opposed to just theoretically having options. This sounds quite intuitively plausible, which is understandable given that most people who give the subject little thought would probably ascribe to such a working definition. Our inherited language and culture are littered with reference to and belief in such a paradigm.

Let us look, though, at what exactly such a belief would entail and why it fails. The above definition I will call libertarian free will (LFW), and the person who believes in it the libertarian free willer (LFWer).

For free will to make much sense at all, it needs to be grounded rationally. To say that one's choice to do an action is entirely irrational or a-rational is to deny any kind of sensible ownership of an action. In other words, it seems not to allow for the kind of responsibility people attach to freely willed decisions and resulting actions.

Imagine a decision. For example, let us take Wendy. She decides at 09:15 to give five dollars to a homeless person she passes in the street. Now imagine that the world continues for any amount of time (say, ten minutes). We then rewind the world back to 09:15. The LFWer believes that Wendy, at 09:15, could just as well have decided not to have given the money to the homeless person, rationally and consciously.

Let us now concentrate on 09:15. This snapshot of time I will call the causal circumstance (CC). A causal circumstance is made up of, well, everything in the universe at that "snapshot." This would include, for any given person:

a) Being born.
b) Their genetic inheritance.
c) Their life in the womb, shaping their genetic self.
d) Their time and place of birth.
e) Their parents, relatives, race, and gender; their nurture and experiences in infancy and childhood.

f)   The mutations in their brain and body throughout life, and other purely random events.

g)   Their natural physical stature, looks, smile, and voice; their intelligence; their sexual drive and proclivities; their personality and wit; and their natural ability in sports, music, and dance.

h)   Their religious training, economic circumstances, cultural influences, political and civil rights, and the prevailing customs of their times.

i)   The blizzard of experiences throughout life, not chosen by them but that happened to them. All the molecules, particles, forces, and wave functions (i.e., the environment).

The point being that these things are in place, immovably so (since they are all variables that are inaccessible by being in some sense in the past, or instantaneously ungovernable) at the time of the decision making. In this causal circumstance (let's call it CC1), Wendy chooses to give the money away (let's call this A). So, in CC1, Wendy does A.

What this means is that Wendy's choice is grounded in and is caused by all of those variables at play at that exact moment. The rational deliberative processes in her brain (themselves variables in the eventual decision) act upon those variables and the outcome is giving the money away.

Now let us return to this idea of continuing to live and then rewinding. This is a thought experiment. It does not matter whether we could, in actuality, rewind the clock. Perhaps quantum indeterminacy would get in the way, or Turing Problems, or some such idea. That is not relevant. What *is* relevant is that Wendy returns to this identical scenario, that being a) to i). *Everything* is the same. We have CC1 in every detail. However, this "second" time, the LFWer believes that Wendy could choose *not* to give the money away, or ~A (not A). This might look like a contravention of the Law of Non-Contradiction, such that it is true that in CC1 Wendy does A and ~A (although there would be philosophical debate as to what truth values are at play here).

The real problem, though, is what would rationally ground this "second," different decision. Since everything in the universe, down to the rationalizing processes and experiences in Wendy's brain would be identical, what possible reason could there be for Wendy choosing to do otherwise? What reason could there be that wasn't there "before"? What would cause a different weighting of deliberation that wouldn't have manifested itself before?

To add to this, random does not help either. Random and consciously rational free will are not good bedfellows, whether induced by quantum mechanics or not. As Saul Smilansky says of the problem:

> The reason that I believe that libertarian free will is impossible, in a nutshell, is that the conditions required by an ethically satisfying sense of libertarian free will . . . are self-contradictory and hence cannot be met. This is true irrespective of determinism or causality. Attributing moral worth to a person for her decision or action requires that it follow from what she is, morally. The decision or action cannot be produced by a random occurrence and count morally. We might think that two different decisions or actions can follow from a person, but *which* of them does, for instance, in the case of a decision to steal or not to steal, again cannot be random but needs to follow from what she is, morally. But what a person is, morally, cannot be under her control. We might think that such control is possible if she *creates herself*, but then it is the early self that creates a later self, leading to vicious infinite regress. The libertarian project was a worthwhile attempt: it was supposed to allow a deep moral connection between a given act and the person, and yet not fall into being merely an unfolding of the arbitrarily given, whether determined or random. But it is not possible to find any way in which this can be done.[2]

Causality is admittedly a complex thing, but given that the term means *something*, contains some *meaningful content or property*, we can safely say that something either has the property of being caused or uncaused (or perhaps a mixture of the two). There is no middle ground or other option here, in the same way we can ascribe the term *mortal* to something, it either has the property of being mortal or ~mortal. So something is caused or uncaused. Uncaused is synonymous with random such that if something happens randomly, truly randomly, it is in effect uncaused.

When something happens, we should always ask *why*. In the case of an event involving an agent and their decisions, the LFWer usually gives up asking the why question when we get to the agent. But, as philosopher Arthur Schopenhauer once said, "A man can do what he wills, but he cannot will what he wills." This shows us that we should still keep asking the why question, even of people's intentions and desires. And if there is an answer to these questions, we invite some kind of causal determinism: effects have a cause, a reason, for being so. LFWers try to ground decisions in the agent, so that Wendy above would become like a miniature version of God, a Prime Mover, an Originator of a causal chain. LFWers cannot have causality working *through* the agent, otherwise reasons for the agent's decision are not grounded (rationally and consciously) in them. In other words, to answer why Wendy chose to give the money away, that reasoning must originate, and not defer to antecedent causality, *in Wendy* (for her to be "morally responsible" for the action, supposedly). Usually, theologians and apologists only ascribe such prime mover skills to God (as can be seen in the Kalam Cosmological Argument [KCA][3]), and yet here they are, smuggling such origination into human beings every time they commit a freely willed action (indeed, LFW and the KCA are incompatible)!

I hope you can clearly see the problems for LFW. A decision in an agent needs to be grounded in the agent with no recourse to antecedent causality. The agent needs to be able to choose A or ~A in any given scenario, and they need to be able to do this with the entire universe being identical (simply the same causal circumstance), thus stripping any differential grounding away from the different decisions. This makes the reason for any difference in decision looking rather like random. There is nothing that could cause the decision, and neither would the LFWer want there to be.

And this is why it makes no sense as a logical position. It can get a little more dry and a lot more complex than this, but, as Smilansky says, this is it in a nutshell.

The remaining positions, then, are *determinism* (often called *hard determinism*), where the universe strictly adheres to cause and effect (or *adequate determinism*, where it adheres to cause and effect on the macro level but has quantum indeterminacy at the micro level) or *compatibilism*, where determinism and free will are compatible with each other. These positions (quantum aside) leave the agent unable to do otherwise, such that compatibilism is often called *soft determinism*. And this is where the equivocation comes in. Determinists deny LFW. It does not exist, so free will does not exist. Compatibilists deny LFW, too. But they take the term *free will* and mold it into something new; they redefine it. Compatibilists usually define it as:

> The ability to consciously and rationally do that which one desires.

Of course, that which one desires will itself be determined, but there is a sense of the agent being the author and owner of their actions. It is worth reminding ourselves here that quantum indeterminacy is no bedfellow to this understanding of free will, since if some of the variables involved in a decision are random, an agent cannot be said to have rational or conscious control over that decision (which is why there is a more recent move to label free will deniers as *hard incompatibilists* rather than *hard determinists*). Other terms, like *volition* and *intention*, are often thrown around, too. We need not concern ourselves with these positions here, because our interest lies in what the average, and not-so-average, Christian thinks.

I hope this shows that the Christian, as an LFWer, is struggling to hold on to a coherent worldview, from a logical and philosophical perspective. Now let us look to the empirical sciences to see what they have to say about the matter.

CORRELATION AND CAUSATION

Data is a funny thing. Science seeks to understand the world, as a method, which means finding out how and why things happen. When we look at the behavior of humans, such that a subset of people are more likely to do X than Y, then we are accepting a correlation, and thus most probably a causation somewhere. The causation might not be direct such that A causes B (this can often be a problem manifesting itself in the *correlation fallacy*) but may involve at least a third party such that C causes *both* A and B.

What would the world's data look like if everyone had perfect, unconstrained free will? Well, the data should look the same as random distribution because if there was no causality acting on decisions, then things would look random. That is what "true" free will would look like. The conclusion to draw here is that any such data that shows a group of people more likely to behave one way than another is an acceptance of determining factors and a mitigation of the sort of free will we were discussing earlier.

Every time, and I mean *every time*, you see such data, bear in mind that this data is a mitigation of free will. It may be the statistic that men are 882 percent more likely to commit violent behavior and crime,[4] or that men born with birth difficulties *and* who are rejected by their mothers are three times more likely to be violent and have antisocial behavior,[5] or that certain people on the autistic spectrum are less likely to believe in a personal god.[6] So on and so forth. Each one of these is a denial of LFW of sorts, or is at least evidence in support of its negation.

Some libertarians will claim that, yes, we are largely determined, or influenced, but that we can overcome this problem with our own volition. This is what I call *the 80/20 Problem*. That is to say, if an LFWer claims that they are influenced, say, 80 percent, then this leaves 20 percent of a decision, making the process open to agent origination (this is often put forward by proponents I have spoken to). The problem is that all the logical issues I mentioned above are now distilled into the 20 percent. In effect, it makes the problem worse. The LFWer here accepts much determination but allows a small window of opportunity and does not escape the grounding objection and any other logical issue with causality previously expressed. The problem of LFW is even more acute, then.

In summary here, bear in mind stats. They don't help the LFWer.

GENES AND GENETICS

Darwin got it right, and he had no idea of the primary mechanism in which to deliver his theory of heredity. A well-calculated estimation of reality indeed. Geneticists, these days, study the very mechanics of how we derive our characteristics. Those who argue for free will are much more likely to accept that genes carry the information for defining our physical characteristics, but argue that our personalities and emotions and decisions that sprout forth from such things are not determined genetically. This appears to be special pleading. Why should genes not define everything about our phenotype? That I generally get angrier than another person, all things remaining equal, or more morose or am prone to depression or what have you, is governed in large part by my genetic heritage. It will usually not be as simple as having a single *gene for* some characteristic or another (even height is determined by perhaps a hundred or so genes interacting) but a set of genes that work in unison to eventuate some characteristic or another. This is not an espousal of genetic determinism such that genes determine behaviors 100 percent. Yes, there are things like being born a certain sex being a genetically determined state of affairs and all the ways this will curtail your free will (I cannot give birth, for example). However, genes interact with the environment, with both determining the outcome. And then we can add the expression timings of genes (epigenetics). It is a complex field. Here are some examples of findings of genetic research (there are some smatterings of genetic points made in other sections, too) that should be of interest:

- Schools are not as important as we think for educational attainment, as genetic heritage is thought to be responsible for 58 percent of attainment in science, and slightly less in other subjects, with environment (home or schools) being only 29 percent responsible.[7]
- Dean Hamer's controversial God Gene is thought to show a spiritual predisposition, which then leads to a greater likelihood of believing in God.[8]
- Autistic people are less likely to believe in a personal God. (Does this mean God hates autistics and doesn't want freely willed relationships with them?)[9]
- Our political predilections and choices are to some degree genetically determined[10] (being also connected to disgust sensitivity, as the work of David Pisarro and others show, and intuitive moral value systems, as the work of Jonathan Haidt shows).

- Male sexuality (female sexuality is seen as more fluid and often more cerebral) has huge amounts of evidence to show genetic, biological, *and* environmental determination to varying degrees.[11]

So on and so forth. I think it unnecessary to list journal after research paper to get a robust understanding of how important genetics is in defining who we are, who we will become, and the decisions we will surely make in a given situation. Cognitive scientist Steven Pinker sums it up with aplomb in his book *The Blank Slate*:

Identical twins think and feel in such similar ways that they sometimes suspect they are linked by telepathy. When separated at birth and reunited as adults they say they feel they have known each other all their lives. Testing confirms that identical twins, whether separated at birth or not, are eerily alike (though far from identical) in just about any trait one can measure. They are similar in verbal, mathematical, and general intelligence, in their degree of life satisfaction, and in personality traits such as introversion, agreeableness, neuroticism, conscientiousness, and openness to experience. They have similar attitudes toward controversial issues such as the death penalty, religion, and modern music. They resemble each other not just in paper-and-pencil tests but in consequential behavior such as gambling, divorcing, committing crimes, getting into accidents, and watching television. And they boast dozens of shared idiosyncrasies such as giggling incessantly, giving interminable answers to simple questions, dipping buttered toast in coffee, and—in the case of Abigail van Buren and Ann Landers—writing indistinguishable syndicated advice columns. The crags and valleys of their electroencephalograms (brainwaves) are as alike as those of a single person recorded on two occasions, and the wrinkles of their brains and distribution of gray matter across cortical areas are also similar.[12]

And:

The genes, even if they by no means seal our fate, don't sit easily with the intuition that we are ghosts in machines either. Imagine that you are agonizing over a choice—which career to pursue, whether to get married, how to vote, what to wear that day. You have finally staggered to a decision when the phone rings. It is the identical twin you never knew you had. During the joyous conversation it comes out that she has just chosen a similar career, has decided to get married at around the same time, plans to cast her vote for the same presidential candidate, and is wearing a shirt of the same color—just as the behavioral geneticists who tracked you down would have bet. How much discretion did the "you" making the choices actually have if the outcome could have been predicted in advance, at least probabilistically, based on events that took place in your mother's Fallopian tubes decades ago?[13]

NEUROSCIENCE: GOING BEYOND LIBET

Benjamin Libet's experiments,[14] and subsequent versions of it carried out by others, are somewhat overused in such debates. Essentially, we have worked out that when someone decided to press a button, it appears that the brain kicks into action sometime before the conscious mind decides to press. In other words, the nonconscious or subconscious brain is doing the decision making. Using fMRI scans, experimenters can also, with some degree of accuracy, predict which hand the subject will use before they are consciously aware (from between three and ten seconds in advance). Scientists have even given transcranial magnetic stimulation to subjects and have made them choose the button on the left or right, to which the subject adds supposedly conscious intention afterward.[15] Variations of these experiments have been repeated to great effect.

Further to these findings, similar investigations into the supplementary motor area (SMA) of the brain have shown that when electrical stimulation has been applied, irrepressible urges to make *voluntary* movements follow. As J. M. Pierre observes:

Doing away with free will does not mean that human beings are automatons in which our motivations, desires and values do not influence action. On the contrary, neuroscience clearly distinguishes between voluntary behavior and involuntary acts or reflexes. With volitional behavior, organisms have choices and make decisions about whether and when to act, what to do, and what not to do. However, in a neuroscientific model, such choices are made within neural networks rather than any immaterial homunculus and often occur outside conscious awareness and before our subjective sense of intention or agency.[16]

In other words, physical areas of our brains appear to control agency and intention, which appears to debunk the idea of a freely volitional model.

Many dispute what the Libet studies actually tell us and whether these automatic types of actions or reactions are the same as more deliberative, rationally derived choices. Let us look, then, at other studies in the area of neuroscience that have created some interesting findings. For example, Chun Siong Soon and his colleagues have taken Libet's study further:

> Here, we show that the outcome of a free decision to either add or sub-tract numbers can already be decoded from neural activity in medial prefrontal and parietal cortex 4 s before the participant reports they are consciously making their choice. . . . Our results suggest that uncon-scious preparation of free choices is not restricted to motor prepara-tion. Instead, decisions at multiple scales of abstraction evolve from the dynamics of preceding brain activity.[17]

Meanwhile, Elisa Filevich and her colleagues looked at finding precursors for *free won't* decisions. Often we think of free will decisions as ones that we choose to do, but often they seem to be inhibitive decisions or decisions *not* to do something. They conclude:

> Our results suggest that an important aspect of "free" decisions to inhibit can be explained without recourse to an endogenous, "uncaused" process: the cause of our "free decisions" may at least in part, be simply the back-ground stochastic fluctuations of cortical excitability. Our results suggest that free won't may be no more free than free will.[18]

Liane Young and her colleagues have astonishingly found that in making moral judgments, a key area of the brain is a knot of nerve cells known as the right temporoparietal junction (RTPJ), and that by sending in transcranial magnetic stimulation (TMS) they were able to change people's moral judgments.[19] The judgments of the subjects shifted from moral principle to verdicts based on outcome (in philosophy we might say from deontology to consequentialism). The ramifications of which are that such judgments are physical in nature or grounding and that physical influences in the brain are likely to have an effect on core moral judgments.

I will now provide you with one of the most important points within this chapter. It concerns Charles Whitman, an otherwise intelligent (138 IQ), "normal" man, who did a very abnormal thing. Whitman went on a mass shooting rampage in and around the Tower of the University of Texas in Austin on August 1, 1966. Three people were shot and killed inside the university's tower and eleven others were murdered (thirty-two were wounded) after Whitman fired at random from the twenty-eighth-floor observation deck of the Main Building before being killed himself. Prior to the shootings at the University of Texas, Whitman had murdered both his wife and mother in Austin. His suicide note reads:

> I do not quite understand what it is that compels me to type this letter. Perhaps it is to leave some vague reason for the actions I have recently performed. I do not really understand myself these days. I am supposed to be an average reasonable and intelligent young man. However, lately (I cannot recall when it started) I have been a victim of many unusual and irrational thoughts.[20]

I would advise looking at the terrible accounts of what took place but also of his own written thoughts and feelings. Suffice to say that, as according to his wishes, an autopsy on his body took place, and it revealed much to explain the man's actions. This former Eagle Scout, marine, bank teller, volunteer, man with good IQ, had (you guessed it) a tumor about the size of a nickel. The glioblastoma had grown beneath the thalamus in the brain, impressed upon the hypothalamus and compressed the amygdala. The amygdalae are involved in emotional regulation, particularly fear and aggression. Damage to this part of the brain, known since the late 1800s, causes emotional and social disturbances; biologists in the 1930s had found that damaging monkey amygdalae caused blunting of emotion, lack of fear, and overreaction. Female monkeys with such damage showed inappropriate maternal behavior, neglecting or abusing their infants.

Whitman hit the nail on the head when he predicted that his behavior, increasing in oddity, was rooted in an issue in his brain. Remember his appeal to give any monies remaining to a mental health foundation. His friends had also noticed changes to his character and personality.

The point I want to make here is that most people think that such a tumor would abrogate moral responsibility in the agent. In other words, Whitman should not be deemed fully morally culpable for his actions since his brain was impaired: he couldn't help himself. I want to look at this claim because it implies that a neurotypical person is categorically different to someone with a brain tumor.

I contest this.

Of course, the tumor makes a person act differently to that which they would have done. But all it actually does is change one form of determined outcome into another. It is not, I posit, a categorical difference. I think people make this mistake too often, such as in the sort of claim that follows: "I think it's not sensible to infer anything about 'normal' cognition from the experience of people who exhibit obviously abnormal cognition."

Behavior X is caused by brain state/neural circumstance/genotype Y. The comment above is separating neurotypical people from non-neurotypical. This is problematic and perhaps entirely subjective anyway. Another oft-stated, highly problematic issue is that non-neurotypical behaviors are uniquely caused by certain circumstances. For example, that autism is caused by X (brains state, genes, brain dysfunction etc.) but that a neurotypical person's behavior is caused by their will, their volition. But, of course, we should be able to infer that since causation is happening in such *abnormal* situations, to claim that mental causation from physical scenarios does not take place in the *neurotypical* ones is special pleading.

In other words, because someone ends up doing something "normal" does not mean they are exempt from causality; rather that "normal" behavior results from "normal" brain states (labeling notwithstanding) that are just as physically grounded in the brain. Brain states and physical phenomena cause mental phenomena. The neurotypical person may have fully functioning rational architecture, but they have no control over choosing that architecture. The outcome is still determined by such processes.

Another way to put it: *neurotypical brains don't suddenly give the agent the special ability to evade logic and philosophy and magically give the agent the ability to do otherwise.*

The above scenario is also exemplified by the recent case of a man suddenly becoming a pedophile, finding out he had a tumor, having it removed, losing the pedophilic tendencies, and then regaining the tendencies, only to find the tumor had returned. It was then removed again, and he again lost those undesirable urges and tendencies.

In simple terms, the brain causes (or some might claim *is* in some way) consciousness. If you contest that brain states cause consciousness or that consciousness supervenes on brain states, then try sticking a fork into your brain and denying an effect on your consciousness.

Again, it is worth mentioning that for every piece of research I have mentioned here, there are hundreds of others that could have been chosen.

PSYCHOLOGY AND PSYCHIATRY

In some sense, there is a huge overlap between the last section and this one, since much psychology is dependent on the neuroscience that underwrites it; the neural structures give rise to psychological events and behaviors. But these behaviors can equally be socially learned, influenced, or constructed.

Priming

Let us first consider the area of priming, which seems to have been gathering evidence apace for its prevalence. This is what is often called, by laypeople, subliminal messaging. The idea is that we are influenced, without our conscious knowledge, in the decisions we make, including the very words we choose. Anecdotally, we all know that when we hear a particular word in conversation that we haven't heard for some time, we end up using it ourselves over the next few days. We know from a study done by John Bargh and his colleagues that people primed with words to do with old age walked down the hall more slowly (in some cases shuffling) compared to the control group.[21] Ap Dijksterhuis and Ad van Knippenberg found that people primed with words associated with intelligence did better in tests than the control.[22] As J. M. Pierre also states:

> A substantial body of similar experiments has demonstrated that sub-liminal primes can activate goals related to a wide variety of higher-order behaviors including social interaction, cognitive performance, moral judgment and decision making. The ability of unconscious stimuli to unwittingly affect the outcome of voluntary behavior therefore does not appear limited to simple motor movements.[23]

The work of David Pizarro and colleagues on disgust sensitivity is really interesting.[24] It turns out that our disgust sensitivity is a good predictor, both in the lab and out in the voting world, of our political worldviews and voting habits.[25] If we are easily disgusted, we are more likely to be right-wing (i.e., Republican) voters, and vice versa. However, within the realms of morality, he has shown that when primed with disgusting sights, smells, tastes, or even "Now Wash Your Hands" signs, people will morally judge actions as being worse (particularly more taboo actions like homosexual sex)[26] or even promote the shunning of certain minority groups.[27] In other words, our moral judgments aren't so rationally derived; or, at least, there is a lot of subconscious influence (added to which William Killgore and his colleagues showed that "sleep deprivation impairs the ability to integrate emotion and cognition to guide moral judgments"—we're not very good at making moral judgments when faced with a lack of sleep[28]). Pizarro's work on many areas of moral judgment and priming are fascinating and well worth looking into. As Pizarro claims himself,

> We have shown, for instance, that people become insensitive to the risks of a gamble taken to win chocolate chip cookies when the smell of freshly-baked cookies permeates the lab, and that men report a greater willingness to engage in risky sexual behavior when sexually aroused.[29]

This builds on Jonathan Haidt's work, whose landmark paper "The Emotional Dog and Its Rational Tail: A Social Intuitionist Approach to Moral Judgment" showed that we morally judge using our intuitions and then *post hoc rationalize* afterward (meaning that we add on our reasoning after already reaching our conclusion from gut feeling).[30] Morality is often not rational.

There are also other priming effects that challenge our sense of ownership over decisions:[31]

- *Mere exposure effect*: This is the effect that merely hearing a name or idea, or seeing a face, etc., will predispose you to favor that idea or person. In other words, the maxim "any publicity is good publicity" is actually quite right! If you were to see a picture, even subconsciously, of someone before viewing a number of pictures of people's faces and were asked to rate them all for attractiveness, you would more likely rate the one you had subconsciously seen a higher attractiveness than if you hadn't already seen their face.

- *Illusion-of-truth effect*: Similar to the last, but with perhaps massive ramifications, you are more likely to rate as true something you have seen or heard before. Think, here, of young-earth creationism, evolution as false, political claims as true, etc., and you will get an idea of its importance to the context of this book. So repeatedly hearing a falsehood (church sermons!) only serves to reinforce its truth value! When Bill Nye debated creationist Ken Ham, this was the rationale behind skeptics who criticized the idea as giving oxygen to the creationist movement.

- *Subliminal pairing* takes place in connecting two ideas. For example, in Bush's TV campaign against Al Gore, "the Gore prescription plan" on the screen was simultaneously shown with the word RATS (which eventually became the word BUREAUCRATS). The idea being that the two concepts would be subliminally linked.

Implicit Egotism

Another aspect to priming is how we look to ourselves in judgments and decisions outside of ourselves. In other words, we are implicitly egotistical. Such findings include conclusions that:[32]

- We have a statistically significant likelihood to get together with a partner of a name starting with the same letter as ours (my previous two partners began with a J).
- If given two supposedly different teas to taste, we are more likely to favorably rate the taste of tea that starts with letters of our name than one that doesn't. This is especially interesting given that both teas came from the same pot in the first place!
- We are more likely to rate Rasputin as a morally nicer person if we are told his birthday is on the same day as ours.
- People named Denise or Dennis are disproportionately likely to become dentists; Laura and Lawrence more likely lawyers, roofers are more likely to begin with R, and so on.

Life Skills and Other Predictive Studies

I could write an entire chapter merely on priming, but alas, there is much more to include. So far, we are beginning to see that we are not as rational and deliberative as we might perceive ourselves to be. Our decisions are victims to the vagaries of environmental influences around us and within us.

One particularly famous study was carried out by Walter Mischel and colleagues, who conducted the famous marshmallow/cookie experiment.[33] Here, the subjects were young children of three to five years old who were given a cookie or marshmallow, but the experimenter promised that they could have two if, after the experimenter returned some fifteen minutes later, the single cookie or marshmallow remained. This was to test delayed gratification: could subjects put off an instant gain for a greater later gain? The interesting findings were not originally planned, whereby the testers looked at the original subjects at several later points in their lives. It turned out that the ability to delay gratification at three to five years predicted general life success skills[34] and SAT scores[35] in later life. Brain scans also showed differences in the prefrontal cortex and ventral striatum of individuals. Further studies that came out of these have shown delayed gratification of preschoolers as predicting body mass thirty years later.[36] A study conducted in 2011, which took brain images of a sample from the original Stanford participants when they reached midlife, showed important differences between those with high-delay times and those with low-delay times. These manifested themselves in two areas—the prefrontal cortex (which was more active in high delayers) and the ventral striatum (an area linked to addictions)—when they were trying to control their responses to the various temptations.[37]

Another similar piece of predictive research (out of many such pieces) is that of Yu Gao and her colleagues, which looked at children who had poor fear conditioning (a lack of showing fear of consequences, for example), finding that at *age three* such poor fear conditioning predisposes the child to adult crime. In other words, one can, with some degree of accuracy, predict criminality twenty years later in a child at the age of three based on their ability to show fear, having its basis in prefrontal cortex dysfunction. As Gao concluded:

> Poor fear conditioning at age 3 predisposes to crime at age 23. Poor fear conditioning early in life implicates amygdala and ventral prefrontal cortex dysfunction and a lack of fear of socializing punishments in children who grow up to become criminals. These findings are consistent with a neurodevelopmental contribution to crime causation.[38]

Kindness is an area that receives a lot of study and analysis. In behavioral studies, kindness is often termed as *prosocial* behavior. It used to be that we thought that kindness was an active suppression of selfishness (in the prefrontal cortex), but work such as that done by Masahiko Haruno and Christopher Frith has shown that it results from activity in the amygdala, a much more automatically functioning area of the brain, controlling intuition and emotion.[39] Conscious deliberation is often not what is taking place in decisions of kindness.

There are simply too many pieces of research concerning kindness, but it is worth considering that twins studies have found that heritability of prosocial behaviors (empathy, cooperativeness, and altruism) have been quantified to be about 50 percent heritable (the other 50 percent being socialized norms and situational factors). Studies abound that show interesting findings, such as that carriers of the COMT gene variant donate twice as much to charity in a given scenario than those without.[40]

Intention

Much research has been dedicated to the idea of intention (for example, as summarized by the late Daniel Wegner in his superb book, *The Illusion of Conscious Will*[41]) and the idea that if intention is missing in the jigsaw of scenario-intention-action, then intention is invented *after the fact* by the mind (post hoc rationalization). The question then follows as to whether, if intention is invented at least *some of the time*, it could be the case that this takes place *all of the time*? This kind of *epiphenomenalism*, whereby consciousness, and in this case intention, is a byproduct of brain states after the fact, is hotly debated both in neuroscience and philosophy. But the evidence does favor, in some sense, an epiphenomenalistic approach.

In the case of inventing intention, this is evidenced by various experiments, including those done with split-brain patients, hypnotized people, and neurotypical people. For example, with left-right hemisphere split-brain patients (having confusion over language and visual stimuli due to the hemispheres of their brains being split, by accident or necessary surgery), certain words shown to the subjects (such as "walk" or "laugh") will cause the subjects to do this, and then invent the reason for doing so. Thus, his left brain would kick in and make up a reason as to why the right brain had made him laugh: "You guys come up and test us every month. What a way to make a living." The word "walk" shown to the right brain would make the patient get up and leave. Upon being asked why, he answered, "I'm going into the house [from the testing van] to get a Coke."[42] This is one of several such examples of left to right brain interaction that leads to the invention of intention.

An example involving hypnosis can be seen in a quote from eighteenth-century German hypnotist Alfred Moll:

> I tell a hypnotised subject that when he wakes he is to take a flower-pot from the window, wrap it in cloth, put it on the sofa, and bow to it three times. All which he does. When he is asked for his reasons he answers, "You know, when I woke and saw the flower-pot there I thought that it was rather cold the flower-pot had better be warmed a little, or else the plant would die [*sic*]. So I wrapped it in the cloth, and then I thought that as the sofa was near the fire I would put the flower-pot on it; and I bowed because I was pleased with myself for having such a bright idea." He added that he did not consider the proceeding foolish, [because] he had told me his reasons for so acting.

Often such invention is as a result of gut reactions or intuition, which brings about non-rationally derived decisions which are then rationalized after the fact. The famous Daniel Kahneman (summarized in his book *Thinking Fast and Slow*) has spent decades looking into this, as have many others. These are decisions that bypass the conscious and rational decision-making processes. Cognitive Dissonance, one of the most widely referenced psychological mechanisms, is very much associated with these sorts of scenarios. In fact, cognitive biases (which I could discuss at great length) are themselves prima facie evidence against a freedom of the will, insofar as they are biases that operate under our radar to work against rational decision-making from the individual.

CRIME AND CRIMINALITY

The research that highlights determination on human behavior is staggeringly large. However, it is worth looking in particular at the drivers for criminality, especially since, like kindness, this is key to notions of judgment from others, like a god.

One fascinating synopsis of the work into the biological and genetic causality involved in criminality is Adrian Raine's superb *The Anatomy of Violence*, and I would thoroughly endorse reading it to understand the causes of violent behavior. Society has always been morbidly obsessed with psychopaths and their behavior, and it is worth noting that they represent only around 1 percent of the population,[43] but they make up approximately 10 to 15 percent of offenders.[44] Probably the most concise way to express many of the findings in his book is merely to list but a few of them:

- According to neuroimaging, most murderers have decreased functioning in the prefrontal cortex, which inhibits volatile emotions like rage and frustration. Psychopaths' brains also differ from normal brains in structure and functioning, with an impaired amygdala and prefrontal cortex.
- Psychopaths have a low-resting heart-rate (thus they are literally cold-blooded!).
- Raine's studies show that omega-3, for example, is critical to brain development, particularly the parts that prevent violence. In one study, children who were provided better nutrition, including increased fish consumption, together with other programs, had a much greater reduction of future conduct disorder than those without it. This would imply that, being overloaded with simple carbs, the diet of most Americans may contribute to increased violence.
- The conclusion of one of his own papers shows that genetic factors on the most reliable measures explain 96 percent of the variance in antisocial behavior.
- Repeat criminal offending should be seen as a clinical syndrome and should be seen as being at the mercy of their biology.

And so on. As Raymond Tallis of the *Guardian* stated of the book:

Raine's key notion that, good or bad, we are the playthings of our brains— "free will is sadly an illusion" (the return of the lumbering robots)— raises the question of why we should stop at the brain in our search for causes. Given that it is a material object wired into the material world, "my brain made me do it" (kill my spouse, write a book on neurocriminology) should translate into "the Big Bang" (ultimately) made me do it.

In fact, the brain is but one player in the complex game of life, not the beginning and end of our destiny.

And Raine seems gradually to accept this. For all his headline-grabbing talk of "murderous minds," "broken brains" and "natural born killers" he ends with "the biosocial jigsaw puzzle," where "the social environment beats up the brain and reshapes gene expression." There is the bit where you say it and the bit where you take it back. He rows back from his initial "biology + genes + brain" thesis towards the kind of "environment (including junk food, toxic metals, maternal rejection, poverty, childhood abuse) + heredity + personal factors" truisms that the rest of us accept. Even so, he is determined to hold on to his brain-centred criminology: "Deprivation makes a big dent on the brain."[45]

I think, without the need to go on, you get the idea of what I am trying to set out here. The notion that we are not as consciously in control of our decision-making as we might like to think is evident. Ezequiel Morsella's "Passive Frame Theory" suggests[46] that the conscious mind is like an interpreter helping speakers of different languages communicate:

"The interpreter presents the information but is not the one making any arguments or acting upon the knowledge that is shared," Morsella said. "Similarly, the information we perceive in our consciousness is not created by conscious processes, nor is it reacted to by conscious processes. Consciousness is the middle-man, and it doesn't do as much work as you think."[47]

Consciousness, per Morsella's theory, is more reflexive and less purposeful than conventional wisdom would dictate. Because the human mind experiences its own consciousness as sifting through urges, thoughts, feelings, and physical actions, people understand their consciousness to be in control of these myriad impulses. But in reality, Morsella argues, consciousness does the same simple task over and over, giving the impression that it is doing more than it actually is.

Conscious will seems to be reactive, the passenger rather than driving the human vehicle. The neural networks and the non-conscious brain make for a powerful mechanism. As far as evidence is concerned, then, I will leave it there. This is a seemingly endless warren of fascinating discovery, and at some point an arbitrary line in the sand must be drawn.

EVIDENCE FOR FREE WILL?

What would evidence for libertarian free will look like? One might think that defenders of LFW would be justified in asking what evidence there would be for it. The problem with this is that if, as I maintain, libertarian free will makes no logical or causal sense, then the concept of having evidence to support it is rendered incoherent. What could possibly act as evidence to something that behaves a-causally? At best, such a-causality would appear to be random. The problem with LFW evidence is the problem with LFW logically.

Some, such as Mark Baker and Stewart Goetz, posit that elementary particles in the brain do not behave in the usual manner, leading to evidence for a soul (theoretically), as supposed by the Standard Model. Fellow contributor to this book, Julien Musolino, does a very good job in taking such tenuous claims to task (here and in his own book).[48]

The only thing that LFW seems to have going for it is that people can *feel* like they have the ability to do otherwise, that, at the point of deciding, one really actually could follow through with the available options.

But then, the world *appears* flat.

RAMIFICATIONS FOR GOD, CHRISTIANITY, AND CHRISTIANS

Calvinists offer themselves to be the odd ones out, certainly in the Christian cloisters of world religion, in that they are deterministic, believing that humans have no free will and everything happens at the predetermined behest of God. Almost all other Christians disagree, and the reason is because they view God as judgmental, with the eventual carrot of heaven or stick of hell being the bribing consequences. For such retributive punishment systems to make any sense, with the notion of *just deserts* driving the punitive cogs, humans have to have the real and actual ability to do otherwise. If this is not the case, then God simply creates people to be condemned to an eternity in hell or rewarded with an eternity in paradise without being able to effect a change to those outcomes.

As I spent some time setting out in my first book, on free will, there are many issues that abound for the classical conception of God, that of an entity with omni- characteristics. A God who is all-knowing, -loving, and -powerful has coherency issues. For example, exactly how does it work that a God can be all-knowledgeable of every single potential counterfactual (an if statement regarding what might happen in a given scenario) in theoretical or actual reality, and that humans have free will to do otherwise? In other words, if God knows that I am going to make a cup of tea at 10:15 on Tuesday morning, then how could I *actually* choose to do otherwise? Some theists claim that just because God knows this eventuality, doesn't mean to say he *causes* it, he just knows my freely willed decision. OK, let's grant that. But this is not quite the point. If I am infallibly, indubitably going to make that cup of tea, then in what possible way am I *able* to do otherwise?

Some theists, such as William Lane Craig, posit that God has Middle Knowledge (also called Molinism) and that God can get around this problem in the way that he created. Middle Knowledge has this order:

Step 1. God's knowledge of necessary truths.
Step 2. God's Middle Knowledge (including counterfactuals).
Step 3. The Creation of the World
Step 4. God's free knowledge (the actual ontology of the world).

Hence, God's Middle Knowledge plays an important role in the actualization of the world. In fact, it seems as if God's Middle Knowledge of counterfactuals plays a more immediate role in creation than God's foreknowledge.

This supposedly ensures that free will is still possible by actualizing a world whereby he knows what will be freely chosen by everyone. God knows all possibilities of what might happen in all possible worlds and decides on which world he wants and creates it (i.e., this one). God then knows the future of this world.

There are many aspects of Molinism that are and have been debated over the years. One such prominent criticism is the "grounding objection," which is based on the following points, looking pretty much the same as my original objection to free will in this chapter: God knows that in CC1, Wendy does A (gives money to the homeless person). God *knows* the outcome. It cannot be A and ~A. Therefore, this choice must be defined or caused by some preceding influence, such as those elements a) through i). Either that scenario (genes, environment, etc.) randomly leads to that outcome, or it causes it. This means that Wendy is not effectively *freely* choosing to go give money. It allows for deterministic values to slip through the door. By saying, "When Wendy sees the homeless person at 09:15, she will give them money," by definition you are denying the alternative possibility of not giving money. Wendy is determined. Defenders of Middle Knowledge claim that God *knowing* Wendy will do this will not *make* her do it, that his knowledge simply "corresponds" with her action. However, this says little about how the causal circumstance affects the decision. God is simply calculating the causal circumstance.

Or, as the *Stanford Encyclopedia of Philosophy* states:

> Even though the theory of Middle Knowledge is a powerful theory of divine knowledge and providence, it is neither necessary nor sufficient to avoid theological fatalism by itself.[49]

One further criticism of Middle Knowledge will serve to segue into my next point—that is, that God's knowledge about which world he would actualize would already be in place due to his foreknowledge. Foreknowledge of all creatures and their possible actions must surely extend to foreknowledge of himself and foreknowledge of what the maximally best choice of universe would be, thus negating the need to "choose" or even to calculate all other "nonevents." It is potentially a chicken and egg situation with what God knows and what he chooses, creating some mind-boggling head scratches. If you believe that God is a-temporal causally before the creation of time, then God can do no choosing, no deliberating, and no calculating—everything is instant and without any type of temporality. God would not have the personhood that we would imagine a "mind" to have.

As a result, God knows every possible outcome for every actualization of every possible world. And God, evidently, chose this one.

First of all, the ramifications are fairly clear for God's own free will. Since he must do what is maximally loving at all times, he cannot do otherwise. One could argue, then, that God does not have free will himself. Without the ability to act contrary to his omnibenevolence, he has only one course of action that he can possibly take, or courses of action that contain equal quantities of "lovingness." A theist could argue that God could do otherwise but chooses not to. This is akin to the taxman analogy, which goes as follows: A taxman assesses your business. He says you have a tax bill for $25,000. He gives you the choice of paying it or not paying it. The free choice is yours. However, by not paying it, you will go to prison or even be sentenced to death. Thus you have a chance to exercise your free will, but one choice will result in your imminent imprisonment or death. What will it be? You can argue, perhaps, that you have free will, but you can also argue that this is an effective denial of free will.

Similarly, God could choose in a way that was not maximally loving, but he never would because it is against his all-loving nature. This is a grey area of free will. There is a debate here as to whether God does not have omnipotence, or whether omnipotence can be a potentiality. If it is a potentiality that can never be made real and existent, then does this equate to it not existing? Surely potential means that in some possible world it *could* happen?

It seems, then, that if God is to keep his omnibenevolent characteristic, then this world must be the maximally perfect and loving world that there can be. If God is perfect, then this must be his most perfect creation. A perfect God could not create something that fell short of perfection, and an all-loving God could not create something that did not fulfil the criterion of being the most-loving creation.

The slightly worrying outcome of this is that a world where 250,000 people and millions of animals are killed in a tsunami, where anywhere between 50 percent and 75 percent of embryos/fetuses are naturally aborted, where cancer and malaria are rife, where a global flood killed all the population of Earth bar eight (and all the animals bar some), where forest fires kill baby deer, is a world where these events are perhaps even necessary for it to be the most loving world. If it is not *presently* the most perfect world, then it must be the most perfect method to getting to that most perfect world. God's foreknowledge and omnibenevolence restrict his decisions to being anything other than that which is omnibenevolent in some manner.

God's free will is also further curtailed, since with his omniscience he cannot be contrary to his own predictions. For example, if you were claimed to be omniscient and omnipotent, and you predicted beforehand that you would make yourself spaghetti bolognaise for supper on Friday, then when it came to making Friday's supper, you would have no choice but to make spaghetti bolognaise. This is because if you decided to be contrary to your own prediction and cook, say, pizza, then your prediction would have been incorrect. This would render your omniscience faulty and would leave you with the characteristic of fallibility.

Likewise, God does not have omnipotence, because he cannot do something that would invalidate his infallible predictions. Or, looking from another point of view, God is entirely constrained by his own fore-knowledge, which, as mentioned, would surely apply to himself. Thus God has no *real* and *actual* ability to do otherwise.

So, logically, God can never be contrary to his own predictions. This has far-reaching consequences: God does not have free will, intercessory prayer is pointless, God cannot change his mind, God's own future and inter-ferences on Earth are determined, and the passage in the Bible where God changed his mind over the fate of Nineveh is patently false (Jonah 3:10).

God gets involved rather a lot in the Old Testament, and he also knows exactly what he will do in the future. These manifest themselves in the mechanism of prophecy. Prophecies coming to pass six hundred years later mean that the world must be micromanaged, or entirely determined, in order to allow for the prophecy to come to pass as predicted. For example, there was a king of Judah (Zedekiah) who is told that God has destined his city to be captured and destroyed; he will be captured and taken to Babylon, where he will die peacefully. But he shouldn't worry too much, as people will surely be sad about his death (Jeremiah 34:2–5). The implications here are that no matter what Zedekiah might do, no matter what stroke of military genius he might have, and despite the fact that his people are supposedly the chosen ones, Zedekiah was going to lose his city, and many people would lose their lives. One assumes that this was destined by God to happen because it all fits into the larger jigsaw puzzle of his intentions for Israel and, as such, the world.

God is decreeing that, no matter what happens, no matter what decisions are made, a good many people will die, and a whole city will burn to the ground. Perhaps God is not actually saying this, perhaps it is not a case of no matter what decisions are made but is more a case that God knows exactly what decisions will be made. This then could potentially lead us to predestination, or determinism. In other words, God has either determined in advance what he wants, and has set all the variables up to achieve these ends, or he has set all the variables up with no particular design for the outcome but can compute what the outcome will be. In either of these scenarios, there is no opportunity for any of the protagonists to exercise anything that resembles a conventional form of free will as already set out here. If Zedekiah had suddenly wanted to go on holiday to Egypt . . . well, he couldn't, since God had decreed exactly what was going to happen to him, and there was nothing anyone could do about it.

This has moral implications for God himself, since this is not just simply a case of determining what Wendy does with five dollars, but it is a case of determining that tens of thousands of people are going to die horrible and painful deaths. Is this the behavior of an omnibenevolent God? Could God have done otherwise? One of the classic defenses that theologians use for the problem of evil (the problem of death and suffering in the world), and of why God shouldn't just simply make there be fewer deaths, is that he has given us free will, and that suffering and death are a result of our mismanagement of that free will. Evil and suffering come from humanity. However, in this case, the death and suffering come from our *lack* of free will; God has determined that this will take place. The only way that this can be justified is by arguing that this is all for the greater good. Now I am sure that God could achieve this greater good by setting up things a little more benignly, so that the tens of thousands of people did not have to die so that Zedekiah would go into Babylonian captivity. Free will is simply not allowed to be exercised here. The future, for this corner of Judah, is in the hands of God, and no one else. If it is in the Judeans' hands, then their hands are tied, and they are acting in the only way they can—and that is a way that results in the sacking of their city and the capture of their king.

As for the many supposed prophecies concerning Jesus, for him to be prophesied God has to ensure that he has the right parents, who have to be, for prophetic reasons and reasons of Jewish authority, in the lineage of David. This is no small organizational feat—the family line must be kept alive throughout the years. In fact, the order is taller than you might think, since it is often not a case of ensuring things *do* happen but ensuring that things *don't* happen. Mary, for example, cannot be bitten by that poisonous snake when she was twelve, must not have injured her uterus when the plough skewed into her abdomen at fourteen, must not have slipped off the wall she was walking along a week later, must not have starved due to a poverty stricken lifestyle, must not have been miscarried, must not have contracted an early form of cancer, must not have . . . the list is tremendous.

And that is just for Mary in her short life. One has to map out the entire history of the world to ensure the rest. It has to be ensured that Jesus doesn't die in some way before his time of preaching and atonement. The entire ancestral line of (one of) his parents must be maintained. The Egyptians must not have been allowed to kill their Hebrew slaves, the surrounding empires must not have obliterated the Israelites in a major conquest, a volcanic eruption must not have wiped out the Middle East, a meteorite must not have hit Earth, man must have evolved in a certain way from the original life-form. And so on, to the point that, in order to ensure that Jesus would come down in the fashion predicted, some six hundred years after the prophecies had been written, God has to micromanage the entire universe, and this smacks, just a little, of determinism. In order for something to happen with any kind of certainty later down the causal chain, God, pretty much literally, has to make the butterfly flap its wings.

CONCLUSION

Therefore, we are left with a concept of free will, this contra-causal, libertarian notion, which makes no sense in philosophical and theological context, and which is not supported by any evidence (nor could it be, it seems). Moreover, the evidence for the fact that free will is an illusion is far-reaching and makes an extraordinarily robust case for the denial of this rather naïve form of free will.

Without free will, however, the Christian God is thoroughly incoherent. God's judgment, and heaven and hell as eternal reward and punishment, are rendered nonsensical when seen in the context of people living in a universe without libertarian free will. Neither God nor humanity have this form of free will, and without this keystone, this fundamental brick, the edifice of Christianity comes tumbling down.

Part 4

SCIENCE AND THE BIBLE

Chapter 10

BIBLICAL ARCHAEOLOGY

Its Rise, Fall, and Rebirth as a Legitimate Science

Robert R. Cargill

Sometime during the twentieth century, "biblical archaeology" became a dirty word in the fields of archaeology and critical biblical studies. The field of study that had given the world so many headline-grabbing discoveries and colorful (mostly male) personalities, and that Christians and Jews pointed to for confirmation of their faith became tainted, disparaged as a field of suspect methodology, questionable motivation, and spurious conclusions. The depiction of biblical archaeology as excavation "with a Bible in one hand and a spade in the other" had fallen out of favor, rightly dismissed as an apologetic enterprise. Too often, confessional Bible scholars, often with little-to-no formal training in archaeological methods, used circular reasoning to find "proof" that their religious beliefs rested on evidence-based, factually verified claims made two millennia ago.

Early biblical archaeology was often little more than treasure hunting done while on expeditions in the "savage" Middle East, attempting to find and identify topographical places mentioned in the Bible. The belief was that if a biblical place like Sodom or Gomorrah could be positively identified as an actual place this would lend support to the veracity of the claims made about it in the Bible. But with the rise of modern science, and with critical methods developed in other fields of study finally making their way into the relatively young arena of field archaeology, traditional "biblical archaeology" began to be viewed with suspicion.

We cannot, however, blame only nineteenth and early twentieth century explorers for this negative conception of biblical archaeology. Interestingly, this practice of making pilgrimages to the Holy Land for the purpose of identifying locations mentioned in the Bible had been going on since at least the fourth century CE, with pilgrims discovering objects they then claimed were proof of the veracity of the Bible's claims, especially those concerning Jesus and Christianity.

EARLY PHYSICAL EVIDENCE

Following the rise of Christianity, as the claims about Jesus' life, teachings, death, and resurrection came to be challenged by non-Christians, the early church sought *evidence* supporting the claims made by and about Jesus. The church had begun the process of collecting (and often editing) the writings of the Apostle Paul, the authors of the gospels, and various other early Christian writings, for the purposes of establishing an orthodoxy and orthopraxy within the church. However, it did not take long for Christians in the early second century CE to notice that the central figure of their faith—Jesus of Nazareth—whom they believed to have altered history following his trial before Pontius Pilate, crucifixion, burial, and miraculous resurrection from the dead (among many other resurrected beings, according to Matthew 27:52), is nowhere mentioned in any non-Christian historical record. The only claims about Jesus from the first century CE stemmed from those few works composed by Christian authors who amassed and reworked the second-hand, oral accounts and traditions of Jesus passed down over the decades following his death. Outside of a small circle of believers, there existed no actual proof of the man, Jesus, whom early Christians claimed changed history.

This posed a problem, especially as Christianity spread north into Asia Minor and west throughout the Roman Empire. In fact, so problematic was the lack of corroborating evidence for the early church that two campaigns of various focus were apparently set in motion in an effort to resolve this problem. The Christians needed *literary* evidence attesting to

Jesus's life and deeds, and they needed *physical* evidence confirming the claims made about him. And two examples show the efforts of early Christians to resolve these dual goals: the efforts of Helena, Empress of Rome, and the Christian tampering with Jewish Roman historian Josephus's historical accounts.

The Empress Helena (250–330 CE), the mother of the Roman Emperor Constantine the Great (272–337 CE), was a devout Christian. In 324 CE, Helena ventured to Jerusalem for the express purpose of identifying the locations of Jesus's birth, crucifixion, and burial and memorializing them with monuments that would serve as official Roman verification of the credibility of the claims made about Jesus.

While in Roman Palestine, Helena dedicated the Church of Nativity in Bethlehem and the Church of the Ascension on the Mount of Olives. She also ordered a temple to the goddess Aphrodite built by Hadrian in the second century CE be torn down, following information she received from local Christians that Hadrian had built the temple over the tomb of Jesus following his suppression of the Bar Kokhba revolt in 135 CE. Helena's son, the Emperor Constantine, ordered the Church of the Holy Sepulcher be built in its place.[1]

During the building of the Church of the Holy Sepulcher, Eusebius of Caesarea records, Helena discovered three crosses, which she was certain were the crosses used to crucify Jesus and the two criminals said to have been executed with him, along with the nails used to fasten Jesus to the cross. Socrates of Constantinople added in his account that Helena also discovered the *titulus*—the wooden sign affixed to the top of the cross that read "King of the Jews"[2]—beneath the destroyed temple to Aphrodite.[3]

In order to determine which of the three crosses actually belonged to Jesus, Helena conducted a rudimentary experiment:

> The sign was this: a certain woman of the neighborhood, who had been long afflicted with disease, was now just at the point of death; the bishop therefore arranged it so that each of the crosses should be brought to the dying woman, believing that she would be healed on touching the precious cross. Nor was he disappointed in his expectation: for the two crosses having been applied which were not the Lord's, the woman still

continued in a dying state; but when the third, which was the true cross, touched her, she was immediately healed, and recovered her former strength. In this manner then was the genuine cross discovered.[4]

Helena brought the nails from the true cross to Constantine, who had the relics fashioned into bridle bits for his horses and a helmet for himself, as he believed these talismans had the protective power of Jesus within them. Thus, it was the pilgrimage of Helena to Jerusalem and the experimental verification used to confirm her discovery that provided "evidence" of the claims made about Jesus to the early church. And it was this Roman endorsement of Christianity at the highest level that served as the unassailable authorization of the veracity of the Bible and its claims.

And as absurd as Helena's test was, with her resulting "confirmation" that a wooden beam discovered near a tomb was, in fact, *the* cross of Jesus Christ, the truth is that nineteenth and twentieth-century biblical archaeology of the Holy Land had not much improved upon this methodology. At its core, early biblical archaeology and Helena's pilgrimage followed the same pattern: A confessional pilgrimage was made to a site identified in the Bible as having affiliation with a biblical character or event. This was followed by an exploration of the site, which inevitably yielded an object or structure that *could* be interpreted as something confirming a claim made in the Bible. The biblical text was then read to inform and confirm the interpretation that the discovery was, in fact, the very "evidence" needed to support the claim made in the Bible. Early twentieth century biblical archaeology was replete with this circular, apologetic reasoning.

EARLY BIBLICAL ARCHAEOLOGY

One classic example of archaeology done apologetically in support of the biblical text comes from Tel Megiddo, biblical "Armageddon," and the case of the so-called "Solomon's stables" in the Stratum IV layer of the dig. Excavator P. L. O. Guy's interpretation was a result of his knowledge of the text of 1 Kings 9:15 and 19, which reads:

> This is the account of the forced labor that King Solomon conscripted to build the house of the LORD and his own house, the Millo and the wall of Jerusalem, Hazor, *Megiddo*, Gezer . . . as well as all of Solomon's storage cities, *the cities for his chariots, the cities for his cavalry*, and whatever Solomon desired to build, in Jerusalem, in Lebanon, and in all the land of his dominion. (Italics mine.)

Guy's interpretation of a number of stone boxes arranged on a flat surface as being "Solomon's stables" was par for the course at the time. Of course, it may very well have been a complex of stables, although several archaeologists since then have disputed this interpretation, opting instead to understand this area at Megiddo as containing storehouses, a small marketplace, or even barracks for troops. Stone troughs do suggest that animals were fed in the area. But even if it *were* a stable-complex, what evidence is there that it was *Solomon's* stables other than a questionable dating to the early tenth century BCE? Recent excavations have suggested that this particular stable-complex—one of *many* stable-complexes on the tell—dates to the ninth century BCE reign of King Omri or King Ahab. So while there is debate about whether this particular structure is a stable for horses, there is still nothing tying it to King Solomon other than the assumption that *if* we dated the stables a little earlier, then it would have been a stable-complex *during the time* of a historical Solomon's reign, thereby *possibly* making them "Solomon's stables." This argument is highly speculative and ultimately not very compelling. All we know is that a major, fortified city lying at a strategic trade crossroads, and requiring a substantive military presence in order to defend the city, had stables for horses. We would *expect* every ruler of Megiddo to have had stables for horses. Nothing makes them *Solomon's* stables.

This "Bible and spade" methodology is frowned upon today—even by very faithful Jewish and Christian archaeologists—because it presumes the Bible is a historically verified, accurate account. Again, this is rightly understood to be circular reasoning because it utilizes the biblical text to interpret archaeological findings and then holds up those same findings as "evidence" of the historical accuracy of the Bible. To be sure, there are many claims made in the Bible that have been confirmed by the Bible.

Then again, there are many that have *not* been confirmed—while there are still more instances where archaeology has completely refuted claims made in the Bible. This has led proper archaeologists to treat the Bible as an *artifact* whose claims must be regularly tested and verified rather than the other way around, as the irrefutable standard of truth to which all archaeology must conform.

And yet, this technique of interpreting *anything* discovered at a site that is mentioned in the Bible, or found to date remotely near to the time period associated with a historical King David or Solomon as something *definitively* associated with the biblical kings, *does* tend to grab the headlines, which in turn sells tickets and books. And yet, at the end of the day, it is often merely speculative sensationalism, as there is often no evidence that links a discovery to a particular biblical character or textual claim other than a *desire* to tie the latest find to some account in the biblical narrative. Of course, once this pronouncement is made, it is then often quickly touted by the faithful as "evidence" of the veracity of the biblical narrative, completing the cycle of circular reasoning: a discovery is made, it is claimed to be associated with something mentioned in the Bible (without any specific archaeological evidence associating the discovery with the biblical character or claim other than the *possibility* that it be so), and then the discovery is presented as evidence for the "truth" of the Bible, leading the next generation of believers to treat the claims of the Bible as "verified" history.

This would be the same as finding a circular object on the floor of the Red Sea (let's say, the steering wheel of a ferry or the wheel of a cart that held goods being ferried across the Red Sea) and then claiming that it is a chariot wheel, and then claiming it is evidence of the biblical Exodus from Egypt, as if at no other time did a wheel roll into the sea or fall off a cargo boat sailing from Sinai to Egypt. Yet this is precisely the claim that pseudo-archaeologist Ron Wyatt claims to have found. His website still touts this as a legitimate discovery and as evidence of the Exodus.[5]

Or it's like finding the remains of a feather underneath the foundation of the Dome of the Rock in Jerusalem, and then claiming it is a feather from al-Buraq, the human-faced, winged steed that is said to have carried the Islamic prophet Muḥammad from Mecca in modern Saudi Arabia to Jerusalem, then on a tour of the universe to meet Jesus and Abraham and Moses and Allah, and then back to Mecca, *all in one night*, and then claiming the feather is evidence that this *Isra'* and *Miʿrāj* (or the "Night Journey" described above, found in Qur'an sura 17) is archaeologically proved to be true. Most readers certainly wouldn't call that "evidence" for the "truth" of the Qur'an, so why do so many Christians and Jews consider a circular object at the bottom of the Red Sea "evidence" of the "truth" of biblical claims?

Perhaps a more realistic account is that of the story of the destruction of Jericho in Joshua 6. Imagine excavating the area where the Bible says Jericho once stood, finding the remains of some destroyed walls, and then claiming that you had discovered *proof* that the destruction of Jericho took place *exactly* as the Bible says it did in Joshua 6.

Actually, you don't have to imagine it, *because that is exactly what happened!* When John Garstang, P. L. O. Guy's boss at the Palestine Department of Antiquities, excavated ancient Jericho in the 1930s, he announced that he had unearthed the *very walls* that had been destroyed during the battle of Jericho. He had read the biblical account, excavated, and "confirmed" the biblical account by finding felled walls where the Bible said he should have found felled walls.[6] However, when subsequent archaeologists like Dame Kathleen Kenyon continued the excavations there in the 1950s, they dated the destruction of the walls to 1550 BCE, which is far too early for them to have been destroyed by a historical Joshua and his Israelite army as depicted in the Bible.[7]

Kenyon argued that the walls were indeed the walls Jericho, but that they *had already been destroyed* by the time that any historical Israelite army would have entered the land. And as we know from Megiddo, ancient cities are destroyed and rebuilt repeatedly on top of one another throughout history. Thus, one can imagine how the biblical story of the destruction of Jericho could have developed over time: the Israelites knew that the city of Jericho had been destroyed, and the city's destruction became part of the Israelite conquest narrative, with credit for Jericho's destruction being transferred to the Israelite army at a later date.

As one might expect, several Christian archaeologists and apologists later challenged Dame Kenyon's dating of the walls. But in 1986, Dame Kenyon's dating was reconfirmed by Piotr Bienkowski, as being the result of a sixteenth century BCE destruction, using objects discovered in the walls of the excavation to date the remains.[8] Furthermore, Drs. Hendrik J. Bruins and J. (Hans) van der Plicht concluded that their radiocarbon dating of organic remains from the same level as these walls also dated to between 1617 and 1530 BCE, yet again confirming Kenyon's dating.[9] Thus, simply finding something that *could* correspond to the claims made in the Bible does not mean that it *does*, and further archaeological investigation often demonstrates that it does not. Jericho had already been destroyed by the time any historical Hebrews entering Cana'an would have come upon it.

The lesson learned by archaeologists dealing with sites directly impacting the claims made in the Bible (that is, "biblical archaeology") is that while the public (and especially those adhering to particular confessional claims and faith traditions) may be predisposed to accept claims made by the likes of P. L. O. Guy and John Garstang as confirming the Bible, this is not necessarily good archaeology. It is certainly not good science. And as archaeology matures as a discipline, the use of the Bible as an indisputable historical reference is being replaced by the more appropriate methodology of seeing the Bible like any other historical document—that is, as an *artifact*, which is a literary account composed and compiled by members of a particular group for the purpose of promoting not only a set of religious beliefs, but their ethnic identity, economic system, and political ideology. In other words, the Bible is a historical record that must *itself* be repeatedly tested to verify the veracity of its claims, and it cannot be relied upon solely to interpret archaeological sites, but rather the archaeological sites must be excavated independently and interpreted on their own merits, with the Bible only *then* being consulted as evidence to determine how one group *understood* the site in antiquity.

EARLY LITERARY EVIDENCE

But the issue of a lack of early archaeological evidence wasn't the early church's only problem. The lack of any non-Christian attestation to Jesus was a further problem, especially considering the public nature of the miraculous claims made about Jesus. The desire for the confirmation of religious beliefs extended to texts as well. And no better example of a religious group's desire to "find" evidence confirming their religious beliefs can be found than the early Christian attempts to insert mentions of Jesus and the Christian faith into first-century historical documents that discussed periods thought to be contemporaneous with Jesus.

Christianity does not exist without the death, burial, and resurrection of Jesus of Nazareth. The Apostle Paul says so himself (1 Cor. 15:14). And while Paul and others preached the death and resurrection of Jesus in Jerusalem throughout the eastern Mediterranean in the first centuries CE, there remained a problem with the message—namely, there was no *evidence* of this supposed resurrection. Jesus's resurrection is a central tenet of the Christian faith, and yet despite it arguably being the most written-about event in history, there is absolutely no archaeological evidence of Jesus's resurrection, or of his very existence!

Now, to be sure, we have ample evidence that Jesus's followers *believed* that he lived, taught, was crucified, and was resurrected, but there is no *literary* evidence from the first century CE speaking about Jesus's life, death, or resurrection outside of the Bible. This is shocking, especially given the fact that the Jewish historian Josephus mentions John the Baptist—according to the Bible, Jesus's cousin—on a number of occasions in his writings, and gives a highly detailed account of the events of the first centuries BCE and CE, and yet *never once* mentions the event that has come to be the best-known event from the first century CE.

The Christians could point to some early *second* century CE non-Christian confirmation of the existence of Christianity, but it wasn't all that flattering. The Roman historian Tacitus mentions a certain "Christus" and "Chrestianos" in his *Annals* (15.44). Tacitus wrote in Latin around 116 CE, almost a century after the events of Jesus's death.[10] However,

while many scholars conclude that this is a reference to Jesus of Nazareth, whom Christians called the *Christos* (Χριστός, the Greek translation
of the Hebrew *meshiah*, מָשִׁיחַ, or "Messiah," meaning "anointed"), it is
noteworthy that Tacitus paints a very negative portrait of the Christians
and does not speak of Jesus's miraculous resurrection, only that he "suffered the extreme penalty during the reign of Tiberius at the hands of one
of our procurators, Pontius Pilatus." Thus, while Tacitus appears to reference Jesus as "Christus" and his followers as "Chrestianos," his portrayal
of them as "a most mischievous superstition" and a lack of any mention
of Jesus's resurrection did not suffice to elevate Jesus to a level of historical significance worthy of the fastest-growing religion in Rome. The early
Christians needed secular attestation from a more trusted source. They
needed Josephus to mention Jesus.

Now, some readers will protest, "But Josephus *does* mention Jesus!"
However, as Richard Carrier and others have pointed out, the *Testimonium
Flavianum* appears to be a later Christian insertion into the text of Josephus, with the intent of making it appear as if Josephus *had*, in fact, at least
mentioned Jesus.

The supposed *Testimonium Flavianum*, or the "Testimony of Flavius
(Josephus)," appears in *Antiquities* 18.3.3 (18:63–64):

> About this time there lived Jesus, a wise man, if indeed one ought to call
> him a man. For he was one who performed surprising deeds and was a
> teacher of such people as accept the truth gladly. He won over many Jews
> and many of the Greeks. He was the Christ. And when, upon the accusa-
> tion of the principal men among us, Pilate had condemned him to a cross,
> those who had first come to love him did not cease. He appeared to them
> spending a third day restored to life, for the prophets of God had foretold
> these things and a thousand other marvels about him. And the tribe of
> the Christians, so called after him, has still to this day not disappeared.[11]

Scholars have several problems with this passage, causing many
of them to conclude that Josephus did *not* write it but rather that it was
inserted into Josephus's *Antiquities* by a later Christian editor, perhaps by
Eusebius of Caesarea, a fourth century CE Christian apologist who relies

heavily on Josephus for his source material and who quotes this supposed passage from Josephus in his *Church History*.[12] Scholars believe that Eusebius (or some other copyist) inserted this claim about Jesus at this point in Josephus's narrative because Josephus was recounting a series of Jewish disturbances that took place under Pilate, and it was a natural place to add a credible Jewish witness to Jesus. Problems with the passage include the fact that Josephus, who himself was a Pharisee (*Life* 12), refers to Jesus as the "Christ" (Heb. מָשִׁיחַ, *meshiah*), which is highly unlikely.

Second, the *Testimonium Flavianum* interrupts the flow of the narrative, which is describing repeated Jewish protests in response to acts done by Pilate, and *not* Pilate working in collaboration *with* "the principal men among" the Jews to execute other Jews as the New Testament and the *Testimonium Flavianum* both suggest. The third and perhaps most compelling reason is the marked lack of sophistication of this disputed passage compared with the surrounding narrative of Josephus, including differences in grammar (like the striking decrease in the use of prepositions Josephus so liberally employs throughout his works), style, and vocabulary. For instance, Josephus only uses the noun ποιητής (*poiētēs*) to refer to poets, but in the *Testimonium Flavianum*, ποιητής is used to describe Jesus as a "doer" of unusual or wonderful works—a usage commonly employed in the New Testament book of James to describe "doers" of good works.[13] In fact, the entire passage reads more like a Christian creed than the typical text of Josephus. As one of the students in my second-year Greek class stated, "I'm no expert, but whoever wrote this passage (the *Testimonium Flavianum* reference to Jesus) didn't write the rest of this book." Well said.

The *Testimonium Flavianum* is also noticeably absent from Josephus' previously authored parallel account about Pilate, in *War* 2.9.2–4 (2.169–77), which recounts the very same story of the standards that Pilate brought into Jerusalem by night.

Finally, Richard Carrier points out that the third century CE Christian theologian and apologist Origen of Alexandria never mentions this reference to Jesus by Josephus, even after he was specifically asked in *Against Celsus* (1.42) to provide evidence of contemporaries of Jesus who could attest to his activities, including his crucifixion and death. Instead of citing the above disputed passage (*Antiquities* 18.3.3), which directly mentions Pilate, Jesus, his crucifixion, and his resurrection, Origen cites only tangential passages from Josephus about John the Baptist (*Against Celsus* 1.47), who *was* mentioned by Josephus, but *without* mention of Jesus (*Antiquities* 18.5.2).[14] Why doesn't Origen mention this reference to Jesus? Scholars argue it is because this passage was not a part of Josephus's account in the third century CE, when Origen was writing. In fact, Origen explicitly states in this very same passage (*Against Celsus* 1.47), "Now this writer (Josephus), although *not* believing in Jesus as the Christ . . . ," which directly contradicts what Josephus supposedly says in *Antiquities* 18.3.3: "He (Jesus) was the Christ." And it is for these reasons that scholars believe that the so-called *Testimonium Flavianum*, Josephus's reference to Jesus, was a later Christian interpolation into the text of Josephus, inserted to remedy the problem of why the one Jewish historian who so meticulously detailed the events of Jewish history, especially those of the first century CE, and who even mentioned John the Baptist, despite the Christian claims of a trial before Pilate, crucifixion, and the dead rising from the grave,[15] *never once* mentioned Jesus of Nazareth.

THE RISE OF SCIENCE AND THE RESURRECTION OF BIBLICAL ARCHAEOLOGY AS A LEGITIMATE FIELD

With the rise of modern science, many aspects of the Jewish and Christian faith slowly came to be challenged, questioned, and ultimately dismissed, save for a small, but quite vocal minority of Christian fundamentalists like Ken Ham, who reject science and cling to debunked notions of a six-day creation (Gen. 1), worldwide flood (Gen. 6–9), and a sun that stands still (Josh 10:12–13) and even goes backward (Isa. 38:8; 2 Kings 20:8–11). Improved techniques in the preservation of archaeological stratigraphy resulting from excavating in 5 × 5–meter squares assisted archaeologists in better understanding diachronic archaeological levels. The science of pottery typography dramatically improved archaeologists' ability to identify and date the residents of a particular archaeological stratum.

But perhaps most important of all was the theoretical shift from using the Bible as a template and a lens through which to interpret archaeological remains to treating the text of the Bible as an *object* that is also subject to critique and examination. Research and developments in the field of biblical textual criticism have made significant contributions to the field of biblical archaeology. As the text of the Bible has been subject to form, source, redaction, and historical criticism, the subjective nature of the Bible has been revealed, as well as its evolution and development from earlier sources. That is, textual criticism has eroded the reliability of the Bible as a historically objective resource, which in turn has eroded the ability of legitimate archaeologists to rely upon the Bible as a lens through which to interpret archaeological discoveries.

By removing this circular self-reinforcing feedback loop, archaeologists are relying on more reliable, better-established archaeological methodology to interpret their findings. As a result, the archaeological excavations in the Holy Land since the second half of the twentieth century have increasingly become more reliable and rigorous in their methodology, and not surprisingly they have yielded more results and interpretations that are *not* congruent with the claims made within the pages of the Bible.

This does not mean that pseudoarchaeologists, who will make sensational archaeological claims for political, financial, and proselytizing purposes, will suddenly cease making ridiculous claims. Simcha Jacobovici will continue to claim that he has discovered the route of the Exodus,[16] the lost tomb of Jesus's family,[17] Atlantis (in Spain),[18] the nails of the cross,[19] the "earliest archaeological evidence of Christianity from the time of Jesus,"[20] supposed "decoded" evidence that Jesus and Mary Magdalene were married and had children,[21] and supposed "encoded" mentions of Jesus and Paul in the Dead Sea Scrolls,[22] and he will sell these books and documentaries to whoever will buy them. But these entertainers, who lack formal archaeological training and yet make sensational archaeological claims pertaining to issues of faith, are now the outliers and no longer part of the archaeological mainstream. And even more encouraging, as legitimate biblical archaeology takes root throughout the Holy Land, the popular media and the public alike are slowly beginning to be able to distinguish between sensationalized pseudoarchaeology and the real thing. The rise of the Internet has only further allowed instant fact-checking from any number of dissenting sources, which has allowed the public to make better-informed decisions regarding archaeological claims. And the increased skeptical nature of the next generation has made it more difficult for pseudoarchaeologists to shill their message and make their money and/ or converts.

The rebirth of modern biblical archaeology is a welcomed one. But today's biblical archaeology is vastly different from our grandfathers' biblical archaeology. For it is now that we've removed the Bible as the lens through which all interpretations must be made that we are engaging in true science. Because critical biblical scholars have been able to demonstrate that the biblical text is by no means an objective history, but rather a subjective history written by adherents to a particular faith tradition who mixed together mythological tales, oral traditions, and *some* historical elements in order to craft a foundation narrative and a history of a people, today's legitimate archaeologists have been able to employ science as the lens through which to interpret our findings. And it is this resurrection of biblical archaeology as a legitimate science that has improved the reputation of the field, and which will continue to provide positive contributions to humanity.

Chapter 11

THE CREDIBILITY OF THE EXODUS
Rebecca Bradley

As stories go, the Exodus of the Bible is a damned fine one. It has everything going for it except comic relief and a love interest: a reluctant but effective hero, magic and miracles, betrayal and intrigue, lots of battles, lashings of violence and disaster, the humbling of the mighty, the liberation of the enslaved, and a reasonably happy ending in a land of milk and honey. But it is much more than that. It is a story that underpins two major world religions—three, if you count the role played by Moses in the legends of Islam. More than any other single narrative from the pre-Christian world, it has conditioned the geopolitics of the modern world; the Abrahamic covenant, redeemed by Moses and Joshua, resounds in the rhetoric of Middle Eastern conflicts and the apocalyptic hopes of far-right fundamentalism.

But did it happen?

To many of the faithful of those world religions, that question is nonsense, even blasphemy. *Of course* it happened—it's in the Bible. The Bible, they say, is a reliable historical resource, recorded not just by eyewitnesses but by the actors themselves, preserved through the ages, richly confirmed by modern archeological research. One can track the footsteps of the Children of Israel across the Sinai, marvel at the mountain where God handed down the tablets of the law, observe the conquest of the Promised Land in the destruction levels of Canaanite cities. Moses and Aaron, Joshua and Gideon, David and Solomon were all flesh and blood individuals and behaved exactly as described. The waters *did* part; the firstborn of Egypt *did* die. And, they say, it is *necessary* for all this to have happened just as the Bible claims—because if it did not, then our religions, our laws, even Western civilization itself, are all founded on a lie.

But did it happen?

The awkward fact is there is a mismatch between the Exodus narrative in the Bible and associated sources and the narrative based on the archeological record. This was not apparent for a long time because the early days of biblical archeology suffered from a peculiar handicap: the Bible. That is, since the Old Testament was assumed to be an honest-to-God history of the Holy Land, the emphasis was on matching the archeological remains to details of the accepted narrative; the match, in turn, was taken as evidence of the accuracy of the narrative. Alternate interpretations were hardly considered, and the narrative itself was not substantially tested until the last third of the twentieth century. Excavated structures were speculatively linked to, say, the building programs of the hyperactive King Solomon; levels of destruction in city tells across the Holy Land were held to be contemporaneous and linked to the Conquest. And so forth. By these standards, the Bible passed the truth test.

Enter more precise dating techniques, more refined chronologies for Egypt and the Near East, more intensive survey and excavation in Israel in the years following the 1967 war.[1] Add to these the fruits of more than a century of documentary analysis, painstakingly parsing the ancient texts. A very different picture emerged of events in that neck of the woods in the final two millennia BCE, a picture that had no place in it for a mass migration of rebellious slaves from Egypt, the watery death of a mighty pharaoh, and the conquest of the Promised Land. The labyrinthine monarchic history of Israel and Judah was somewhat supported, but with hugely significant divergences from the events recounted in the chronicles of the kings and prophets.

But not only did the picture change—it was placed in its wider context. To the faithful, the biblical narrative *was* the canvas, the entire canvas. The Children of Israel loomed large on the world stage, as would befit Jehovah's chosen people and pet project. Archeology and the historical disciplines, however, painted a sweeping great mural of the past, on which the concerns of the minor kingdom of Judah were not much more than a vignette in the clash of empires. Furthermore, what became known of that larger picture made it impossible for the Pentateuch to be regarded as an accurate historical account. An engaging and powerful mythos—yes. Brilliant propaganda from a distant era—certainly. Veridical history—not so much.

To put it baldly, there is *no* archeological evidence for the Exodus and the conquest of Canaan as described in the Bible, and there is significant archeological evidence ruling it out. That bold statement, no longer news, is the archeological consensus, despite continued resistance in fundamentalist quarters. Generally speaking, we will be looking at the positions of two opposing camps.

- The book of Exodus (and the rest of the Old Testament) is a divinely inspired eyewitness account of past events, accurate by definition. The Exodus *did* take place, as described in that inerrant document. If the document conflicts with the archeological evidence, then the problem is with archeology. Predictably, belief in a literal Exodus overlaps with the creationist agenda, and goes along with the search for Noah's Ark, the Ark of the Covenant, and other biblical relics.
- The book of Exodus (and the rest of the Old Testament) is a compendium of folktales, myths, annals, laws, and literature, compiled and redacted in the seventh to fourth centuries BCE for reasons that were as much political as religious, and of great interest in themselves. The Exodus did *not* take place as described in the document, though the folktales and myths may be rooted in historical events. If the document conflicts with the archeological evidence, then the problem is with the document.

There is a large literature dealing with miraculous details of the narrative, attempting naturalistic explanations of the Plagues, the slaying of the firstborn, the parting of the Red Sea, manna from heaven, and so forth. In my view, these are beside the point if the basic narrative has no evidential foundation; for example, if no Israelites crossed the Red Sea, it makes no sense to seek the place where the waters parted. Therefore, this chapter will focus on the credibility of a large Semitic population migrating en masse from the Nile Delta to Canaan, via Sinai and Kadesh-Barnea, and eventually founding the kingdoms of Israel and Judah.

THE STORY

The bare bones of the Exodus as recounted in the Pentateuch, plus its patriarchal backstory and its sequels in the Promised Land, are well enough known that they can be presented almost telegraphically as a series of narrative set pieces:

- God makes a covenant with Abraham.
- Jacob/Israel and his twelve sons live as herders in Canaan.
- Joseph, sold into slavery, makes good in Egypt.
- The Israelites settle in the Land of Goshen, in the eastern Egyptian delta.
- The Israelites, procreating mightily, are enslaved by a new pharaoh.
- Moses, the baby in the bulrushes, is adopted by Pharaoh's daughter.
- Moses, exiled in Midian, receives God's commission through the burning bush.
- The Ten Plagues culminate in the first Passover and the slaying of Egypt's firstborn.
- The Exodus takes place, with the parting of the Red Sea and the destruction of Pharaoh and his army.
- God gives Moses the Ten Commandments at Mount Sinai.
- The Israelites sojourn in the wilderness for forty eventful years, most of them in Kadesh-Barnea.
- The death of Moses.
- The conquest of Canaan.
- The era of the Judges.
- The kingdom of David and Solomon.
- The kingdoms of Israel and Judah.
- The rediscovery of the holy texts.
- Disaster and Exile.
- The return from Babylon and the building of the Second Temple.

Less well known to the Christian public are the additions and embroideries found in the rabbinical writings, the Islamic tradition, and

Josephus's *Antiquities of the Jews*. The details found in these sources, many of them used to fill out the towering figure of Moses himself, make the Pentateuch sound rather like a Reader's Digest version of events. Though these sources are noncanonical to Christians, their pages have occasionally been mined by the faithful for clues to some of the problems posed by the biblical narrative. Of particular charm and interest are diverse legends regarding Moses's life as an exceptional child in the court of Pharaoh and as the heroic conqueror and ruler of Ethiopia before his escape to Midian.

A number of the episodes above are untestable—but others do allow predictions to be made about the archeological record: archeological patterns, remains, and textual references that would be expected if the narrative were true. We would expect to see a homogeneous Semitic population with a distinct culture in the Eastern Delta in the mid-second millennium BCE; the numbers would increase gradually over a period of about four centuries, but the material assemblage would show a decline in markers of wealth and prestige in the later levels. This distinctive culture would then vanish abruptly from the Delta, but be traceable briefly in Sinai and Kadesh-Barnea, in the wilderness of Zin; meantime, the Nile Valley would display economic upheaval, weakness, and catastrophic depopulation of both humans and livestock. The distinct Delta culture's disappearance from Kadesh-Barnea, in turn, would be followed by its sudden, dramatic florescence all over Canaan, immediately overlying the destruction of a number of named cities. We would expect to see marked urbanization and monumental building projects in Jerusalem, with attributes of full state organization in both Jerusalem and Samaria by the eleventh century BCE. And we would expect to see extensive mention of these events and that culture in the records of the highly literate surrounding societies, linked as they were by trade, competition, and complex political relationships.

What does the archeology show? As it happens, there was indeed a strong Semitic presence in the Delta in the stipulated millennium, which then vanished from Egypt. A new cultural complex did indeed enter the Levant at the end of the Bronze Age, atop the ruins of some Canaanite cities. Urbanization and state-level organization did crop up at Jerusalem and Samaria. Egypt did suffer a period of upheaval and disaster—several, in fact. These data, cherry-picked from the insanely complicated archeological record of Egypt and the Levant, have frequently been used to validate the biblical history. This, though, is like trying to put together a picture using pieces of several different jigsaw puzzles. Satisfying as the picture may be, the pieces did not all come from the same box.

We will approach this first through some issues of chronology, and then by looking at what the archeological record suggests was happening in the three areas critical to the Exodus story in the relevant periods: the departure point in Egypt, the areas en route, and the ultimate destination in the Promised Land.

CHRONOLOGY

An ocean of ink has been spilt on speculations about when, exactly, the Israelites settled in Egypt, when they were enslaved, and when they departed so dramatically. Extreme believers tie these events into a young-earth creationist timeline calculated partly by totting up the life spans of the Patriarchs. Oddly enough, this has a partial parallel in one of the strands of Egyptian chronology—king lists recording the names and length of reigns of the pharaohs, preserved in several inscriptions and fragments of papyri. Oceans of ink have been spilt on this exercise, too, starting with the Egyptian historian Manetho in the third century BCE, who devised the basic dynastic framework that is still in use. But, whereas Egyptologists have been able to expand, refine, and partly validate Egypt's timeline with a wealth of other evidence—everything from seal impressions to structures, acres of inscriptions, astronomical observations, archeological deposits dateable by scientific methods, pottery sequences, cross-referencing data from Egypt's contemporaries, and so forth—those seeking to date the Exodus, finding themselves on the wrong side of the archeological evidence, have little concrete to go on outside the pages of the biblical account. Unfortunately, the internal evidence from the book of Exodus creates more problems than it solves.

Most of the believers' schema cluster around two dates for Exodus:

an "early date" of 1450 BCE, in Egypt's early Eighteenth Dynasty, and a "late date" of about 1290 BCE, in the Nineteenth Dynasty. The early date is founded largely on 1 Kings 6:1, which explicitly dates the construction of Solomon's temple to 480 years after the Exodus; the foundation of the temple, in turn, has an accepted (among believers) date of 966 BCE.[2] The "late date" attempts to pay more attention to the archeological record, and to some major anachronisms embodied in the "early date," notably place names in the Delta that first appear in the Ramesside period, at least a couple of centuries later than 1450 BCE. Each of these proposed time periods has some wiggle room, allowing a few decades either way, and allowing several pharaohs to be proposed as the one whose chariots came to grief in the Red Sea.

This is not all the fault of the literal-Exodus believers. Egyptian chronology itself is a work in progress, with a fair amount of wiggle room and many heated controversies, though some recent attempts at major revisions[3] gained no traction. But even with that flexibility, there is no way to reconcile dynastic history with the Exodus account; the biblical details are strewn all over the timeline derived from archeological and other evidence. That basic mismatch and such fatal anachronisms as camels and Philistines (more on them later) are occasionally recognized by believers, but they are dealt with mostly by finagling the text and the archeology into some kind of uneasy truce. A couple of examples will suffice.

In an extreme example that departs from the mainstream of believers, Dr. Gerald Aardsma, young-earth creationist and self-styled biblical chronologist, handily summarizes some of the problems with both the early and late conventional dates and proposes a dramatic solution: a scribal error that knocked a full thousand years off the history of the Israelites.[4] Instead of the 480 years mentioned above, he proposes 1,480 years elapsed between the Exodus and the founding of Solomon's Temple, pushing the Exodus back to 2450 BCE and adding several centuries to the era of the Judges. The collapse of the Egyptian Old Kingdom at the end of the Sixth Dynasty could then be tied in with the chaos resulting from the Ten Plagues and the destruction of the army in the Red Sea. As a bonus, his theory would enable Joseph to be identified as the famous Imhotep, vizier to the Pharaoh Djoser, though Aardsma does not go so far as to claim that Hebrew slaves built the Old Kingdom pyramids. Among many other problems, it does not dismay him that the Old Kingdom collapse can be reliably dated more than two hundred years after his proposed date, tied in with the major climatic event that knocked over most of the ancient world at the end of the Early Bronze Age. Apparently, "the difference of 247 years between these two dates is close enough for such ancient times to regard the dates as the same. Uncertainties of a few hundred years in historical/ archeological chronologies are normal at such early times in the history of civilization."[5] In a timeline as well-populated as Egypt's, this is nonsense.

Aardsma may be regarded as something of a crank, even by other Exodus-believers, but he is not alone in proposing convenient reinterpretations of the data. For example, William Shea, supporting the traditional "early date" of 1450 BCE, proposed Amenhotep II as the pharaoh of the Exodus; but, to get around the awkward fact that Amenhotep II reigned for many years after the putative Red Sea disaster and is still with us in mummified form, Shea made a novel suggestion. The pharaoh in question was actually two pharaohs with exactly the same name; when the first died ignominiously in the Red Sea, the second was installed secretly and seamlessly, as part of a cover-up of Egypt's shame. Shea attributed Amenhotep IIB's military campaigns in Syria to the need to refill Egypt's treasury after the Israelites walked off with so much of Egypt's riches, though one would expect more difficulty waging war when one's army had recently drowned in the Red Sea.[6]

But even those who are not so far out in left field have an out when the evidence does not match the Bible. The logic is summed up in the following comment on the early/late controversy among believers, by a proponent of the early date:

> Those who opt for the late date of the Exodus do so primarily on the basis of archeological evidence. And yet that evidence is always colored by the presuppositions and prejudices of those interpreting the raw data. On the other side are those who opt for the early date of the Exodus. They do so primarily because of the biblical data. So what is the answer? All truth is God's truth; yet the only truth which can be known absolutely is

that truth which God chooses to reveal in His Word. Thus the biblical evidence must be the primary evidence. For this reason the writer accepts the early date of the Exodus as being the better alternative.[7]

That is to say, when archeology says one thing, and the Bible says another, believe the Bible. But what are the kinds of data that must then be thrown out to make the Exodus chronology work? There are many, but we'll focus on just a couple. Cue the camels and the Philistines.

CAMELS

Forget those movies about Ancient Egypt where Pharaoh's subjects lead heavily laden camels out of the desert. Though the beast is now virtually a symbol of Egypt, bedizened in Giza with tourists on top, it was a relatively late addition to the faunal assemblage of the Nile Valley. The pack animal of choice through most of dynastic history was the donkey; the other livestock represented in faunal remains and artwork were cattle, sheep, goats, and pigs. Horses only became common after about 1700 BCE, in the Second Intermediate Period. Camels, however, were so conspicuously absent from Egypt that there was no word for them in the Egyptian language, not even a hieroglyphic sign in the extensive sign list. Most likely domesticated in South/Central Arabia in the Early Bronze Age, they did not feature significantly in the Levant until the early Iron Age, in the tenth century BCE, and not in Egypt until the Persian period, starting in the sixth century BCE—at which point a Semitic loan-word for camel also entered Egyptian Demotic texts.

Their relevance to the Exodus question lies in several anachronistic mentions in the Pentateuch. Joseph is sold by his jealous brothers to Midianite/Ishmaelite spice merchants, on their way to Egypt with camel-loads of "aromatic gum and balm and myrrh" (Genesis 37:25). On the believers' timeline, this is dated to the late nineteenth century BCE; unfortunately, this is about nine centuries before camel caravans revolutionized the spice trade and opened up new overland caravan routes. Of course, it is just barely credible that occasional camels passed through Canaan or reached Egypt prior to their significant introduction in the Iron Age, but it is a stretch. The incense-bearing camel caravan mentioned in Genesis is equally a stretch; the trade described is more characteristic of the Iron Age products of Arabia. But it is the next mention of camels, in Exodus 9:3, that blows the game entirely:

> Behold, the hand of the LORD is upon thy cattle which is in the field, upon the horses, upon the asses, upon the camels, upon the oxen, and upon the sheep: there shall be a very grievous murrain.

This is a reference to the fifth plague, whereby God would smite the Egyptian (but not Israelite) livestock with a deadly epidemic. Including camels among the herds implies that camels were being bred in Egypt at the time, which in turn would mean that their bones, hair, and dung should be turning up in significant quantities in Egyptian deposits. The fact that they do not is an irreconcilable difference between the document and the archeology. Equally damning is the mention of Job's herd of six thousand camels (Job 42:12) at the impossibly early date of 2100 BCE according to conventional Bible chronology. Details like these, however, tie in very well with mainstream scholars' contention that the narratives were written down and compiled no earlier than the seventh century BCE, when camels had become commonplace.

THE PHILISTINES

According to the Bible, the Philistines were the Israelites' *bête noire* (chronic pain in the arse) in Canaan during and after the Conquest, but they are first mentioned in connection with the Exodus. Moses is instructed to lead the Israelites out of Egypt going the long way round: not the short hop along the coast to Canaan but a turn south into Sinai and a long—*very* long—digression through the wilderness. The reason, according to Exodus 13:17–18, is the Philistines:

And it came to pass, when Pharaoh had let the people go, that God led them not through the way of the land of the Philistines, although that was near; for God said, Lest peradventure the people repent when they see war, and they return to Egypt. But God led the people about, through the way of the wilderness of the Red sea: and the children of Israel went up harnessed out of the land of Egypt.

This implies that, whatever the date chosen for the Exodus, the believers must posit that Philistines were already established as a powerful presence on the Mediterranean coast. This is a problem for even the "late date," since the Philistines' arrival was part of a dramatic process that only began at about 1200 BCE: the general collapse and chaos that marked the end of the Late Bronze Age and the segue into the Iron Age. All over Anatolia, the Aegean, and the Levant, cities, states, and even empires crumbled, over a period of about fifty years; Egypt and parts of Mesopotamia survived but were shaken; the infamous Sea Peoples swept like a scythe along the Mediterranean coast, cutting down cities as far as the borders of Egypt, where Rameses III managed to hold them off. A good case can be made for another climate event setting off what turned into this perfect storm of misfortune—an extended drought similar to but not as severe as the one that put a finish to the Early Bronze Age, nearly a thousand years before. But whatever the proximal cause, this rolling catastrophe accounts for a number of the destruction levels previously credited to Joshua and the Conquest of Canaan.[8]

The Philistines were also part of this process. As one of the named ethnic groups comprising the Sea Peoples, they were defeated in the epic battle with Rameses III in about 1180 BCE and then were allowed to settle along the southern coast of the Levant. From the late twelfth century BCE, they were a strong presence in the form of the Philistine Pentapolis, until they came under Assyrian control in the eighth century BCE, along with most of their neighbors. Biblical references to them controlling the coast at the suggested dates for the Exodus pose yet another irreconcilable anachronism; worse, Genesis 21 and 26 portray Abraham and Isaac visiting Philistines, a good nine centuries before the Philistines were there to visit. Attempted rationalizations are pretty feeble, often involving an indefensible claim that "Philistine" was a generic name for the Sea Peoples.[9] Archer simply insists that the Philistines must have been present in Abraham's time, because scripture says so.[10] Aardsma suggests the earlier and later Philistines were entirely different groups, for which the Israelites rather inconsiderately used the same name;[11] again, the only evidence offered is that the Bible says so. Certainly, the name does not occur in any extra-biblical texts before the reign of Rameses III.

In summary, the chronological issues seem to rule out any easy decision as to when the Sojourn and the Exodus could be slotted into Egypt's busy schedule. The internal evidence points to an Iron Age origin for the documents, composed by writers and redactors whose world included camels and Philistines.

ARCHEOLOGICAL ISSUES: THE LAND OF GOSHEN

The eastern Nile Delta, known in the Bible as the Land of Goshen, is the flank that Egypt exposes to the Levant, and it not surprising that it was frequently home to West Asiatics wandering down along the Gaza coast. Through much of the Middle Bronze Age, in fact, a mixed bag of Canaanite/West Semitic populations predominated and even ruled in the Delta, engaging in a complex relationship with the indigenous Egyptian pharaohs who ruled in the south. In the traditional Egyptological narrative, a significant—and peaceful—infiltration began during the nineteenth century BCE, when the Pharaohs of the powerful Twelfth Dynasty allowed Canaanite traders and herders to settle and prosper in the Delta. Eventually, aided by a weakening of the southern regime, the Canaanite element grew strong enough to set up an independent line of rulers with a capital at the important trading center of Avaris (Tell el-Daba), in reasonably peaceful coexistence with their southern neighbor. They were enshrined in Egyptian history as the Fourteenth Dynasty, which overlapped chronologically with the Thirteenth Dynasty in the south. Then a period of famine and possibly epidemic weakened both regions and opened the way for a more violent Canaanite/West Semitic invasion of the Delta in about 1650 BCE, by a people known to history as the Hyksos.

The name "Hyksos" is derived from the Egyptian *heqa-khase*, "rulers of foreign lands," though Josephus and others mistranslated it as "Shepherd Kings." Josephus, writing in the first century AD, speculatively identified the Hyksos with the Israelites, and the same line is followed by some who favor taking the Bible as a reliable history. According to ancient Egyptian traditions, the warlike Hyksos invaded the Nile Valley and conquered Egypt as far as Thebes, before consolidating their power in the Delta as the Fifteenth Dynasty. More recent research casts doubt on that traditional narrative and the previous Egyptological consensus, calling into question whether the Hyksos arrived as hostile invaders; it now seems more likely they were an in-place development of the existing Canaanite regime. In any case, the Hyksos kingdom prospered in the northern section of Egypt until about 1540 BCE, when a more ambitious power arose in Thebes, strong enough to mount an offensive against the northern regime and reclaim the Delta. The Hyksos were routed and driven back to the Negev, leaving the triumphant Egyptians free to turn their attentions toward annexing Canaan and Nubia. The New Kingdom, Egypt's golden imperial age, began with a bang. In later Egyptian retellings, the Hyksos were stigmatized as vile usurpers, but that seems to reflect a strong element of anti-foreigner propaganda in the woes of the Bronze Age collapse.

This, then, is the background to the Canaanite presence in the Delta that has been taken as evidence for the Exodus narrative's validity. Literal-Exodus believers identify the nineteenth-century influx as the coming of Jacob and his progeny; Joseph is identified with one of the Canaanite kings of Avaris; various rulers of the Twelfth through Seventeenth Dynasties are identified with the nice pharaoh who elevated Joseph, or the nasty pharaoh who enslaved the Hebrews. The Hyksos, largely by being Asiatic in origin, are identified with the Israelites. These attempts to match a literal Exodus narrative with the archeological record, however, paper over the cracks of some major issues:

1. The Bible speaks of a large ethnic minority in the Delta, enslaved for generations after a new Pharaoh came to the throne and set to building cities of brick—cities that, in fact, are not attested until later

in the New Kingdom. The Canaanites in the Delta show no sign of ever being enslaved; in contrast, they show every sign of being well integrated through the Middle Kingdom and independent of the southern pharaohs during the Hyksos period, in full control of the "Land of Goshen." The Hyksos were never in bondage in Egypt.

2. The Canaanite population in the Delta were diverse in origin, not at all the genetically and culturally homogeneous population one would expect if they were all the descendants of a single patriarch.

3. The Children of Israel in the biblical account left Egyptian territories peacefully, though under something of a cloud. The Hyksos, in contrast, were expelled by force.

4. The dates are hopelessly mismatched. Conventional Bible timelines that adhere to the "early date" for Exodus place the bondage of Israel at 1600 BCE to about 1450 BCE. Hyksos domination is dated to ca. 1650–1540 BCE. That is, for the first portion of the proposed period of bondage, the Hyksos were anything but slaves; for the second portion, they were already gone. The mismatch is even worse for the "late date" in the thirteenth century; by that point, the Hyksos had been gone (though not forgotten) for upward of two hundred and fifty years. Donald Redford has made an interesting case that widely circulating legends and memories of the Hyksos episode may have some relevance to the later Hebrew Exodus mythos—but as the source population for a literal Exodus, the Hyksos are a nonstarter.[12]

ARCHEOLOGICAL ISSUES: EGYPT IN GENERAL

There are other serious problems in finding traces of a literal Exodus from the Nile Valley, as described in the Old Testament. For one thing, in a literate culture with a mania for recording, one would expect at least some mention of these dramatic events to show up in Egyptian annals. An argument commonly used, that the pharaohs would suppress the memory of such a humiliation, does not really hold water—the disappearance of a whole army and such a huge source of labor would surely leave ripples in the financial accounts, if nowhere else.

The single appearance of Israel in the multitudinous texts of Egypt is a brief and equivocal reference dated to the end of the thirteenth century BCE, on a victory stele of the pharaoh Merenptah of the Nineteenth Dynasty. The mention of Israel is a literal one-liner, part of a postscript to an account of victories in Libya, where the action shifts at the end to a kind of *meanwhile, back in Canaan*: "Ashkelon has been overcome; Gezer has been captured; Yano'am is made non-existent. Israel is laid waste and his seed is not." In context, it is not clear whether "Israel" refers to a region or an ethnic group, or even a generic term for a pastoral group. Similarly, the word "hapiru" in Egyptian texts, often interpreted as "Hebrew," was almost certainly a generic term for rootless nomads, wanderers, bandits, and people generally living on the margins of settled society.

In any event, the Children of Israel do not appear to have loomed large on the Egyptian horizon; this is ironic, as the sheer size of the Exodus described in the Bible is also something of a problem. Literalist believers defend the number of men claimed in Exodus 12:37–38 and extrapolate it to a total figure of 2–2.5 million, plus herds:

> And the children of Israel journeyed from Rameses to Succoth, about six hundred thousand on foot that were men, beside children. And a mixed multitude went up also with them; and flocks, and herds, even very much cattle.

This is a hefty slice of a total Egyptian population, estimated to peak in the New Kingdom at around three million.[13] It would be roughly equivalent to the entire populations of Wyoming and both Dakotas upping stakes and heading for Utah. Also, given that the Egyptian livestock were wiped out again and again in three of the Ten Plagues, while the Israelites took theirs with them, Egypt must have been a very quiet place after Pharaoh's army set off for the Red Sea. Needless to say, no such massive depopulation nor die-off of livestock is either recorded or shows up in the archeological record of Egypt. In fact, it is quite the reverse: the dates proposed by believers land squarely in the middle of Egypt's greatest period of prosperity and empire-building.

ARCHEOLOGICAL ISSUES:
THE SINAI AND THE WILDERNESS OF ZIN

As noted above, the Exodus narrative pushes an estimated two-million-plus people, with extensive herds, out into the Sinai wilderness—agricultural peasants and (reputedly) construction workers, who were several generations away from any expertise in nomadic herding and life in the desert and were moreover heavily laden with children, the elderly, and the treasure "borrowed" and pilfered from the Egyptians. The Sinai is not a welcoming place for a mass migration: stretches of desert, broken backside-of-the-moon rockscapes, winding escarpments, fractal wadis that are dry much of the year. No wonder we're told the Children of Israel complained.

How would they subsist on their journey across this forbidding landscape? Believers point to the miraculous: God provided manna and quail for the Israelites' sustenance, plus water conveniently laid on when Moses struck certain rocks with his staff. But even if that were verifiable, subsistence does not end with food and water; nothing in the narrative indicates any miraculous provision was made for feeding the livestock, which would be a serious problem in such a marginal area, nor for the fuel that would be needed to keep two million people warm through the cold desert nights. As for hygiene issues, one hardly dares to imagine. Perhaps it is no miracle we're told the Israelites were prone to epidemics.

Anyway, the question is moot; no archeological remains suggesting a mass migration of Asiatics across Sinai have ever been recorded, though several archeological explorations have been carried out. *But*, say the believers, *nomads do not leave archeological remains*. Actually, they do, often in patterns that reflect seasonal transhumance, clustering around certain resources at certain times of the year and in places where they interact with sedentary populations.[14] Of course, because pastoralist or foraging groups tend to be small and mobile, the archeological footprint is more subtle than those of sedentary populations, but nomads are by no mean invisible in the archeological record. As for the Exodus, however, there would be nothing subtle about a great mass of people and animals shuffling across the landscape. The equivalent of a fair-sized city on the move—say, Greater Cleveland, or Calgary *plus* Edmonton—should have left a large, clear footprint, including a good scatter of artifacts.

Further, scholars have a reasonably clear idea about what *was* happening in the Sinai in the Chalcolithic and Bronze Ages: it has been designated the Timnian Complex, a sequence of hunter-gatherer groups that added some pastoralism to their subsistence by about 5500 BCE and segued by the Early Bronze Age into full nomadic pastoralists with an additional, critically important, resource: The Sinai Peninsula is the site of very ancient copper and turquoise mines, an activity that the Egyptians took over as early as the Third Dynasty and maintained (with the odd hiatus) right through to the mid-twelfth century BCE, toward the end of the Late Bronze Age.[15] During this long period of Egyptian mining, pastoralist camps continued but tended to be scarcer, clustered around the Egyptian mines and interacting regularly with the Egyptians, who referred to them as *shasu*. Egypt was clearly in control, maintaining a strong military presence in a string of forts along the coast, though Elizabeth Bloxam also suggests a certain integration between the indigenes and the Egyptian mining authorities, rather than "forceful domination."[16]

The bottom line is this: at the times when Exodus-positive scholars propose the Israelites were in a holding pattern in the wilderness, (1) the Egyptians maintained a strong presence in the Sinai peninsula due to the valuable copper and gemstone deposits, (2) pastoral groups displaying a high degree of cultural continuity from predynastic times continued to subsist in the Sinai, and (3) Kadesh-Barnea, where the Israelites were said to have hung out for much of their forty years in the wilderness, was archeologically barren at the time—though known for its Iron Age fortress dating from the seventh century BCE.[17] All these factors rule out a migration across Sinai on the scale described in the Bible and give no support for a sojourn in Kadesh-Barnea.

ARCHEOLOGICAL ISSUES: THE PROMISED LAND

Canaan, the endpoint of the Exodus, the promised land of milk and honey, is where archeological predictions based on the biblical narrative are most glaringly falsified. No conquering multitudes suddenly burst out of the Wilderness of Zin; no new Israelitish culture suddenly explodes in what would become the kingdoms of Israel and Judah. The full story is outside the main scope of this chapter, but here is the bottom line: the area that would become the Judahist heartlands was a marginal zone that experienced several waves of settlement from 3500 BCE through to the beginning of the Iron Age.[18] In each case, the settlers were not incomers but indigenes responding to changing climatic and trade situations by shifting emphasis from pastoralism to cultivation and back again—a common strategy of ecotonal peoples. The evidence points to in-place development rather than incursions, on a background of overall cultural continuity. Israel, it seems, did not come out of Egypt; it appears they were in Canaan all along.

NARRATIVE SOURCES

The conclusion strongly suggested by archeology—that the Exodus narrative is an origin myth rather than veridical history—is further supported by examining the bare bones of the story itself. A good case can be made that its composers relied heavily on reworking folk memes that were floating about the ancient Near East at the time, which can also be glimpsed in other legendary-historical traditions. The striking similarity of Moses's backstory to that of the much earlier backstory of Sargon of Akkad is a case in point, and also the tip of a huge folkloric iceberg.

Sargon, usually monikered "The Great," was the king of the city-state of Kish in the Mesopotamian uplands, late in the Early Bronze Age. He is remembered as the architect of the first great empire, stretching as far north and west as the Mediterranean Sea and Anatolia, and lasting from the mid-twenty-fourth century BCE to the general collapse in about 2200 BCE. But Sargon, according to his legend, began as a nobody: the bastard (or virgin-born) son of a high priestess, who bore him secretly and then made a little boat of reeds or rushes, sealed it with pitch, and surrendered the child to the river. Baby Sargon floated downstream for a bit, until a palace gardener found the little boat and brought the foundling up in humble circumstances to be a palace gardener himself.

He was not destined to stay in lowly circumstances. The King of Kish saw him and, struck with his fine qualities, appointed the lad to be his cup-bearer; but once he became part of the king's court, Sargon was troubled with prophetic dreams, which in turn troubled the King of Kish, because the dreams were about him (the king) drowning. To avert this doom, the king sent Sargon to another city to deliver a message, which boiled down to "kill the messenger." Fortunately, with the help of the love-smitten goddess Ishtar, Sargon was able to navigate the hazardous shoals of royal Kish and end by supplanting the king.

There is an obvious parallel to Moses being placed by his mother in a little ark among the bulrushes of the Nile, to save him from pharaoh's murderous edict about drowning male Hebrew babies. We see similar details in the story of Romulus and Remus, the scandalous bastard offspring of Mars and a vestal virgin, cast adrift on the Tiber in a basket; or Karna of the *Mahābhārata*, whose little ark is drawn out of the Ganges. These, and other baby-in-a-basket stories, are part of a more general "Hidden Child" folkloric motif, where the birth of a hero is accompanied by dire prophecies—often, that the child will cause the downfall of the city or the reigning king. The king orders the child to be killed as a precaution, often by exposure, but the child survives. The underling entrusted with the murder may take pity on the child, as in Oedipus or the fine folkloric example of Snow White and the Seven Dwarves. The prophetic element is not clear in the biblical version of Moses's backstory but is made explicit in the later rabbinical literature: Pharaoh's soothsayers warn him that a Hebrew child will be his downfall, which leads to the order to drown all male Hebrew babies at birth. The rabbinical literature even throws in a touch of marital irregularity in the relations between Moses's parents. In all cases mentioned, the future hero/liberator grows up in obscurity, hidden behind a false identity until his true exceptional nature is revealed. Moses, in a neat reversal, is presented as a son of slaves raised by a princess, but the story arc is unmistakably part of the same tradition. (Parenthetically, the story of Jesus's birth presses quite a few of the same buttons.) The Judean composers of the Pentateuch may or may not have drawn *directly* on the Sargon legend, but the character of Moses draws on the same widespread archetypal memes.

Echoes of other widespread memes can also be heard in the Exodus narrative. For example, elements and snippets from Egyptian stories— Sinuhe, the Tale of the Two Brothers, various dream narratives, the parting of the waters and other magical feats recounted in the Westcar Papyrus— were common memes that could have been borrowed by the biblical composers. More interesting is the possibility that folk memories of verified historical events may have been incorporated into the narrative. Redford argues convincingly for a folk tradition among the Levantine Canaanites preserving the memory of the Hyksos' glory days in Egypt and their ultimate expulsion, a tradition that survived into Phoenician and Hellenistic lore and the origin stories of various Semitic groups—including the upland Judeans, who were not themselves related to the Hyksos.[19] Even the great take-home lesson of the Exodus, the Ten-Plus Commandments, sits squarely in a widespread legal genre of great antiquity and cannot be viewed as seminal, or even particularly original.

CONCLUSION

So where does the captivating story of the Exodus come from? Its details do not fit the time periods in which it was supposedly set; its exposition of Israelite origins runs counter to the archeological record. And yet, it overwhelms the reader with detail and incident, long-running plot lines, a wealth of characters, local color, and local history, some of which does correspond to extra-biblical material—for periods in the Iron Age, at least. And that is the key.

Converging lines of evidence point to the Pentateuch, including the Exodus narrative, being a product of the kingdom of Judah; its compilation begins in the seventh century BCE, continues through the Babylonian

Captivity in the sixth century, and culminates in post-Exilic times, from 539 BCE on. A pattern of anachronisms—camels and Philistines, spice caravans, place names and personal names—points to Iron Age authors and redactors describing the distant past in terms of their contemporary world. What legends and folktales they may have drawn upon is a much wider study, but a good case can be made for the reworking of folk memories of the Hyksos expulsion, the chaos of the Late Bronze Age collapse, and the semi-pastoral origins of the main players.

Indeed, could one of the legends woven into the Exodus narrative be based on a factual small-scale escape from Egypt? Not the cast-of-millions of the Bible story, but a few dozen or a few hundred enslaved prisoners of war slipping out of the Delta and making their way across the Sinai, avoiding the forts along the Mediterranean coast? Of course it is possible—but unknowable. If, however, there is one aspect of the story that rings true to me, it is the messianic motif embodied in Moses: a charismatic leader who persuades a pack of marginalized people to follow him into the wilderness on the promise of a better life, and then subjects them to increasingly despotic behavior and draconian rules, the Jim Jones of the Wilderness of Sin. The narrative of the Exodus and the sojourn in the wilderness is one of blood and terror, hideous punishments, and ruthless intolerance of dissent, a fitting prelude to the genocidal horrors of the Conquest. As Redford puts it:

> One final irony lies in the curious use to which the Exodus narrative is put in modern religion, as a symbolic tale of freedom from tyranny. An honest reading of the account of Exodus and Numbers cannot help but reveal that the tyranny Israel was freed from, namely that of Pharaoh, was mild indeed in comparison to the tyranny of Yahweh to which they were about to submit themselves. As a story of freedom the Exodus is distasteful in the extreme . . . and in an age when thinking men are prepared to shape their prejudice on the basis of 3,000-year old precedent, it is highly dangerous.[20]

Chapter 12

PIOUS FRAUD AT NAZARETH

René Salm

M ost of the excavating in Nazareth over the past century has taken place in a small area of the northern hillside known as the "Venerated Area," land owned by the Catholic Church since 1620. From the time of Constantine forward, Christian tradition has located the ancient settlement there, following the famous account in the Gospel of Luke where Jesus's sabbath reading in the synagogue incited the crowd to fury: "They got up, drove him out of the town, and led him to the brow of the hill on which their town was built, so that they might hurl him off the cliff" (Luke 4:29). The fact that no cliff suitable to such a scene exists in or around the Nazareth basin is but the first indication that, perhaps, the Lucan account is fictive. A second indication is that at the turn of the era synagogues did not yet exist in the Galilean backcountry. Thirdly, the synoptic gospels describe Nazareth as a *polis* (Gk. "city"), yet no evidence of a substantial settlement existed there at the turn of the era. In fact, an astonishing conclusion of my 2008 book, *The Myth of Nazareth*, is that "not a single post-Iron Age artifact, tomb or structure at Nazareth dates with certainty before 100 CE."[1] My second book, *NazarethGate*, published in 2015, reaffirms that conclusion.

The above untoward observations (and many more are possible) are indeed troubling to a Christian tradition embattled today on multiple fronts. We may wonder whether the historicity of its iconic founder, Jesus of Nazareth, can survive without any *demonstrable* existence of his hometown. Perhaps recognizing this, the Christian tradition (Catholic, Protestant, and Greek Orthodox) has recently filled any perceived archaeological lacuna with an unprecedented spate of excavations and religious developments in Nazareth, all bolstering the traditional story.

The closing years of the twentieth century also witnessed a major shift in Israeli archaeology, one still very much in effect today. That shift is perhaps best described as an awareness—at a fundamental level—of the commercial potential of Israeli archaeology. That awareness is not itself new, but only in the last generation does it seem to have become an integral part of the government's ubiquitous involvement in the field. The commercial exploitation of contemporary archaeology in Israel appears to be top down—deliberate and made at the highest levels.

In 2014, Israeli tourism garnered $11 billion, some 4.5 percent of the country's nominal GDP. Tourism employs 100,000 workers, two-thirds of them in the hotel industry. A large part of that tourism is, of course, specifically Christian, in the form of pilgrims from the West visiting sites allegedly associated with the life of Jesus. A 2002 article in the *Washington Post* describes a concerted effort to reach out to these potential visitors:

> In an effort to solidify its relationship with American evangelicals, the government of Israel has launched initiatives that include expense-paid trips to the Holy Land and strategy sessions with the Christian Coalition and other conservative groups. The objectives: to revive Israel's sagging tourism industry and strengthen grass-roots support in the United States. The target audience is the estimated 98 million U.S. evangelicals, but especially a subset of that group, Christian Zionists.[2]

The author goes on to state that the Israeli government hired an American PR firm and approved a "multimillion-dollar marketing plan . . . with certain aspects dependent on funding by the Knesset, the Israeli parliament."[3]

The first stage in this commercialization of Israeli archaeology took place with preparations for the millennium celebrations, which included an historic visit by Pope Paul II to the Holy Land in March 2000. Nazareth was an important stop on the pope's tour, and in preparation the city undertook a major renovation and expansion initiative (the "Nazareth 2000" tourist development project), which included considerable rebuilding. In Israel, the erection of new structures often uncovers antiquities that, by law, must be vetted by the Israel Antiquities Authority (IAA). Thus, many salvage excavations were initiated by the Nazareth municipality and undertaken by the IAA at the close of the millennium, including those at Mary's Well (which we will discuss below). New hotels and one major Christian pilgrim destination (the Nazareth Village Farm) were success-fully completed in time for the pope's visit.

It can be noted that, in the decades before the Second World War, archaeological digs in Nazareth were fairly unprofessional affairs con-ducted by untrained researchers—mostly seminary-trained priests. The science of archaeology had not yet come of age. Many of the recovered artifacts from those decades were lost, while those that have survived in boxes or small displays are often not labeled or are mixed with materials from several venues. Stratigraphy was not used, and chronologies were only vaguely known.

In those early days, the Catholic researchers were surprised to find so many tombs in the settlement that ostensibly existed in the time of Jesus. They noted a number of tombs even under the Church of the Annunci-ation—the spot where the Virgin Mary allegedly lived.[4] Apparently, the pre-war researchers were not aware of Jewish purity laws, particularly the proscription against living in the vicinity of tombs.[5] They noted the many tombs in and around the Venerated Area with alacrity, even hoping perhaps to discover the burial of a member of the Holy Family.[6] This accumulated early information would later prove damning, for the "traditional village" of Nazareth lies in the midst of a Roman cemetery, as Father C. Kopp observed in 1939.[7]

The Second World War brought immense advances to all branches of technology, including to biblical archaeology. New insights also brought an acute awareness of the Jewish proscription against living in the vicinity of tombs. The primary Christian answer to this dangerous information at Nazareth was to simply ignore all evidence of tombs collected in pre-war decades. Thus, Jack Finegan's much-referenced *The Archaeology of the New Testament* (written in 1969) includes a map of the tombs of Naza-reth—*yet it omits all tombs within 250 meters of the Venerated Area!*[8] We also look in vain through Fr. Bellarmino Bagatti's magnum opus on Naza-reth, *Excavations in Nazareth* (1969), for any mention of tombs—with one exception. That exception regards a tomb under the Church of the Annunciation that Bagatti was unable to ignore as it had been repeatedly noted and even diagrammed in the pre-war years. Therefore, Bagatti attempts to redate the tomb to Crusader times. I discuss this desperate stratagem in my 2008 book, noting that the macabre custom of medieval Christians burying their dead under the house of the Virgin Mary is otherwise thoroughly unknown.[9]

It can also be noted that the lion's share of movable artifacts from Nazareth excavations (pottery and oil lamps) have been found in kokh-type tombs,[10] a type of tomb dating no earlier than ca. 50 CE in Galilee—that is, *after* the time of Jesus.[11] This is according to the detailed work of H.-P. Kuhnen and other specialists—work that is still assiduously ignored by the tradition, for it is obviously too threatening.

We will now consider the most important recent develops in the archaeology of Nazareth.

THE NAZARETH VILLAGE FARM

The Nazareth Village Farm (NVF) is a fifteen-acre development half a kilometer west of the Church of the Annunciation. Popular with children and adults, it features "a carefully researched re-creation of Jesus' hometown" complete with "exact replicas of first century houses, synagogue, mikveh (a bath used for the purpose of ritual immersion in Judaism), and olive presses" (from the NVF website). Activities include a Parable Walk, museum, study center, and biblical meals. The NVF is funded by an international consortium of Christian groups called the Miracle of Nazareth International Foundation. Since the project's inception, the consortium has raised over $60 million toward the venture. High-profile contributors in the United States include former president Jimmy Carter, singer Pat Boone, and Rev. Reggie White, the former Green Bay Packer football star.

Planning for the NVF began in 1996 and the doors opened to visitors in 2000. "Scholarly research led by experts from the Jerusalem-based University of the Holy Land underpinned the project," according to the NVF website. Those scholars claimed that the site was "a working terrace farm in Jesus' time." In 2007, Stephen Pfann (director of the University of the Holy Land), Ross Voss, and Yehudah Rapuano authored a sixty-one-page report in a scholarly journal that claimed that a substantial amount of pottery found at the NVF dates to Hellenistic and Early Roman times.[12] However, my careful examination of the report revealed multiple errors in its data and also showed that the pottery in question reduces to eleven small shards whose dating rests not on scientific grounds but on one archaeologist's (Rapuano's) *unsupported* opinion. Furthermore, all the NVF pottery shards whose descriptions were accompanied by scientific parameters (such as typological parallels to artifacts of similar form, color, composition, and decoration) in every case could be dated after 70 CE. In sum, upon careful scrutiny there is no evidence from the NVF of human activity "at the time of Jesus."

My written rebuttal of the above NVF claims, published in 2008, in turn provoked Rapuano to publish an amendment to his original findings, one twice the length of his original pottery report.[13] Unfortunately, the amendment was a step backward, not forward, in that it abandoned verifiable parallels for the early shards in question in favor of unpublished comparisons that no one can even check.

Of course, in an elaborate project geared to Christian tourists such as the NVF, millions of dollars are at stake. It is not surprising that the claim of "a working terrace farm in Jesus' time" would be vigorously defended by the "experts" and by the backers of the enterprise. Noteworthy, however, is that the archaeologist charged with providing a pottery rationale for that early dating—Yehudah Rapuano—is a Jewish archaeologist employed by the Israel Antiquities Authority. This shows the close cooperation between Christian and Jewish interests that now obtains in large Christian enterprises such as the NVF.

MARY'S WELL

As mentioned above, salvage excavations took place at Mary's Well in the context of the Nazareth 2000 development project. The Israel Government Tourist Authority funded the excavations, carried out in 1997–98 and directed by IAA archaeologist Yardenna Alexandre.[14] Ms. Alexandre has been professionally active for several decades, either assisting or directing a number of excavations in the vicinity of Nazareth. Some years ago, during the lengthy research for my book *The Myth of Nazareth*, she and I engaged in an email correspondence in the course of which Alexandre made the claim of having found "Hellenistic" evidence at Mary's Well. She sent me an attached document that would later prove of considerable value: her one-page IAA report on the Mary's Well excavation. I was surprised to see that the document had only just been prepared—it was dated 2006 though the excavation had taken place eight years before. Alexandre's report only mentioned the word "Hellenistic" once, in the final line and in a vague context: "some occupation here in the Hellenistic, Crusader and Mamluk (1250–1517 CE) periods."

Hellenistic evidence was entirely inconsistent with the data that I'd been collecting from the Nazareth basin. Strangely, it was also inconsistent with the rest of Alexandre's 2006 report, which mentioned no evidence from the Hellenistic era at all.

If present, such early evidence would effectively scuttle my case against Nazareth's existence at the time of Jesus. In our email exchanges, I pressed Alexandre on this provocative claim. However, she was either unwilling or unable to produce documentation. Nor did she offer any description at all of the "Hellenistic" evidence involved.

Information surfaced a year later on the nature of the "Hellenistic" evidence from Mary's Well. Astonishingly, the details came not from Alexandre but from Stephen Pfann and his colleagues, in their 2007 report on the NVF, a site located over one kilometer from Mary's Well. Why apparently irrelevant details on Hellenistic evidence from Mary's Well would be disclosed in an NVF report is curious. The difference in venues was sufficiently confusing that it misled Bart Ehrman. In Ehrman's 2012 book *Did Jesus Exist?*, the latter described the alleged Hellenistic evidence as if it were found at the NVF and not at Mary's Well.[15]

What was this evidence? Pfann and his colleagues wrote that Alexandre, in 1997 and 1998, had uncovered 165 coins at Mary's Well.[16] They continued, "The coins were overwhelmingly Mamluk, but also included a few Hellenistic, Hasmonean, Early Roman, Byzantine, Umayyad and Crusader coins (Alexandre, forthcoming)." Remarkably, Alexandre had mentioned nothing in her 2006 IAA report (previously emailed to me) regarding "Hellenistic, Hasmonean, [and] Early Roman" coins.[17] It was also more than passing strange to me that vital evidence for Nazareth's early existence was appearing in such a roundabout way—eight years after the excavation and disclosed not by the excavator herself but by third parties!

There is insufficient space here to detail the convoluted history of this early coin claim, one that has engaged scholars far and wide but that has no basis in fact, and I detail it in my 2015 book. The chapter on coins in Alexandre's 2012 Mary's Well monograph was not written by her but by Ariel Berman. That chapter is remarkably flawed, with false references, a double-dating (the same coin dated in two places to widely separate eras), and an illustration where all five captions are in error.[18]

The question in this case is ultimately not who wrote what, but the nature of the coins. To resolve the matter, in 2015 I personally engaged a professional photographer to visit the IAA archives in Jerusalem and to photograph these bronze coins that have been at the center of so much discussion. The results were stunning. The photographs reveal that these "early" coins are so water-worn as to be thoroughly unreadable. On one coin, for example, Berman identifies an "anchor" on the obverse and a "star" on the reverse. Yet one side is concave, having been completely worn away from water wear, and both sides are so pitted that not a single feature (line or letter) is preserved, much less any design.[19]

Even more damning than the complete unreadability of these coins, however, is the fact that the water channels at Mary's Well were—by Alexandre's own determination—*built in Middle Roman times*![20] The alleged early coins are extremely water-worn and were dredged out of those very channels. This being the case, their deposition could not predate the existence of the channels themselves. This automatically eliminates the coins from Mary's Well as evidence for pre-Roman (and even Early Roman) activity at the site. In turn, we can be quite sure that those unreadable coins have clearly been both misread and falsely assigned to their alleged early periods.

THE "HOUSE FROM THE TIME OF JESUS"

Four days before Christmas 2009, IAA archeologist Yardenna Alexandre gave a press conference in Nazareth, Israel, at newly excavated ground across the street from the Church of the Annunciation. Major news agencies from around the world were present, including AP, UPI, Reuters, and Agence France Presse. Alexandre had just finished excavating the site, which was situated under the International Marian Center (IMC), an ambitious Roman Catholic venture that opened to the public in 2012 and encompasses five stories, and includes cinemas, a restaurant, boutique, chapel, offices, and a gift shop.

Alexandre announced to the global media that she had uncovered for the first time a house "from the time of Jesus" in Nazareth at that precise spot. It was a modest dwelling of two rooms, with a courtyard and cistern for the collection of rainwater. The recovered pottery was from the first and second centuries CE. In addition, Alexandre reported that she excavated a refuge pit at the site, "probably hewn as part of the preparations by the Jews to protect themselves during the Great Revolt against the Romans in 67 CE."[21]

The above information quickly circled the globe and has since been used in both scholarship and the popular press as clear evidence for the existence of Nazareth at the turn of the era. However, Alexandre's foregoing claims are in fact false.[22] The site was clearly that of a Middle-Late Roman wine-producing installation, with numerous distinctive features, including a plastered treading floor and down-sloping channel leading to a plastered collecting vat with several characteristic steps in one corner. Most egregiously, in her attempt to validate a dwelling, Alexandre apparently invented a critical wall where none even exists in the evidence.[23] Furthermore, the ground slopes in the wrong direction for the cistern to have been used for the collection of water. It would also be far too large to have served the modest purposes of a humble family (as portrayed in the press). On the other hand, the cavity perfectly suits both the characteristics and dimensions of a cellar used for the storage of filled wine jugs.

Alexandre claimed that another cavity on site served as a "refuge pit" during the First Jewish Revolt.[24] This claim added a touch of drama to the ensuing news stories, but it is untenable for several reasons. First of all, the cavity is far too modest in size for that purpose, measuring only one meter in diameter. Only a couple of people would have been able to squeeze into it, and they would have had to do so without provisions or even room to lie down. Scholars such as Mordechai Aviam and Yuri Tepper have shown that Jewish hiding places in antiquity were large, complex, and well hidden.[25] Usually, they were multi-chambered affairs reachable through narrow underground passageways. This layout discouraged both intruders and discovery, for to reach the hideout people had to crawl on their belly in single file for some distance.

The cavity at the IMC denominated a "refuge pit" by Alexandre is something entirely different. It is small, single-chambered, and at the surface. It probably also served as a storage cellar for the nearby wine-producing installation. In any case, hiding complexes were typical of the Bar Kokhba Revolt, not of the First Jewish Revolt.[26] Furthermore, the people of Sepphoris (a town located six kilometers north-northwest of Nazareth) did not side against the Romans during the First Revolt (engendering considerable animosity among the Jews), and thus it is unlikely that hiding places in that time and in that area were hewn at all.

It can also be noted that Alexandre's various statements on the IMC site contradict one another at important points—as regards, for example, the basic size of the "house" (small or large). No final report on this excavation has been published. Perhaps one is planned for the future, but as of this writing we are left with an "official" interpretation of the site that is tradition-friendly and strongly enhances the commercial attractiveness of the new mega-complex known as the IMC, while at the same time it flies in the face of the obvious evidence and is empirically untenable.

THE SISTERS OF NAZARETH CONVENT

Over a series of several summers beginning in 2006, the British archaeologist Ken Dark worked at the site of a well-known kokh tomb located under the Sisters of Nazareth Convent (SNC), scarcely one hundred meters to the west of the IMC site described above. A large rolling stone at the door of the SNC tomb has been a minor, but noteworthy, pilgrim attraction for generations. The Catholic order of the Sisters of Nazareth (not to be confused with a London-based order of the same name) purchased the site in 1881, land that local residents claimed had once been the site of a "great church" and the "tomb of a saint."[27] As we noted in the opening section of this chapter, the pre-World War II Catholic tradition was keen to identify tombs in and around the Venerated Area, in hopes that one or more of them might be the burial places of members of the Holy Family. Indeed, as recently as 2009 the nuns at the SNC were solemnly informing visitors that St. Joseph, the father of Jesus, was buried there.[28] However, as the kokh-type of tomb did not reach the Galilee until the middle of the first century CE—per the work of H.-P. Kuhnen (see above)—that pious tradition is most unlikely, though it could perhaps be argued that St. Joseph was an old man when he died and that he was buried in one of the first kokh tombs hewn in the valley.

Ken Dark, however, has now published radically different conclusions regarding the SNC site.[29] Dark proposes that a *dwelling from the time of Jesus* existed at the SNC site! In fact, he claims to be the first archaeologist to uncover such remains,[30] thus competing with Alexandre for that distinction, for we have seen that in 2009 she announced to global media the discovery of the "first" house from the time of Jesus.

However, my research has shown that Dark's conclusions regarding a dwelling at the SNC are as specious as those of Alexandre at the nearby IMC site.[31] First of all, we are confronted with Dark's astounding proposal that a Jewish dwelling once existed *above* a tomb! Of course, he is quite aware that this scenario is more than a little unusual, not to mention abhorrent in a Jewish context. Thus Dark further proposes a two-phase chronology in which the tomb did not yet exist when the dwelling was inhabited.[32] In order to date the dwelling precisely to the first century CE, Dark appears to adopt a series of hypothetical conclusions, none of which are tenable and all of which, when viewed in their totality, border on the absurd. We cannot enter into the details here, but Dark proposes, in sum:

(1) that a dwelling at the SNC site was constructed, inhabited, *and* abandoned all during the first century CE; (2) that the kokh tomb was also used and abandoned before ca. 100 CE; and (3) that somehow the timespans of dwelling and tomb did not overlap.

The question is not whether the above venturesome scenario is possible—even remotely. The problem is that *it has no basis in the material evidence* at the SNC site. The structures Dark interprets as domestic are, in fact, agricultural and funereal, as witnessed most dramatically by the fact that the extant early walling is *rock-cut*, low, and coarsely chiseled—as one might expect in tombs and agricultural installations (but not in dwellings). Furthermore, many of the walls of Dark's proposed Early Roman "courtyard house" are hypothetical—they do not exist at all in the remains.[33]

A forthright analysis of the remains will have no difficulty positing (1) a funerary phase followed by (2) partial destruction of the kokh tomb (in Byzantine times), and (3) many centuries of agricultural work (including filtration basins, a possible wine-making complex, and storage of liquids and grain in cisterns and silos). Dark's proposal of a pre-existing dwelling has no empirical basis. Since it is false for the SNC site, it is much more so for the archaeologist's attempted extension of this dwelling-tomb sequence to other sites in Nazareth! In addition, Dark proposes that a substantial water source (i.e., spring) once existed at the SNC site.[34] All these bold proposals lie in the realm of the fabulous and may be diplomatically described as "irresponsible."

Unexpectedly, Dark's arguments also contain elementary errors that are surprising in a scholar writing and publishing on the archaeology of Nazareth. Such errors include the misapplication of Jerusalem chronologies to the Galilee (with critical consequences for Dark's dating scheme), as well as a fundamental misunderstanding of kokh tomb typology.[35] The British archaeologist also seems to have an unfortunate penchant for misrepresenting the work of his peers when it serves his purposes.[36] The latter is, of course, entirely indefensible on scientific and ethical grounds.

THE CAESAREA INSCRIPTION

The "Caesarea Inscription" (CI) purports to be a marble plaque displayed during Byzantine/Medieval times in a synagogue of Caesarea Maritima. The alleged plaque consists of three small fragments, which, together, make up less than 10 percent of a lengthy inscription listing the twenty-four priestly families (or "courses," 1 Chr. 25:7–18) in order, together with each course's domicile in the Galilee in Middle Roman times (or later). The list of courses and their Galilean domiciles is mostly known from poems and chants (*piyyutim*) dating to Byzantine times. Inscribed stone plaques containing the list were sometimes also placed in synagogues. Rare fragments of such plaques have been recovered in the remains of a few synagogues from Yemen to Syria.

Since discovery of the first fragment of the CI in 1956 (under very dubious circumstances, however—see below), interest in it has been high among both Jews and Christians. A second excavation season in 1962 uncovered two further fragments (also under dubious circumstances), one containing the word "Nazareth." The reconstructed eighteenth line reads, "The 18th course, Hapises, Nazareth." Hence, this discovery purportedly represents the earliest non-Christian epigraphic witness to the existence of Nazareth. Evangelical and conservative Christians have gone so far as to date the migration of priestly courses northward to immediately after the First Jewish Revolt. Thus, they have found in the CI evidence for the existence of Nazareth already ca. 70 CE, a scant forty years after Jesus' alleged death.

However, more liberal scholarship today doubts that a northward migration of priests ever took place.[37] It views the priestly courses' inscription as myth, primarily self-validating propaganda authenticating the rights of priests in towns and areas *after* they established their new domiciles in the Galilee—a way of retroactively establishing claim to the land as far back as Hasmonean times. In any case, Jerusalem was largely spared by the Romans in the First Revolt, and there would have been no reason for priests to leave. Even though the temple had been destroyed, they were hoping for its eventual rebuilding and would not have been willing to forsake the holy city unless compelled.[38] If a northward migration of priests occurred at all, it would have taken place after the Bar Kokhba Revolt, during which Jerusalem (and, indeed, all of Judea) was devastated, and after which Emperor Hadrian forbade Jews to reenter Jerusalem, renamed Aelia Capitolina. Of course, if the priests were indeed expelled from the city in the 130s CE, then the priestly courses' inscription is no longer relevant to the question of Nazareth's existence at the turn of the era.

Thus, recent historical considerations argue against the conservative Christian position that the CI witnesses to a first-century Nazareth. But such scholarship was as yet unknown in 1962 when the fragment containing the word "Nazareth" in Hebrew letters (fragment A) was discovered. Immediately, the discovery was hailed as proof for the existence of Nazareth in the first century CE. Since then, the CI has remained a fixture in the Nazareth evidence, with scholars disagreeing principally on whether the Hapises migrated up to Nazareth after the First or after the Second Jewish Revolt.

However, my researches have now shown that none of the above positions can be correct, for the CI is itself a forgery.[39] In brief, my suspicions were aroused in 2012 when a colleague in Italy, Enrico Tuccinardi, alerted me to his article on the CI entitled "*Nazareth, l'épigraphe de Césarée et la main de Dieu*" (Nazareth, the Caesarea Inscription, and the Hand of God).[40] The article points out some surprising anomalies, including that the underlying inscription (as discovered) seems to contain errors in spelling, that the 1956 fragment was unprovenanced and is known only through a single, poor quality black-and-white photograph[41] (no scholar on record claims to have actually seen the artifact), and that the discoverer of the "Nazareth" fragment had a proven history of fraud.

The director of both Caesarea excavations (in 1956 and in 1962) was Michael Avi-Yonah. Subvention was largely provided by the Fund for the Exploration of Ancient Synagogues. This graphically betrays an interest of Avi-Yonah and his backers to uncover evidence of ancient synagogues in the land. The 1956 discovery of a priestly courses' fragment in Caesarea hinted that a synagogue might be in the area of ancient northwest Caesarea. In 1962, Avi-Yonah returned to that area and was only too pleased to uncover two further fragments, allegedly of the same plaque.

However, they were not from the same plaque. In the summer and fall of 2012, Tuccinardi, a very helpful Israeli researcher by the name of Evgeny Oserov, and myself verified that the three fragments of the CI do not match one another in orthography (size of letters and line spacing). This immediately signaled that no single "Caesarea Inscription" exists. It is a *myth*. We are dealing, rather, with *three different plaques*, fragments from which have been falsely combined into one plaque.

At this point, another scholar enters the discussion. Marylinda Govaars participated in subsequent excavations at the Caesarea site in 1982 and has co-authored the most comprehensive review of the archaeological remains to date.[42] Her 2009 book was extremely critical of Avi-Yonah's sloppy excavation methods, and it also took the latter to task for not properly documenting his finds and not publishing them. (Avi-Yonah never published a final report on the Caesarea excavations.) Most interesting for our purposes, however, is that Govaars was unable to find any reasonable evidence of a synagogue in the area that Avi-Yonah excavated. Evidently, in his haste to find what his financial backers wished him to find, Avi-Yonah had jumped to conclusions regarding structures in the earth and had all-too-quickly pronounced them a "synagogue."

Tuccinardi, Oserov, and myself immediately saw that, if Govaars was right, then the authenticity of *all three marble fragments* was questionable. After all, priestly courses' plaques are known to have been placed *only* in synagogues. How, then, could any such fragment have been found in an area of Caesarea *where no synagogue existed?*[43] Furthermore, the reader is reminded that we are dealing with three incompatible fragments. The upshot is that whoever would defend the case for authenticity must posit the ancient existence of *three different synagogues in the same small area of Caesarea where archaeology has failed to locate even a single synagogue!*

At this juncture, my suspicions regarding the discoverer of the critical "Nazareth" fragment became acute. Upon rereading Tuccinardi's article, I was flabbergasted to register his name: Dr. Jerry Vardaman. This evangelical archaeologist-scholar (d. 2000) is remembered today mostly for his notorious claim of having found microscopic Christian lettering ("micro-lettering") on ancient Roman coins. Vardaman never presented the coins themselves, and his claim has now been completely discredited and Vardaman exposed as a fraud.

Further research revealed that Vardaman had embarrassed himself in excavations at Shechem, that he had narrowly escaped arrest in Jordan for bribery in the field, and that he was actually arrested by the Israeli authorities—during the 1962 Caesarea excavation season, and *on the same day that the Nazareth fragment was found!*

Vardaman's arrest occurred on August 14, 1962, within hours of his "finding" the critical marble fragment bearing the word "Nazareth" in Hebrew. The excavation would last another week, but Vardaman did not return to the field.

No charges were pressed against Vardaman. In fact, his find was soon hailed as monumentally important. The Jewish establishment was pleased to have such clear evidence of a synagogue in the northwest of the land. Christians, too, were thrilled that evidence for Nazareth's early existence had now been produced. Vardaman became something of a celebrity in archaeological circles. The following year he sat next to Avi-Yonah at an awards ceremony in Jerusalem's King David Hotel, during which Avi-Yonah told Vardaman that "our joint discovery of this [priestly courses'] inscription resulted in [my] recent promotion from associate professor to full professor at the Hebrew University."[44]

However, it is very possible that all three fragments of the CI are forgeries. Each is attended by a long list of problems. As regards the first fragment found in 1956, it was apparently discovered not in a controlled excavation but by an amateur collector of antiquities, and its provenance is unknown. It could have come from anywhere. Indeed, its relevance to a putative synagogue in the excavation area (and hence to the CI) is entirely unsubstantiated. Its writing does not match the other two fragments. Furthermore, scholars have apparently only seen a photograph of it—no one claims to have actually seen it. Finally, the present whereabouts of this first fragment are unknown. It has been "lost" since the late 1950s.

The second "Nazareth" fragment of the CI was found by Vardaman during the second Caesarea excavation, in a wheelbarrow containing debris destined for the dump. This is curious, for such a substantial piece of marble would, under normal circumstances, have been noticed on excavation from the earth and would never have entered the debris pile. This raises the suspicion that the fragment was not found in the earth but planted in the wheelbarrow.

Vardaman was absent in the early weeks of the 1962 excavation for days at a time and is known to have been in Jerusalem, not in Galilee, on at least one of those trips. His activities in Jerusalem are unknown. Vardaman was one of the co-directors of the Caesarea excavation. Why was he absent while digging continued? Why was he in Jerusalem for days at a time prior to discovery of the "Nazareth" fragment? Evidence is only circumstantial, but if a *forged* fragment of the Caesarea inscription were fabricated, it would almost certainly have been produced by a black market supplier working in Jerusalem.

In any case, the most incriminated circumstance relating to the "Nazareth" fragment must be Vardaman's arrest by "Israeli authorities" (presumably the police) on August 14, 1962—the same day the "Nazareth" fragment was found.[45] We do not know the reason for the arrest, but it surely has something to do with Vardaman's astonishing discovery of that afternoon. One can speculate that somebody tipped off the police to foul play in the excavation. Indeed, we know that Vardaman made enemies in the earlier Shechem excavation, and that his activities were watched with apprehension by at least some of his peers.[46]

Serious questions also attend the third CI fragment. It was found also in the 1962 excavation season, a few days after Vardaman's arrest. One might suppose, then, that Vardaman had nothing to do with this fragment. But its strange nature and the strange circumstances of its discovery lead to other conclusions, as described in detail in *NazarethGate*.[47] Briefly, the third fragment contains only the first letter of three consecutive lines. The letter is a *mem*, as it should be in the priestly courses' inscription. However, while the top *mem* is expertly chiseled, the other two were accomplished in a very different and coarse manner—perhaps quickly. In any case, it is clear that the three letters do not go together. In addition, the line spacing and size of the letters in this fragment do not match those of the other two fragments. Finally, the reports contradict one another regarding the find spot of the third fragment ("area E" vs. "area F").[48]

Today, over fifty years after the fact, forgery is all but certain when we consider that authenticity is virtually impossible. Firstly, the fact that no synagogue was in the putative discovery area of any of the three fragments is the most incriminating element. It means that the fragments were *brought in*. Secondly, the fact that the three fragments do not match one another shows that something is seriously amiss. Thirdly, Vardaman's arrest the very day of his discovery of the "Nazareth" fragment is a smoking gun. Finally, Vardaman's known history as a lawbreaker must keep suspicion upon him.

We can now be certain that the "Caesarea Inscription" is a hoax. The CI cannot be used to substantiate the existence of a synagogue in Caesarea, nor to substantiate the existence of Nazareth in Roman times. Simply put, it must be expunged from scholarship.

CONCLUSION

The lack of ancient literary attestation for Nazareth, the embarrassing presence of tombs in the Venerated Area, the dating of *all* the recovered post-Iron Age evidence (including the tombs) to after the time of Jesus—these major problems for the tradition are endemic to the Nazareth evidence. In addition, the Christian tradition has systematically treated that evidence with equivocation, suppression, and even fraud. Published claims have been made for Hellenistic evidence at Nazareth that is not present in the ground (Mary's Well), for houses dating to the time of Jesus (the IMC, the SNC), and even for an inscription attesting to the early existence of the settlement, an inscription that never existed (the CI). The extent of archaeological dissimulation has been broad, systematic, and breathtaking. Today, it is a juggernaut that encompasses Jewish as well as Christian archaeologists. Millions of tourist dollars, as well as the integrity of the Christian tradition, depend upon the truth of Nazareth *not* becoming known.

Part 5

SCIENCE AND THE CHRIST

Chapter 13

THE BETHLEHEM STAR

Aaron Adair

About two centuries ago, there was a major transition in the way
scholars were approaching the stories of the Bible, both the Old
and New Testaments. There was a greater attempt to look at the historical context and formation of the holy book and its stories, and the tales
of Jesus were a major issue for critical scholars and theologians. It was
also at around this time that the acceptability of wondrous stories became
unpalatable, at least for the educated, for whom a deistic god was more
ideal—one that did not perform miracles and was consistent with the universe of Newtonian mechanics. A naturalistic understanding of the world,
inspired by the success of the physical sciences, along with inspiration
from Enlightenment thinkers, changed the way people looked at the world,
and this caused a significant reassessment of the spectacular stories of the
ancient world. What was one to do with the miracle stories of Jesus if miracles don't happen? The solution was a series of rationalizations, none seen
as terribly plausible but preferable to claiming a miracle or a myth. For
example, Jesus walking on water was a mistake on the part of the disciples,
seeing their master walk along the beach shore on a foggy morning and not
actually atop the water. Even the resurrection of Jesus was thus retrofitted
into scenarios that are unlikely, to say the least, but at least not impossible.

This sort of reconfiguring of the stories of the New Testament is significantly less common among modern biblical scholars, but a major strand
of this project exists in the sciences today: an astronomical (or astrological)
explanation for the Star of Bethlehem as described in the second chapter of
the Gospel of Matthew. This Star, which rises in the east, leads the Magi
("Wise Men" in the King James Bible) from Jerusalem to Bethlehem, and

then stops over a particular locale where the infant Messiah rests with his family. For centuries, the story was interpreted by Christian theologians as a story involving a star that does the impossible, as was the opinion of as varied a group as Augustine of Hippo, Thomas Aquinas, Martin Luther, and John Calvin. Even astronomers of the early modern period, including Tycho Brahe and Johannes Kepler, said the Star was truly out of this world. But in the early nineteenth century, the Christmas Star, like the other amazing tales of Jesus, was forced from its context to fit into the modern square hole of what made sense to modern readers, no matter the actual shape of the tale.[1]

Now, it is not necessarily the case that this is a poor way to approach the Bible. After all, there have been excellent scholarly works to show how the strange and weird tales of ancient peoples can be explained by completely natural phenomena and observations. The best example is perhaps the discovery of fossils, often from ancient pachyderms and dinosaurs, and various ancient sources tell us that they were thought to have been the bones of giant humans or fantastical creatures. For example, to the untrained eye an elephant thigh bone looks much like a human thigh bone, just amazingly large.[2] So perhaps some event in the sky was misinterpreted by ancient sources that made it into the Gospel account or we need a more charitable reading of the story of the Star.

However, there are two major points to consider that make this way of approaching the story of the Bethlehem Star problematic. First is the issue of finding anything that can fit the description of the Star in the original Greek text. There have been over two hundred years of hypotheses trying to fit this, and the success is less than stellar, as will be shown below. Second is the problem that, unlike the case of the discovered elephant bones, we have good reason to doubt the events happened in the first place. Not only might we be trying to fit a scientific theory to something not scientific, but we might to trying to fit to something that never happened nor was written to say what did actually happen at around the time of Jesus' birth. While many have argued in recent years that the genre of the Gospels is that of Greco-Roman biography,[3] there is the issue that this era produced biographies of people that never even existed, such as the founder-king Romulus. Too often, the questionable arguments about genre[4] seem to bypass the fact that the stories in the gospels, canonical and non-canonical, are theological first and perhaps completely so. Treating the books that happened to get into the Bible as true biography and automatically carrying historical information is to privilege the texts and potentially do apologetics rather than critical history. In fact, the entire project of the Star of Bethlehem seems to be largely motivated by the preservation of the Bible as a historically reliable document and source of faith traditions. For example, astronomer David Hughes writes, "If Matthew invented the star, what else did he invent? Is his whole gospel full of mis-truths? I do not think so. To me, the Gospel of Matthew rings true. All of it."[5] Others are even more explicit that their project is apologetic, attempting to prove the historicity of the entire Bible. From statements like this, an all-or-nothing stance exists, that if one part of the story is shown to be invention, the whole thing falls apart, and what seems to "ring true" is all in the mind. However, resting the approach to this story based on what is called the affect fallacy (it feels real, so it must be true) promotes bias to evidence. While some in biblical studies try to promote the Gospels as attempting to write the history of Jesus,[6] our document in question looks far more like a piece of theological writing than a sober, critical account once we place the tale into the context of how stories were written in antiquity.

This is hardly a new point; it was argued back in 1835 by the young Bible scholar David Strauss, and his way of approaching the Gospels as works of theology first and not history has been pretty much the main-stream for the last century and a half among critical scholars. This isn't to say that everyone in New Testament studies is skeptical of all that is written in the Gospels, and at the recent conference on the Star of Bethlehem at the University of Groningen a diversity of such views in the field were seen. The results of that conference are included in this chapter, though much of it is consistent with my previous work on the subject. One key point is that comparing the nativity story in the Gospel of Matthew with other such accounts, along with literary practices of the period, does not bode well for the historicity of the tale. The nature of what kind of story the natal account is should force us to realize that we might as well be trying to use science to explain the origins of the lightning scar on the forehead of Harry Potter—you just have to ignore all those magical parts. If that sounds like an odd way of doing history, then you are not alone. In this chapter, these two points will be made: no astronomical or astrological event or configuration can fit what the Gospel of Matthew describes, and the story cannot be historical and likely was not intended as history in the first place.

I have covered the details of the arguments about the myriad attempts to explain the Star in my book, *The Star of Bethlehem: A Skeptical View*. Many of those details about ancient astrology and Zoroastrian priests, along with others, were consistent with what was said by the various experts at the Groningen conference. The story that needs to be explained is this: During the final years of King Herod the Great of Judea (ca. 6 BCE), Magi from the East came to find the newborn king of the Jews as indicated by the rising of "his Star." Herod tries to figure out what he can about the birth of the Messiah, inquires of the Magi, and sends them to find the child in Bethlehem with the ultimately hope to destroying his infant rival. The Star then, after apparently not being visible, "goes before" the Magi until it arrives and stops "over where the young child was." The Magi then prostrate themselves before the Christ child, give expensive gifts, and then leave by another route because of an angelic dream message. What phenomenon could fit the details of this Star?

The major ideas presented over the years to explain the Star have been a comet, a supernova, the conjunction of planets, or a particularly auspicious horoscope. One way or another, these stellar circumstances signaled to the Magi, living in the Persian Empire (the East, Matt. 2:2) that a king of the Jews had been born. Then the Star had to "go before them" and then stand "over where the child was" (Matt. 2:9). With ancient records and modern computer models of the solar system, we can reconstruct what was in the skies on any given night; our confidence in the positions of the sun, moon, and planets is very high, and our records of notable comets from this period is reasonably good, primarily coming from China. For example, we know there was a comet in 5 BCE, not long before the death of King Herod and thus fitting the basic timeline indicated by the Gospel of Matthew. There is no record of a supernova around this time, and given their rarity, unless a lost record comes to light, it is probably the case that no such event happened in the skies around this time. There were many different conjunctions of the planets in this period, and the most discussed ones in the scholarly literature are those between Jupiter and Saturn three different times in 7 BCE. As for a horoscopic approach, the most thorough attempt at this looks to April 17th of 6 BCE.[7]

From what we know about how comets were viewed in antiquity, the comet makes for an implausible hypothesis for the Bethlehem Star. With very few exceptions, comets were seen as omens of disaster, especially for a king. There are only two historical records from the ancient world where they were used as a more positive sign, but there were political motivations for those interpretations. The comet seen at the funeral games of Julius Caesar was at first not even thought to have been a comet by the supporters of Octavian (later known as Augustus), and instead it was seen as a star, and the soul of Caesar rising to heaven; when it could not be denied as being a comet later on, it was still used as a sign of Caesar's apotheosis and divine favor for his adopted son, Octavian. The enemies of the Julian dynasty who eventually lost out did their best to say it was an ominous sign, but failure to win control of Rome also meant failure to control the narrative, as Augustus would mint coins with the Julian star on them, saying *Divus Julius*, and a temple to Caesar with a star on his forehead was established in the Eternal City. It was more politics than prophecy that made the comet into a favorable sign.[8] The same can be said for the only other case in the historical record with an auspicious tailed star, the comets of Mithridates VI of Pontus.[9] From Babylonian, Assyrian, Persian, and later Arabic works, as well as antique Jewish sources, no examples of fortuitous comets have been found, and it is from this broad time and place that the Magi are supposed to have come from. On top of comets almost always being found to be ominous, they are relatively common, so it is hard to understand how such a celestial sign was seen this one and only time as the birth of a Jewish king and inspired a trek to the Holy Land. There does not seem to have been any belief that a comet was to herald the birth of the Messiah, or any king for that matter, as indicated by the ancient Christian writer Origen.[10] Usually, then, the comet hypothesis relies upon other astrological theories.

As for supernovae, there is no indication from the historical record that astrologers distinguished between such stellar explosions and comets, and some Chinese records of certain comets may be novae or supernovae.

At best, we don't know how novae would have been interpreted, and more likely they might have been viewed in the same way as comets, making them no better a candidate for the auspicious Star of Bethlehem. These objects also do not move against the background of stars in the sky, so it is even harder to explain how a nova could have "gone before," stopped, and then "stood over" anything. This, along with a lack of a record of such an event from the time of Jesus's birth, makes this a faulty premise to hang a scientific hypothesis on when explaining the Star.

As for conjunctions, they received the greatest amount of interest in the literature explaining the Star, with papers and books speculating on how ancient astrologers would have interpreted them. One thing to first note about conjunctions is that they are relatively common. The conjunctions of Jupiter and Saturn, for example, happen about every twenty years, though the triple conjunction in 7 BCE is much more uncommon. However, the question of rarity is only a part of the equation when it comes to explaining the Star; the astronomical or astrological circumstances need to be uncommon enough to make sure the Magi were not running around the Middle East every other year, but we also need some reason to think that these or any other conjunctions would have indicated a king was born, Jewish or otherwise. Extant Assyrian and Babylonian records provide no indication of how Jupiter/Saturn conjunctions were interpreted, so astronomers have looked to disparate sources from a much later time. In particular, some look to what an Iberian rabbi in the fifteenth century had to say about what constellation represented the Jewish people,[11] even though around that same time other rabbis contradicted each other.[12] Worse, the astrological theory that we do know of that gives high importance to these sorts of conjunctions was not invented until centuries after the time of Jesus.[13] Out of speculative desperation, other conjunctions have been cited, perhaps the most common secondary candidate being those of Jupiter and Venus in 3 and 2 BCE, even though they are outside the time of Herod the Great's life in most modern chronologies of Judea.[14] No ancient astrological evidence is cited for this, and it rests merely on the imaginative exegesis of the Old Testament. We simply have no ancient evidence to indicate what conjunctions would have been seen as particularly auspicious and indicative of

Jewish royalty. In fact, we have no reason to think anyone even should have had such an idea in mind in the first century; the speculations about what was auspicious to ancient stargazers is built upon even more unjustified premises run by modern Christian bias, that what is important to a contemporary believer was important to non-Jews and non-Christians in the beginning of the era. This is a dubious way to do history, to say the least.

Horoscopes, on the other hand, have the advantage that we possess many Greek and Latin books about how a given configuration of the planets could be interpreted. The best recent attempt to do this focuses on the works of the influential astronomer and astrologer Claudius Ptolemy (second century CE), especially for arguing that the constellation (or more accurately, astrological sign) that was indicative of Judea is Aries the Ram. On the other hand, a broader look at the given reconstructed horoscope, along with other ancient sources on natal astrology, shows there is nothing less than a sea of contradictions. Depending on your source and interpretation of a given writer's geography, either Judea is not mentioned at all or it is within some other territory (such as Syria province or Phoenicia) under the influence of some other constellation. No two sources can agree.[15] Moreover, the reconstructed horoscope for Jesus, while possessing several features that can be said to be auspicious and similar to those found in the horoscopes of emperors Augustus and Hadrian, has other features that would indicate the newborn should have been a sex slave with elephantiasis-like symptoms and epileptic seizures.[16] In actual fact, any given horoscope can have "good" and "bad" features just by looking for them. Perhaps it is no wonder to the scientifically literate that modern astrologers cannot predict any better than chance the outcomes of personalities of people based on their horoscopes, but it is more astonishing that they also fail to agree on what any given horoscope means better than chance.[17] Ancient sources agree that interpreting a horoscope is extremely difficult and highly prone to error, as expressed by the most influential of the ancient Hellenistic astronomers and astrologers, Claudius Ptolemy.[18] If professional astrologers like Ptolemy cannot tell you what a given horoscope means, then how could any modern historian do any better? Conversely, if a horoscope could potentially mean anything, how were the poor Magi supposed to have known about the birth of Jesus any better than lottery-levels of chance? On top of this unavoidable ambiguity of interpreting a horoscope, they are not used to predict births, but instead the time and place of birth are used to construct the horoscope. In other words, using a star chart to predict when someone will be born is doing astrology back-ward. At this point, such a method of explaining the Star might fall into the category of "not even wrong." In actual fact, that was the opinion of Albert Schweitzer, the famous scholar and philanthropist, when discussing the same approaches to rationalizing the Star of Bethlehem a century ago.[19]

The main problem with all of these attempts to say what would or would not have been auspicious to a sky watcher in antiquity is that, without explicit indication from someone in that period, we really don't know. As the expert historian of astrology Franz Boll put it a century ago, trying to find what was auspicious to the Magi in order to search for the Messiah is like trying to solve the equation $A = x + y$ with x and y unknown—an impossibility no mathematician would consider.[20] The only instance where we can guess with confidence is in the case of comets—but not in a favor-able way. Along with this matter of any interpretation of ancient skies by modern astronomers being speculative at best, the fact is we do not have a historical precedent of such signs making anyone think that a king had been born, let alone one of such importance that it was worth trekking for untold miles in order to worship an infant. Unless we get amazingly lucky in the future and discover the testament of the Magi themselves or other astrologers of the period telling us about auspicious conjunctions at about the time of Jesus's birth, there is no way to justify any astrological theory for the Star of Bethlehem. Without the ability to establish anything with positive evidence, that would make the project untenable, though we may always be left wondering and considering it just possible something hap-pened even if nothing can be proven. However, what is the most damaging to all of these attempts is the simple fact that they fail to conform to the description of the Star in the Gospel account.

The first detail about the Star is that it rose up (Matt. 2:2, 9). The exact meaning of the Greek phrase is debated, but, no matter which is best, there is nothing implausible about stars, planets, comets, etc., rising in the east, as do all stars except those close to the celestial poles (for example, Polaris). There is more difficulty, however, getting the word "star" to mean a planet or comet without additional philological evidence; on its own, especially when referred to as "his Star," this does not suggest a planet, let alone a conjunction or horoscope, and instead it is more like a special and personal star, something all souls were assigned according to classical sources.[21] Yet this is a quibble compared to other issues. The details as provided in verse 9 of the story are where the true problems lie. What is said there is that the Star "went before them." The verb used by the author is one that in classical usage is one of leading forward or being in the front of a procession (cf. Matt. 21:9), and that is an astronomical problem. The Magi see the Star leading them to Bethlehem from Jerusalem, which is toward the south. Stars do not move south but from east to west, so already the Star is doing something unnatural. Then the Star stops as it arrives at its destination. There is no way for a normal star to have come to rest in the sky. The straightforward, literal meaning of the words paints a picture of something contrary to astronomy.[22]

To avoid this, various attempts have been made to have the Greek words mean something else. However, in no case is anyone able to competently show that anyone in antiquity talked about the motions of the heavens in the way the author of Matthew does. Moreover, the inventive readings have to ignore important details in order to work. For example, some have tried to argue that this "going before" and "standing" of the Star was about retrograde motion and stationary points of planets, totally normal motions the planets appear to make in the sky. However, the measurable change from retrograde motion to stationary cannot happen in a single night but instead takes at least several and is imperceptible to the naked eye; the story implies the trip took little more than the walking time from Jerusalem to Bethlehem, a distance of a several miles and perhaps two hours of walking or riding. Also, the context is rather clear that the Star is going before and standing relative to the Magi and objects on the ground, not the stars in the sky. The text says the Star "went before them," and the only "them" mentioned is the Magi.

Another way of trying to understand the movements of the Star is to say that the light was simply in front of the path of the Magi ("before them") and it was above the town of Bethlehem as seen by an observer in Jerusalem (or perhaps even over a particular house in the little town from a certain point of view). In this way, the Star was "before them" and it "stood over" the appropriate locale. However, the Gospel verse says that the Star "went before," indicating motion, not merely being stationary. That is reinforced by additional verbiage saying how the Star arrived and stopped. Moreover, the Star went before the Magi "until" a certain point in time. That means the Star cannot be "before them" and "stand over" the place at the same time; one happened and then the other. Since no Star can move and then stop in place in a matter of hours, let alone in a moment, as the context of the story would suggest, this interpretation also fails.

However, this is still not the worst part of such rereadings of the tale. The most problematic point of the description is the last bit of the verse, where the Star is said to stand "over where the child was." There, the preposition and grammar used is that of close proximity rather than up in the sky. The same sort of wording is used for the angel that sat *upon* the stone door to the empty tomb (Matt. 28:2), and there was the sign that was nailed *above* Jesus on the cross (Matt. 27:37). Astronomical and astrological texts never used the same language for describing stars in the sky above the ground. Perhaps then it is no wonder why the many ancient and medieval commentators on this story all agree that the Christmas Star was not a part of the natural world but instead a light that came down, guiding the Magi to a particular house and standing right above it as an ancient GPS unit, as some modern Bible scholars put it.[23] Unlike the case of the misidentified elephant bones, this belief about the Star was not out of ignorance of the science of astronomy and astrology. Augustine, for example, was well-versed in astrology since it was a part of the Manichean religion that he belonged to before converting to Christianity. So he is of considerable authority when he says the Star was not of the natural order but a special creation of God, divorced from astrological speculations or calculations.[24]

On the other hand, there is one naturalistic object that could move and stop in the way described: an alien spacecraft. This has been soberly proposed by several people, including one educated theologian.[25] We may wonder why the Magi would have referred to a spaceship as a star rather than a ship, and we have to speculate how they connected a flying saucer to Jewish royalty, but the most obvious problem with the extraterrestrial explanation is the fact that we don't even know if such creatures exist, let alone visited the earth, let alone would have buzzed around the Palestinian countryside to guide a few Persian scholars around. The alien explanation is little better than the miraculous explanation, a modern *deus ex machina* to make any story possible.

However, the E.T. hypothesis is, in a way, the uncovering of the id (or unconscious drive) behind these sorts of speculations: miracles are seen as implausible to the modern thinker; even a fundamentalist Christians will initially doubt the claims of the paranormal from another religion (and perhaps even those of another denomination). However, any sort of naturalistic explanation has two features in its favor: it already comes across as more plausible, and it now has the authority of science behind it. The idea that a Bible story can be scientifically proven is indeed a powerful tool for an apologist, and one such apologist has admitted just that in the case of using natural events in the sky as explanations for the Star, even if the Star is supernatural.[26] Other scientists may be interested in the subject because of their own curiosity in explaining the ancient world by scientific means, but my own survey of the literature finds neither secular nor non-Christian writers on the Star of Bethlehem favoring an astronomical hypothesis. Far more then, this approach to the Bible is Christian apologetics rather than secular history.

This begs the question: can we say the events of the story didn't happen? For many, that the tale is about a miraculous Star is enough to make it just as doubtful as the stories of impossible trips to the heavens on the backs of flying animals. Some will look for any way to rationalize things, including impugning the intelligence of the author of the story for not understanding what happened. For those who accept supernatural tales, additional reasons to doubt the story are demanded. And there are several. For one, there is no independent corroboration of this tale by any source, Christian or otherwise. Our only source, the Gospel of Matthew, comes from an unknown author writing at least a full lifetime after the described events, and the author provides no details about how he knows about what he writes or that he even has a source, let alone a reason to believe such a source is reliable. By the time of composition, all of the characters in the story are long dead—father Joseph is already gone and presumably dead by the time Jesus's ministry begins in the Gospels, Jesus is obviously dead (and undead but gone nonetheless), while Mary was likely dead by the middle of the first century; the Magi returned to their own land and would have been deceased by the time ink reached parchment. Not only is there no living witness to the story when it was written down sometime in the late first century or so, but the only other canonical nativity story, from the Gospel of Luke, contradicts key points of the story. In particular, the census mentioned in Luke 2:2 that brings the Holy Family to Bethlehem where Jesus is then born takes place at least a decade after the time depicted in the Gospel of Matthew (during the reign of Herod).[27] Luke also fails to mention anything about a Star, Magi, or an attempt on Jesus's life. He either knew nothing of such a story (in which case the source for Matthew is all the more mysterious and unverifiable) or didn't think the story belonged in his attempt to tell the story of Jesus (in which case it is seen as a dubious tale).

Even on its own, the story is filled with details that make neither logistical nor contextual sense. Consider these points of character when looking at the Magi and the king of Judea: The paranoid Herod, believing he is threatened by an infant, completely trusts the people who undermined his authority (the Magi) to report to him after finding the child. One spy tailing them would have been enough to change history, and yet Herod does not do what would have been expected by a ruler desperate to hold onto power. While Herod is perhaps overly concerned about a baby, the interest of the Magi is contrary to history. These people would have been a part of the Zoroastrian priesthood, so they would not have been worshipping a Jewish king. In fact, later inscriptions showed that the head of the magi gladly persecuted Jewish *and Christians* in the lands of the Persian Empire.[28] This is highly unlikely if several members of this priestly caste went and literally worshipped at the feet of the founder of Christianity. The magian interest in the stars is also anachronistic; while Hellenistic authors erroneously believed that the magi were astrologers and Zoroaster was a founding figure of the practice, in actual fact there is no evidence the magi were methodological astrologers at the time and significant evidence against it. Only centuries later do we begin to see any Persian interest and influence in astrological methods.[29] What this shows is that the author of the story is building a narrative built upon a common Greco-Roman stereotype and not what would have been accurate for the magian priesthood, a classic sign of crafting fiction.

Perhaps the most glaring problem with this story is that not only is there no independent attestation of the tale, but that there certainly should have been because what is described is an international incident. The magi were a part of the Persian government, both in the sphere of religion as well as in who controlled the empire. One Christian source even says the Magi were "almost kings."[30] This means having magi coming into Judea, a Roman territory administered by Herod the Great, means that not only are the Magi usurping the authority of King Herod but also Caesar Augustus. This should have been a major event in east-west relations, similar to other cases where the Romans and Persians wanted different people in control of territories between them. In particular, disagreements about who was the appropriate king in Armenia led to diplomatic showdowns and wars in the first century CE, recorded by several Greco-Roman historians, including the Jewish historian Josephus.[31] The complete lack of any mention of an even more brazen dispute about who controlled a Roman satellite territory is exceedingly implausible, as if the communist takeover of Cuba never even made it to the newspapers, let alone never having a response by the US president.

On the other hand, instead of looking at the story of the Bethlehem Star as a poorly sourced biographical account that fails to conform to what we know about history, it is far more productive to see if it fits into a different category of tales from the ancient world: the birth of the legendary hero. It has been noted that the Jesus story, especially the nativity, fits amazingly well to the general archetypal structure of the birth of the hero, from the miraculous conception, to his father being a god, to the attempt on the life of the child who is spirited away to safety.[32] By the first century, this sort of narrative was well known and worn out; in particular, the Jesus nativity in Matthew's account has many similarities to the natal account of Moses as retold by Josephus (around the same time as Matthew wrote). That story included angelic dreams on the part of Moses' father, a prophecy to the pharaoh that a Jewish leader was to be born, and an attempt to kill the infant and future rebel by slaughtering all male Hebrew babies.[33]

As for guiding stars, there are literary precedents for that. The most important is the story of the blazing star of Aeneas. After Troy had been destroyed by the victorious armies of Greece, Aeneas and other survivors of the city plan to find a new home, and prayers for a sign are answered by a great star, flying over at treetop level and acting as a guide to their new kingdom in Italy. This story is told by the poet Virgil,[34] and his blazing-tailed star is to make it more like the comet seen at the Caesarean funeral games, believed by many to have been the soul of the Julian dictator rising up to heaven. The star is thus a guide and a sign of the greatness of Rome and its divine leader. This was an extremely well-known epic throughout the Roman world, and commentaries provide other interesting details. In particular, the star that guided Aeneas and company was said to have been Venus as the morning star (*lucifer*),[35] which is not unlike the Bethlehem Star, which may also be described as a morning star when it rose (Matt. 2:2). We can now make sense of the Christmas Star story with this context; the adoption and transvaluation of this powerful symbol of the Roman Empire by the author of the Gospel of Matthew tells the audience how Jesus is the true king and son of God, that the great morning star leads to him and not to Rome and its founding figure.

Along with the Aeneas story, there is an equally important text that likely influenced the nativity story: the Star Prophecy from the Old Testament (Num. 24:17). This was a well-known and widely cited prophetic passage, found in the Dead Sea Scrolls and the Talmud to that end. The messianic claimant Simon bar Kokhba was said to have changed his name to fit this prophecy, his name meaning "son of the star." Early Christian commenters widely cite the passage from Numbers as prophecy for Jesus's stellar birth—the messiah was supposed to have been heralded by a star. We can see then why it would have been so potent to have Jesus in his birth story fulfilling this important prophecy in the eyes of Jews and Christians. Additional exegesis of the Old Testament can explain other details of the story of the Star. For example, Isaiah 60 talks about kings coming to the rising light of

Israel, bringing gifts of gold and frankincense—two of the gifts of the Magi, brought to the new king in Jerusalem via a rising light. This sort of reading of the Old Testament and creating new interpretations and narratives was common practice among Jews and Christians of the time, so categorizing the Jesus nativity story along with this sort of writing is far more logical than acting as if it were a historical account, either of good or poor quality.

There is considerable precedent for this sort of creating exegesis from the scriptures, even in stories about the birth of Jesus that are found in later sources.[36] The most interesting of those sources must be *The Gospel of Pseudo-Matthew*, written in Latin along with a forged letter from St. Jerome to give the fake gospel significant authenticity. This document adds many details to the original Nativity story, such as how an ox and donkey were beside the manger holding the Christ child, a common feature in modern nativity scenes. The additional information is not mere window dressing but supposedly there to fulfill prophecy about how the ox knows its owner, the donkey its master's feeding trough (Isa. 1:3) and that between two animals the Lord is made known (Hab. 3:2) as explicitly quoted by the author of this gospel (*Ps-Matt* 14). Such citations of the Old Testament are following the style of the Gospel of Matthew itself, which often cites a verse from the scriptures to say how such-and-such was in fulfillment of prophecy. This goes to absurdity in the pseudo-Matthew gospel, where Jesus is able to tame dragons (*Ps-Matt* 18), fulfilling the prophecy of Psalm 148:7 that dragons will praise the Lord. Even though it seems fantastical, this gospel proved to be influential in the medieval period, finding it tales retold in the *Golden Legend* (thirteenth century), Christmas carols, various works of art, and other apocryphal infancy gospels, all influencing the modern version of the Nativity. If the author of *Pseudo-Matthew* was able to concoct fanciful stories about Jesus by inventive renovations to the Old Testament and be believed, then so could the author of the Gospel of Matthew. The Star of Bethlehem looks far more like a product of borrowed tales and/or theology, quite unlike a product of biographical inquiry.

Putting this all together, the weight of the evidence strongly favors the idea that the tale of the Star of Bethlehem was deliberate theological fiction. The stellar object is described as something astronomically impossible, as recognized by all of the ancient commenters, and none of the scientific rationalizations of the story can find a plausible way to make it conform to ancient descriptions of astronomical or astrological situations. Our story, told to us by someone who fails to provide any reason to think he has done a careful retelling of history, makes little sense when put into the context of what we know of the Zoroastrian priesthood, King Herod, and Roman-Persian relations. In general, the story is highly improbable, with all sorts of legendary embellishments, from the attempt on the life of the infant to angelic announcements and virgin births. On the other hand, the story is in great conformity with antique legends of heroes and standard literary practices by Greek-speaking writers, especially Jewish authors practicing theology with creative writing, not unlike the various Jewish novels contemporary with early Christianity. When the Star fits so well within the category of ancient fiction, while fitting so poorly to history and astronomy, continuing to explain Christmas with science can only be as dubious as trying to explain how the magic words "Open Sesame" unsealed the cave door in *Ali Baba and the Forty Thieves*.[37]

But what if someone wanted to be forceful and not accept that Matthew could have "just" made up the story, that there had to have been a source. Could there be something to the claim that there is something else behind the tale? Perhaps so, but it is not as expected by anyone looking for a historical event. Consider that there are other texts that connect a star with Jesus, but in a different way. Three times in the New Testament (2 Pet. 1:19; Rev. 2:28, 22:16) we find that Jesus is called a star, specifically the morning star. Note that this is not the same as there being a star at Jesus's birth but rather that Jesus is actually some sort of celestial light. This is also found in the second-century writings of the bishop Ignatius of Antioch (*Eph.* 19) in which Jesus is said to be a bright star. Ignatius's letters are figured to be independent of the Gospel of Matthew, and Revelation and 2 Peter are not using Matthew 2 as a source either. In which case, perhaps there is a common source, but one that has Jesus as a star rather than there being a star at the Nativity. In particular, the Jesus-as-star idea seems to be related by Ignatius to his resurrection, and the morning star has been so interpreted by early Christian commentators. There is also the larger dying-and-rising god mytheme, starting with the Sumerian goddess Inanna and later changing into the Babylonian goddess Ishtar, both deities related to the planet Venus, the morning star. This may suggest then that the Star of Bethlehem was originally about the resurrected Jesus.[38] While this is somewhat speculative, this line of inquiry has far more going for it than the last two hundred years of Star of Bethlehem astronomy. In the end, if we must continue to dig into the story, we will find myth, not history.

Chapter 14

IF PRAYER FAILS, WHY DO PEOPLE KEEP AT IT?

Valerie Tarico

Nothing fails like prayer. This simple phrase, coined in 1976 by secular activist Ann Nicol Gaylor, has replicated across the Internet for two reasons.

One, it's true. In aggregate, research on prayer shows no overall effect, or one so weak that the most that can be said for God is that he, maybe, operates at the margins of statistical significance—not a very impressive claim for an omnipotent, interventionist deity. The more carefully constructed the research, the less likely that prayer has any measurable effect on those prayed for. Put it this way: a pharmaceutical company that made similar claims and had similar results would be sued out of existence.

Ask and ye shall receive? Knock and it shall be opened unto you? Not unless the implacable laws of physics and biology are already trending in your favor. But don't take it from me. We'll come back to this point.

The other reason former Christians and other non-theists keep repeating Gaylor's adage—the reason it decorates bumper stickers, posters, and even an annual award by the Freedom from Religion Foundation[1]—is this: Despite the stack of evidence that God is either deaf or dead (or otherwise unaffected by human supplication), theists keep sending requests heavenward. In a 2010 Pew Survey of 35,556 Americans, over half said they prayed daily.[2] Even among Millennials (born 1982–2002), 48 percent said prayer is part of their daily lives. Prayer is far too pervasive simply to be dismissed or ignored by anyone who wants to understand either the lived experience of individual human beings or the complexities of modern society.

Many people only pray quietly alone or else in the company of co-religionists. But some insist on displays of prayer in public places and in public roles, like football players who kneel on the field and in front of the cameras or elected officials who appeal at the microphone for God's guidance. When faced with "acts of God" like hurricanes and tornadoes, or acts of humans like terrorist bombings and mass murder, millions respond with words like, "Please pray for [Haiti-New Orleans-Paris]" or "Our prayers are with the victims."

Given prayer's lack of efficacy and religion's causal role in mass violence, offering or requesting prayer in response to a natural disaster or terrorist assault may seem particularly cynical or cruel. But these requests and offers often are made sincerely—by smart, kindhearted people who seemingly should know better. What is going on? Why do intelligent, compassionate adults—folks who would laugh if you suggested they carry a lucky rabbit's foot or cross the street to avoid a black cat—still pray?

Humankind has outgrown millions of superstitious beliefs our ancestors treated as knowledge. We have discarded countless rituals for currying supernatural favor and cherished talismans for warding off harms, each of which failed to precisely the same degree as modern prayer requests. But millions of people still prostrate themselves toward Mecca daily, while millions of others light candles and incense in front of icons and billions more implore the favor of one or another god in some less formal fashion. How should we make sense of this?

In this chapter, we will explore the nature of prayer as a human endeavor, then look specifically at how the Bible and Christian tradition treat prayer, then review the disappointing results of research on the power of petitionary prayer—meaning prayer that makes a request of some sort—which is central to biblical Christianity. We will discuss what people *do* get from prayer, which—I think—answers the question of why prayer remains so popular, and then ask how some of these positive benefits might be carried forward without the baggage of supernaturalism and superstition.

WHY BOTHER TALKING ABOUT PRAYER?

Prayer is the very soul of religion.
                                        —August Sabatier, 1897

If you've read this far in this book, you probably have some inkling that prayer doesn't work—not in the way your Sunday school teacher or Bible study mentor or Christian college professor said it would—at least not if your experience is at all similar to mine. So, why bother reading a whole chapter about prayer?

Because prayer is as much a part of humanity's past and present as virtually any other enterprise you could name, short of eating, sleeping, and sex. It was part of the everyday life of our Bronze Age ancestors, and it is part of everyday life for many of your Information Age neighbors. In the 2008 report of the General Social Survey, a national survey spanning the thirty-four years from 1972 to 2006, 97 percent of Americans said they prayed at least on occasion, and 57 percent indicated they pray daily.[3] Preschoolers pray and octogenarians pray, as do folks at all points in between. Soldiers and civilians pray—including atheists in foxholes. People mutter prayers when they don't believe they are actually talking to anyone but themselves, and possibly the only thing more bizarre than that is that most people who mutter prayers think they *are* talking to someone other than themselves.

I would argue that prayer is so endemic to our species that it's impossible to understand humanity fully without making some attempt to understand the behaviors we call *prayer*. Although it is strongly shaped by cultural and religious traditions, prayer reflects the thought patterns and yearnings of those who pray. If we assume there's nobody on the receiving end, then prayer is in some ways like an ink blot test. It's us projecting ourselves into the universe. It offers insight into our deepest fears and highest hopes and most transparent self-absorption and silliest gullibilities. It reveals how we relate to each other by showing how we relate to an *other* who is woven from the fabric of our own neural architecture. And in coming years, the study of prayer may offer insights into that neural architecture itself.

WHAT IS PRAYER?

> *A man who prays is one who thinks God has arranged matters all wrong but who also thinks he can instruct God on how to put them right.*
>
> —**Christopher Hitchens**

When Christopher Hitchens not so subtly pointed out the arrogance of prayer—that the pray-er thinks he knows better than God how to run the world—he was both right and wrong. Hitchens was right that *petitionary prayer*—the kind that attempts to influence God—may be considered the most archetypal kind of prayer. In fact, I suspect that if human beings became convinced that we can't influence God, not only prayer—but religion itself—might go away. But he was wrong in that prayer is much more than simply a set of requests that God change or fix something.

Scholars for centuries have tried to categorize and define prayers, and in recent decades they have applied investigative methods from fields including psychology, sociology, anthropology, medical research, and neuroscience to this task.

At its heart, prayer is a form of communication, which makes it a social activity. But some forms of prayer are more like conversations than others, depending on how the person praying understands God. Western religious traditions describe God as a person who is external to the self and the universe as a whole. By contrast, Eastern traditions tend to see the universe and ourselves within it as manifestations of the divine—not that the division is clean. A Christian mystic or Sufi imam may hold a view of God that is far more like that of a Buddhist monk than might the average Christian or Muslim. And reformers within Christianity have tried to move fellow believers away from the concept of a person-god. Anglican Bishop John Shelby Spong, for example, refers to God as the Ground of all Being, and has said that, instead of being born again, Christians need to grow up. Nonetheless, broad differences persist. These differences in god-concepts determine whether a person treats prayer as half of a dialogue or more like a way of centering the self within the structure of a universe infused with the sacred.

This point is important, because as atheists and other non-theists move to develop wholly secular spiritual practices that eschew any form of supernaturalism, we may find that some forms of prayer are fundamentally archaic or superstitious but that others can be adapted for purposes very like those that motivate prayer on the part of believers: values clarification, centering, and even listening to the quiet internal voice that we now recognize as our own. The secular value of prayer depends in large part on the intent. Are prayers meant, in the words of theologian J. Harold Ellens, "to persuade, inform, beguile, suggest, inveigle, ingratiate, encourage compliance with our wishes, or merely reflexively benefit us psychologically and physiologically?"[4] The latter set of goals is perfectly compatible with superstition-free living, and in fact is highly desirable.

Psychologists Bernard Spilka and Kevin Ladd have written a book, *The Psychology of Prayer: A Scientific Approach*, that attempts to summarize the current state of research on the topic of prayer.[5] Their top take-away is that prayer is a broad, complex, multidimensional human enterprise. A number of researchers have created taxonomies to help make sense of this complexity. They cite theology professor Richard J. Foster, who divided prayer into three types, which face "inward, outward, and upward."[6] Inward-facing prayers are those seeking personal transformation, either by creating some form of transformative insight/consciousness or by requesting God's assistance with change. Outward-facing prayers are those seeking to influence the world external to the person praying. They may be public prayers, for example, or requests that God fix some problem situation. Upward-facing prayers are those that seek to alter the believer's relationship with God himself.

Ladd and Spilka analyzed a number of prayer studies and then used factor analysis to look at Foster's three categories. What they found was that prayer can be hard to categorize. Nonetheless, themes emerged—including "personal examination, tears, sacramental, rest, radical, suffering, intercession, and petition."[7] In their analysis, "radical" meant seeking God's help with boldness or radical personal change. "Intercession" included requests made on behalf of other people, while "petition" meant requests made on behalf of the "self."

GETTING WHAT WE NEED AND WANT

Why did I say earlier that petition may be considered the archetypal prayer? Because our interest in God is instrumental. It is a means to an end. As *humans*, a huge part of our energy goes into trying to figure out the cause and effect relationships that govern our lives and wellbeing. To that end, we need a god who cares how we think and feel and behave because *otherwise we have no way to manipulate what God does*. That is what I mean when I say our interest in God is instrumental. It is not actually about God, per se; it is about us.

Our ancestors generated a whole host of ideas about who the gods are and then rules about how humans can relate to them in ways that get us what we want: health, children, enduring prosperity, protection from our enemies, bountiful crops—and more esoteric desires like a sense of tribal superiority and individual righteousness—and perhaps most importantly the ability to delay or avoid death, or at least make it not permanent.

Not everybody wants the same thing from God, either here on Earth or in the afterlife. At the most simple level, people who preach and practice Prosperity Gospel, like Joel Osteen and his followers or Creflo Dollar, may basically want money. Believers in this tradition may treat God like an investment of sorts. Put money in the offering plate, and it will return to you tenfold, even a hundredfold, some say.

But most believers want something more complex, more like what one Bible writer called the fruit of the spirit—love, joy, peace, long-suffering, gentleness, kindness, meekness, temperance—and of course faith or hope. Worship can trigger a powerful experience of transcendence or joy or wonder or one-ness, a deep pleasure that is hard even to explain.

Similarly, images of the afterlife can be pretty crassly material—mansions and streets of gold, gemstones and crowns, eternal youth and white robes signaling no need to work. When people are desperate and powerless and poor, the symbols and trappings of wealth can be very appealing. But again, the afterlife that people yearn for can be something much simpler or more complex, something that includes submission and selflessness, maybe an eternal variation on that incredible worship experience that is such a powerful feel-good.

Regardless of the specifics, when we commit our lives, our money, and our energy to a god, we expect something back. Without the hope that our devotion can change our lives and afterlives for the better by winning God's favor, then the question of whether God exists simply isn't interesting to most people.

If we knew for sure that God was the god of Thomas Jefferson—a prime mover who put the universe in motion and then disengaged—or if we knew for sure that God was the god of Albert Einstein, best understood as a set of mathematical intricacies frankly incomprehensible to most of us—if we knew that God's predestined plan was going to play out no matter what we did, then people would simply get on with their lives: trying to take care of their kids and pay their bills and maybe occasionally practice random acts of kindness and senseless beauty.

What people *wouldn't* do is this: They wouldn't spend time trying to cultivate a relationship with God—whether that means prayer without ceasing or church on Sundays. And they wouldn't spend time trying to win converts, to get other people to cultivate the same kind of relationship with God. The word cultivate and the word cult, meaning religious practice, have the same root word—the Latin *cultus*—which literally meant the care and feeding of the gods. We cultivate the ground to get crops out. Salesmen cultivate clients. Nonprofits cultivate donors. Prayers, even prayers of praise and thanksgiving, are forms of cultivation. Our ancestors used to take care of the gods so the gods would take care of them. In our own twenty-first-century way, we do the same.

WHAT THE BIBLE AND CHRISTIAN TRADITION
SAY ABOUT PRAYER

The texts of the Bible were written over hundreds of years, and they contain a wide variety of exhortations, prescriptions, and stories about prayer. Some of these provide models for meditation and other types of contemplative, reflective spiritual experience. Others offer examples of celebratory rituals, with the power to induce and express spiritual euphoria either alone or in community. But most, at some level, teach believers how to communicate with God for the purpose of avoiding his wrath and cultivating his favor so the believer can avoid natural and unnatural disasters and instead receive blessings, including health, wealth, joy, love, and eternal bliss.

In the earliest texts of the Hebrew Bible, communications between the god Yahweh and his people are modeled on the relationship between an Iron Age warlord and his subjects. This includes hymns of praise and ritual displays of allegiance, often conveyed to Yahweh by a professional intermediary, a priest. As was common in surrounding cultures, the Torah prescribes sacrifice of innocent life as a high-cost display of loyalty, producing burnt-offering smoke that both "pleases God" and helps lift prayers to the heavens where God resides, literally. In fact, textual analysis of the Old Testament suggests that the early Hebrew religion included human sacrifice, which later writers seek to reframe or repudiate.[8]

Christianity emerged after hundreds of years of evolution in the Hebrew religion. Even so, the idea that God is pleased or soothed by the sacrifice of innocent life persisted, as it does today in Christian sects that treat the crucifixion as a form of substitutionary atonement. The model of a God who wants and needs constant adulation from his subjects persists clear through the New Testament and Christian history and provides the basis for modern Christian images of an afterlife in which angels and believers spend eternity singing God's praises.

This image of heaven raises fascinating questions about human social psychology and, in particular, our powerful attraction to dominance hierarchies—but that is a digression. A more relevant question here is whether Christianity's myriad efforts to relate to God—prayers of praise and confession and repentance and adulation and supplication and so forth—actually produce the desired effect. Does prayer get people what they want from God? Some parts of this question are hard to answer because people want things like forgiveness and communion and peace and joy, which are difficult to measure, and because Christianity teaches that many fruits of prayer won't be seen until sometime after death, which makes them even harder to measure.

But the Bible makes some promises that are much more concrete. Specifically, it promises that prayers will affect the here-and-now of the material world. These claims, if they are true, should have effects that are tangible and measureable. And whether these claims actually pan out should offer some evidence regarding the trustworthiness of the other claims that are less testable. If I promise that I will cure your illness or give you money today, and then I don't, why would you believe me when I promise that I will make you healthy and wealthy after you're dead?

Consider the following verses from the New Testament (New International Version):

- Which of you, if your son asks for bread, will give him a stone? Or if he asks for a fish, will give him a snake? If you, then, though you are evil, know how to give good gifts to your children, how much more will your Father in heaven give good gifts to those who ask him! (Matthew 7:9–11)
- I tell you the truth, if you have faith as small as a mustard seed, you can say to this mountain, "Move from here to there" and it will move. Nothing will be impossible for you. (Matthew 17:20)
- Again, I tell you that if two of you on earth agree about anything you ask for, it will be done for you by my Father in heaven. (Matthew 18:19)
- If you believe, you will receive whatever you ask for in prayer. (Matthew 21:22)
- I urge, then, first of all, that petitions, prayers, intercession and thanksgiving be made for all people—for kings and all those in authority, that we may live peaceful and quiet lives in all godliness and holiness. (1 Timothy 2:1–2)
- Is any one of you sick? He should call the elders of the church to pray over him and anoint him with oil in the name of the Lord. And the prayer offered in faith will make the sick person well; the Lord will raise him up. If he has sinned, he will be forgiven. Therefore confess your sins to each other and pray for each other so that you may be healed. The prayer of a righteous man is powerful and effective. (James 5:14–16)

Note that none of these verses says, as did my youth pastor, that God will always answer with "yes, no, or wait." They don't suggest that a person must first figure out what God already wants and intends, and then ask for that. They don't have asterisks suggesting that anyone signing the prayer contract must first read a long list of fine-print exclusions and qualifiers. There is no indication that something as acutely and personally important as healing for a child dying of cancer or protection against rape and stabbing or the salvation of a loved one—or something as magnanimous and beneficent as food for the starving or world peace—might be outside the scope of these promises.

Furthermore, from the time of the Early Church to the present, Christian leaders have always taught that prayers of petition or intercession have real-world consequences. In the first century, Ignatius of Antioch urged Christians to pray for nonbelievers: "Only you must pray to God for them, if by any means they may be brought to repentance, which, however, will be very difficult. Yet Jesus Christ, who is our true life, has the power of [effecting] this."[9] In his letters, he repeatedly asks for prayers of intercession for himself (eight times), for the Christian church in Syria (seven times), for persecutors, heretics, and all people generally (once each).[10] In December 2013, Pope Francis called for a "wave of prayer" around the world, to address the problem of hunger. Christian leaders from every pulpit and position of power continue, to this day, to send the message that God intervenes in human lives and in our natural world in real time in response to prayer.

EMPIRICALLY MEASURING
EMPIRICAL CLAIMS FOR PRAYER

Extraordinary claims require extraordinary evidence.
—**Marcello Truzzi**

If prayer actually worked, everyone would be a millionaire, nobody would ever get sick and die, and both football teams would always win.
—**Ethan Winer**

How might one test the claims made by the New Testament writers and modern Christians? On the surface, it seems rather easy to point out that if prayer worked, Christians wouldn't have needed to convert Europe, and later Latin America, to Christianity at sword point. Likewise, it seems easy to point out that after December 9, 2013, patterns of world hunger declined and increased precisely as might be predicted had the Catholic "wave of prayer" not occurred. But apologists for Christianity have always found ways to explain away historical events and actuarial statistics—for example, the fact that the Black Plague devastated the most devout medieval communities or the fact that modern Christians, on average, live no longer than Muslims or Buddhists of comparable socioeconomic status.

Since the birth of the social sciences, serious academic researchers have suggested that we can come at this question through a different approach, by applying the scientific method to claims about the power of prayer. Sir Francis Galton, statistician and founding father of modern psychometrics, published his first prayer study in 1872. Galton pointed out that royal sovereigns are the most prayed-for of any public figures. *God Save the Queen*. But when he compared the longevity of kings and queens to eleven other groups of privileged people, he found that "the sovereigns are literally the shortest-lived of all who have the advantage of affluence. The prayer therefore has no efficacy."[11]

In recent decades, research on petitionary prayer has focused primarily on prayers for healing. Many prayers seem trivial or self-serving in a way that makes God's lack of response easy to dismiss: please help me find a parking spot, or please help me get an "A" on this test, or please let my team win. But we generally assume that God cares about illness and suffering. Jesus was called the Great Physician, and many of the miracle stories of the New Testament are stories of compassion and healing. So a natural place to examine the efficacy of prayer is in the field of medicine.

Attempts to assess the healing power of prayer have varied in their results, as do individual studies in most scientific fields—especially when human beliefs, behaviors, or relationships are a part of the equation. As a consequence, many believers have heard of one or another study suggesting that prayer made some kind of intriguing difference in health outcomes, a difference that seems inexplicable except by appealing to the supernatural. Why haven't these studies led to further research on how to better tap into the power of prayer? They have. And the best evidence available suggests that there's no *there* there.

The highest quality study to date on this topic was the "Study of the Therapeutic Effects of Intercessory Prayer," also known as the STEP study funded by the faith-friendly Templeton Foundation. The decade-long study applied a rigorous clinical research design including double blind with random assignment, and followed 1,802 patients who received coronary artery bypass surgery at six hospitals.[12]

At the conclusion, Templeton, which had spent $2.4 million on the project, issued a press release:

> This project applied a large-scale controlled randomized research model to contribute to a growing number of scientific studies about prayer. Previous studies had attracted widespread public attention and discussion due to claims of positive health outcomes for distant intercessory prayer in which patients were unaware of being prayed for in the context of a research study.
>
> Analysts, however, had pointed to methodological weaknesses calling these results into question. In view of both the empirical uncertainties and the potential significance of a non-null result, the Foundation's advisory board advocated that substantial resources be put forth in order to advance methodological rigor in the design and execution of a new "blue ribbon standard" study.
>
> . . . [T]he null results obtained by the methodologically rigorous STEP experiment appear to provide a clear and definitive contrasting result to an earlier published finding (Byrd study) of a positive effect for patient-blind distant intercessory prayer in a prayer experiment involving recovery of patients in a cardiac care unit. Result: The STEP project did not confirm these findings.[13]

Scientists have a set of tools for wading through contradictory research results in order to identify the best evidence available and then synthesize that evidence. This means conducting a "meta-analysis" that combines results from relevant studies, often weighting them based on how rigorously variables of interest were defined, controlled, and measured, and how carefully alternative explanations were ruled out.

In 2009, Leanne Roberts, Irshad Ahmed, and Andrew Davison conducted a meta-analysis titled, "Intercessory Prayer for the Alleviation of Ill Health."[14] In 2014, their review was edited and republished by the leading publisher of medical meta-analyses, Cochrane Reviews, without any change to conclusions. After discarding lower quality research, the authors included in their review ten randomized trials "comparing personal, focused, committed and organized intercessory prayer." The studies included 7,646 patients. Overall, they found no clear effect of prayer on either general clinical state or death. Four studies looking at heart attack patients specifically found no difference in readmission to the coronary unit. Two that reported on re-hospitalization more generally, likewise found no significant difference between patients receiving standard care and those receiving standard care plus prayer. It is worth noting that the authors were not hostile to the possibility that prayer might have a measurable effect and, in fact, Roberts was an employee of the Anglican Church. But they concluded that:

> These findings are equivocal and, although some of the results of individual studies suggested a positive effect for intercessory prayer, the majority do not and the evidence does not support a recommendation either in favour or against the use of intercessory prayer. We are not convinced that further trials of this intervention should be undertaken and would prefer to see any resources available for such a trial used to investigate other questions in health care.[15]

BUT, BUT, BUT

This is the kind of language that consigns a drug or surgical procedure to medical history. But when prayer studies produce null results—or weak results at the margins of statistical significance—researchers tiptoe, reassuring the public that their findings aren't the last word or that results have little relevance outside of the specific conditions of the research. When the STEP results were released, one coauthor—a chaplain at the Mayo Clinic—made assurances that the study had no bearing on the efficacy of personal prayer or prayers offered for family and friends. Bob Barth, director of an intercessory prayer ministry in Missouri expressed optimism that future research would pan out: "We've been praying a long time and we've seen prayer work, we know it works, and the research on prayer and spirituality is just getting started."[16]

Some religious leaders and theologians take another tack. They argue that prayer is inherently exempt from evaluation. The Bible contains warnings against "putting God to the test," which in the minds of some literalists makes any scrutiny of prayer sinful. Of course research on prayer doesn't work! Theist philosopher, Richard Swinburne dismissed the STEP results by arguing that God answers only those prayers offered "for good reasons."[17] Columbia behavioral medicine professor Richard Sloan told the *New York Times* that "the problem with studying religion scientifically is that you do violence to the phenomenon by reducing it to basic elements that can be quantified, and that makes for bad science and bad religion."[18] In a form of special pleading that should be embarrassing, apologists like these argue that prayer, uniquely, has an effect on the natural world that is at once enormous, important, and unmeasurable. God heals people, but only if we aren't watching and measuring.

Many also hasten to reassure people that seemingly unanswered prayer is actually a good thing. In the words of evangelist Ken Collins, one reason God might not answer prayers immediately is that "if He did, you'd stop praying! So He delays His answers to give you something better: fellowship with Him through persistent prayer."[19] Even if unanswered prayer results in suffering unto death, Swinburne insists that what might look and feel bad is actually just one more way God shows his goodness: "Although of course a good God regrets our suffering, his greatest concern is surely that each of us shall show patience, sympathy, and generosity, and thereby form a holy character. Some people badly NEED to be ill for their own sake; and some people badly need to be ill in order to provide important choices for others."[20]

Taking such obfuscation to its logical extreme, Bernard Spilka and Kevin Ladd argue that "scientists must be willing to acknowledge the distinctly nonscientific possibility that prayer operates 'as advertised' in a realm that is both nonlocal and nonphysical."[21] They are absolutely right about this. Faced with assertions about a realm that is completely outside of the natural order, scientists must plead ignorance. However, *so must all human beings, including theologians.*

Spilka and Ladd further point out that "some may argue that explanations involving invisible, unmeasurable actions of a supraphysical Deity are equivalent to no explanation at all."[22] In this, they are equally correct, since the act of explanation itself is about positing cause-effect relationships that have some pattern of relevance to human experience. As I said earlier, humanity's interest in prayer is fundamentally an interest in ourselves, a desire to manipulate the world around us in order to get what we need and want. Actions of a supraphysical deity that have no discernable pattern of impact on the actual lives of physical beings are, from a human standpoint, simply irrelevant. Humanity's prayer habit persists because people believe, thanks to faulty reasoning and sloppy evidence, that they can and do perceive patterns in which prayer affects this physical world and their own lives.

Atheist Sam Harris argues that prayer apologists have cut themselves far too much slack even before wading into the argument that prayer claims are uniquely exempt from the scientific method. To his mind, the evidence against prayer is incontrovertible even without double-blind randomized studies: "Get a billion Christians to pray for a single amputee. Get them to pray that God regrow that missing limb. This happens to salamanders every day, presumably without prayer; this is within the capacity of God. I find it interesting that people of faith only tend to pray for conditions that are self-limiting."[23]

Since the year 2000, the US government has spent over two million dollars on prayer studies, without producing any result that is remotely congruent with the bold claims made by the authors of the New Testament. And yet the bold claims of those authors are quite honestly a reasonable set of assertions to make about an all-powerful and all-loving interventionist deity. God the Almighty shouldn't operate at the margins of statistical significance. He shouldn't be most evident when the evidence itself is of the poorest quality, fading into invisibility as the light of scientific rigor becomes brighter. He shouldn't need defenders who are willing to tie their reputations to expensive research that they then dismiss as irrelevant when results are disappointing.

Even more so, God shouldn't need defenders who engage in rabbit hole reasoning, who insist that he moves in our world and in our lives, but only as long as we aren't looking; or who insist that despite all evidence to the contrary bad is actually good because it must be good, because by definition God is good and he's in charge.

PETITIONARY PRAYER FAILS CONCEPTUALLY

> *To pray for particular favors is to dictate to Divine Wisdom, and savors of presumption; and to intercede for other individuals or for nations, is to presume that their happiness depends upon our choice, and that the prosperity of communities hangs upon our interest.*
> —**William Paley, eighteenth-century philosopher**

Let's concede for the sake of argument that the effects of prayer are uniquely impossible to measure. Even in this case, one might argue that intercessory prayer is a fatally-flawed concept. As Adam Lee at Daylight Atheism put it, "If God is omniscient and omnipotent, nothing can happen if he does not desire it to happen. Nothing can happen against his will. Therefore, whatever state of affairs a theist seeks to change through prayer must be God's will, and by seeking to change it, they are in essence saying that what God has done already is not good enough, or just plain wrong."[24]

Lee is far from the first or last to point this out. "How can he—who "changeth not"—alter circumstances he foreordained in response to prayer?" asks former evangelical Christian, Marian Wiggins, who once wrote Christian curricula.[25] The lack of evidence that petitionary prayer actually works is only one of many layered problems—empirical, rational, and moral—that mitigate against the whole endeavor.

THE PSYCHOLOGY OF PRAYER

Night follows day, and day night. The seasons preserve their
succession. . . . We may not hope to suspend their operation by
our prayers. . . . And yet notwithstanding all of this, we hold in
an undoubting faith the doctrine of the efficacy of our prayers,
or to use the language of another, "of an influence from above
as diversified and unceasing as are the requests from below."
—William Peabody, nineteenth-century theologian

Any doctor seeking evidence-based medical interventions would have to conclude that weak bleach solutions do a better job of saving lives than does petitionary prayer. Why, then, do millennia of experience with illness and death, followed by more than a century of disappointing research, so utterly fail to persuade humanity that we are better off promoting bleach than petitionary prayer?

Alternately, if we look at petitionary prayer outcomes from an interpersonal standpoint, one might argue that kindergarteners and chimps respond to plaintive pleas for help more often than God does. Why, then, do we spend so much more time and effort seeking assistance from a seemingly unresponsive absentee than from chimps and tots? Why is humanity virtually indifferent to the fact that, as Gaylor put it, *nothing fails like prayer*?

One simple answer, of course, is that human beings are broadly superstitious. We see patterns in all sorts of random phenomena and engage in wishful thinking that knows few bounds. Given sufficient ambiguity, we perceive what we want to perceive and believe what we want to believe. The scientific method is powerful precisely because it erects barriers against this pattern, forcing us to ask the questions that could show us wrong. It has been called, "what we know about how not to fool ourselves." We don't have to lower the standard of evidence far before all manner of pseudoscientific hogwash seems real.

But why is it that this particular thing—the idea that some dominant supernatural being with magical powers both exists and actively intervenes in our lives—so garners our emotional devotion that we defend it with arguments ranging from "I just know" to complex theologies?

The answer appears to lie in the cognitive, emotional, and social architecture of the human mind, in hard-wired patterns of thought and feeling that neuroscience is only just beginning to understand. Human beings are social information specialists, meaning a large portion of our brain space is optimized for processing information from and about other beings with minds much like our own. Consequently, we make gods in our own image psychologically. As anthropologist Pascal Boyer lays out in his book *Religion Explained*, humanity's gods all have human psyches with minor modifications, and the God of Christianity is no exception.[26]

As hierarchical social animals, we are predisposed to seek authorities and allies. Our ability to survive and thrive depends on our ability to relate. Fortunately, we can represent other minds as "introjects" within our own, which allows us to anticipate the thoughts and preferences of others, and to act accordingly. Children construct imaginary friends who meet important developmental needs, and then develop relationships with these imaginary companions, sometimes carrying on elaborate conversations with, essentially, parts of the self. They also create "introjects" or virtual copies of their parents, which allow them to draw on parental guidance when a physical parent is not present. Subvocal self-talk remains important on through adulthood. The same cognitive capacities that enable these important, adaptive processes also allow us to develop relationships and conversations with supernatural beings. Is prayer, like self-talk, adaptive? Or is it simply an artifact of our makeup? The answer to that isn't entirely clear. Either way, the habit of prayer comes quite naturally to our species.

ISN'T ASKING FAVORS FROM GOD HARMLESS?

> *Prayer makes us feel good. It gives comfort. It's a way to feel like we're doing something important with minimal effort.*
> —Seth Andrews, The Thinking Atheist

It is important to remind ourselves that not all prayer takes the form of soliciting favors from gods. Prayers of meditation, thanksgiving, and so forth may provide benefits and pleasures that are outside the scope of this critique, but even petitionary prayer may have benefits that don't require supernatural intervention—in particular, first-person benefits for the person doing the praying. For example, expectancy and placebo effects may be operative even when divine intervention isn't. Formulating a request may help a person clarify his or her own desires. Prayer may strengthen resolve or lend confidence to an endeavor. Handing off anxieties or burdens to a deity provides relief; praying for others may offer comfort at times that we otherwise feel powerless to help. So, shouldn't we let prayer and prayers well enough alone? Isn't asking God for favors harmless, especially if driven by an impulse of altruism, generosity, or compassion?

While I am sympathetic to this perspective, I believe that petitionary prayer is far from harmless:

- Petitionary prayer undermines agency. *Let go and let God*, say evangelicals. *Que sera, sera*, say Latin American Catholics. If God is in charge and whatever happens after prayer must be his will, then there's not a lot a person can do but work on acceptance.
- Petitionary prayer undermines dignity. Singing the praises of a powerful person who requires underlings to ask for favors, even though he already knows what they need, and who grants or denies these requests in some inscrutable pattern is not love. It is groveling on the part of the subordinate and abuse on the part of the master.
- Petitionary prayer suppresses critical thought. Seth Andrews of *The Thinking Atheist* asks, "Why is everyone praying instead of asking why Yahweh . . . would require one of his children to be peeled out of a flaming school bus wreck with the Jaws of Life to endure excruciating skin grafts?"[27] According to the Bible writer, prayer manifests faith. But "faith means the purposeful suspension of critical thinking," says Bill Maher. "It is nothing to be admired."[28]
- Petitionary prayer promotes a habit of self-deception. Even those who claim they believe in the power of prayer at some level believe otherwise. As physicist Lawrence Krauss asks, "If you are choking next to me and either I could perform the Heimlich maneuver or I could pray for you, which would you choose?"[29]

- Petitionary prayer distracts from more promising endeavors. Whole industries have grown up to promote petitionary prayer, which is one of the most lucrative products sold by missionaries and televangelists. What else might that energy and money do in our world?
- Petitionary prayer promotes victim blaming, including self-blame. If a perfectly loving God grants the requests of the faithful, what does that say about those whose prayers evoke no response?
- Petitionary prayer replaces compassionate action. In her monologue, *Letting Go of God*, actress Julia Sweeney gets hit with the realization that we are responsible for each other: "'Wait a minute. What about those people who are like . . . unjustifiably jailed somewhere horrible, and they are like . . . in solitary confinement and all they do is pray . . . this means that I . . . like I think they're praying to nobody? Is that possible?' And then I thought, 'We gotta do something to get those people outta jail!' Because no one else is looking out for them but us; no God is hearing their pleas. And I guess that goes for really poor people too."[30]

BEYOND PRAYER

> *For although I had long ceased to believe in the efficacy of prayer, I was so lonely and so in need of some supporter such as the Christian God, that I took to saying prayers again when I ceased to believe in their efficacy.*
> **—Bertrand Russell at age 22**

I could conclude this chapter by reiterating how dramatically prayer fails. That would be easy, but also incomplete. Prayer wouldn't be such a profound and pervasive part of human experience if it simply was a package of non-responses and harms. If we hope for humanity to move beyond faith and superstition, we need to treat prayer seriously, meaning we must seriously examine the hypothesis that prayer provides real benefit. Blogger Adam Lee offers[31] a warning to fellow atheists:

Human beings have always been, and still are, at the mercy of a complex and often frightening world. It is only natural that people in such circumstances would be eager, even desperate, for a way to calm their fears and give themselves confidence, and this is what prayer provides. It gives believers a "direct line" to the highest power in the universe, the one whom they are told is on their side and will make sure everything turns out all right for them. This ability to cope has always been one of the major perceived benefits of religious belief, and atheists who seek to make inroads against theism would do well to remember it.

Some forms of prayer are relatively compatible with the emergence of wholly secular forms of spirituality and may be borrowed, largely intact. Where the Abrahamic religions treat God as an external "other," many Eastern traditions treat the divine as an internal center point. Contemplative prayer practices can help to draw a person into this center, sharpening consciousness and clarifying values so that they can shape lived experience. Sam Harris discusses psychological benefits of these practices in his book *Waking Up*. Embracing ancient wisdom may be essential if secular communities want to offer humanity a real alternative to tired dogmas and text worship.

Even forms of prayer that are laced through and through with superstition may provide enough natural benefits to those doing the praying that they will be broadly abandoned only when something else takes their place. For example, counselor Michael J. Formica talks about petitionary prayer as a form of resolution or setting intentions.[32] A variety of studies looking at the personal effects of prayer on the believer suggest variously that more prayer is associated with increases in hope, attachment, and forgiveness, resistance to addiction, feelings of unity, and decreased marital conflict. To date, these studies document a correlation between prayer and measures of wellbeing, but without establishing causal relationship. It is to our benefit that prayer has now become a serious topic of study, with dedicated communities of scholars within the Society for the Scientific Study of Religion, the Religious Research Association, and Division 36 (Psychology of Religion and Spirituality) of the American Psychological Association.

If humanity is ever to benefit from the best that religion has to offer, we must unpack the box and sort the contents so that we can separate handed-down rubbish from that which is timeless and useful. In the Bible, prayer takes many forms, ranging from prayer "without ceasing"—meaning an everyday mindful connection with something bigger than the self—to "forty days in the wilderness"—meaning a retreat from everyday routines to re-center on that which matters most. In the Christian tradition, prayer can be a social activity, an orchestration of shared values and intentions. Alternately, it can be a solitary cry of anguish or confession of that which feels too dark to share. Each of these types of prayer expresses a core part of the human spirit—what it means to us to be who we are, where we are: mortal women, men, and children, keenly aware of our fragility, seeking to live well and die well in community with each other.

These dimensions of prayer are too meaningful to be left, to quote Christopher Hitchens, "in the dustbin of history," even though many aspects of religion must be. They are too important to be ceded to the traditional purveyors of institutional superstition and patriarchy. They are part of humanity's inheritance, a finely evolved product of millennia of human suffering, joy, wonder, and yearning. They belong to us all, and for the sake of our children we must begin the long, complicated process of cleaning and claiming them.

Chapter 15

THE TURIN SHROUD

A Postmortem

Joe Nickell

At the middle of the fourteenth century—a time rife with alleged holy relics—the linen cloth now known as the Shroud of Turin first appeared. This took place in about 1355, at a little church in Lirey (in the diocese of Troyes) in north-central France. Its owner was a soldier of fortune named Geoffroy de Charny. Unlike other alleged Holy Shrouds, de Charny's cloth bore the front and back images of an apparently crucified man, together with the red markings of his blood.

Unfortunately, de Charny never explained how he—a man of modest means—had acquired the most holy relic in Christendom, although it was later alleged to have been received as a "gift" or suggested that it was a *butin de guerre* (a spoil of war).[1] If it were indeed a consequence of the crusades, and had actually come from the Holy Land, might it really be the True Shroud of Christ?

PROVENANCE AS A RELIC

Three of the four gospels vaguely refer to Jesus's shroud (e.g., "a clean linen cloth," says Matthew 28:59); however, John's gospel reports that Jesus was *wound* with *multiple* cloths and spices, myrrh, and aloes, "as the manner of the Jews is to bury" (John 19:39–40). A separate cloth, the "napkin," covered the face (John 20:7). In contrast, the Turin Shroud is a single fourteen-foot-long cloth with its two images head to head (as if half the cloth was placed under the body and the remainder, folded at the top of the head, was draped over the front). And whereas Jewish practice required washing of the body, the figure on the shroud was not washed, as evident from the dried "blood" (as on the arms). And neither myrrh nor aloes is found on the cloth. In other words, the Turin Shroud is both incompatible with John's gospel and contrary to Jewish burial practices, as described in the Mishna (the first part of the Jewish Talmud).[2]

Moreover, there is no provenance for any of the alleged relics of Jesus before the fourth century, and then only references to legend. In 326, Helena, the mother of Roman emperor Constantine, went to Jerusalem where, supposedly aided by divine inspiration, she discovered Jesus's lost tomb and with it a treasure trove of relics. (In early Christianity, a relic was some portion of the remains of a holy person—a piece of skin or fragment of bone, for example, or some object that that had been contact with the body, such as a swatch of cloth from his clothing.[3])

The tomb contained, it would be alleged, the cross of Jesus, as well as the *titulus* (cross title board), the crucifixion nails, crown of thorns, crosses of the two thieves crucified with Jesus (Mark 15:15–28), and much more. Subsequently, Cyril, the fourteenth-century saint, wrote that "the whole world is filled with fragments of the wood of the cross," and Protestant reformer John Calvin quipped that there were enough pieces to "form a whole ship's cargo."[4] Just one piece allegedly from this "True Cross" has been radiocarbon dated, along with the titulus, yielding the dates of 1018–1155 CE for the cross fragment and 980–1146 CE for the titulus. These dates prove the objects are fakes, consistent with known medieval trafficking in bogus relics.[5]

Some of the fake or at least doubtful relics were also attributed to St. Helena—such as fragments of the bones of the Three Magi (who heralded the infant Jesus [Matthew 2:1–12]) that I visited in Milan. There are countless other relics. At Bruges, Belgium, I held in my hands the reliquary alleged to contain the blood of Christ, although the substance is suspiciously still red and the earliest document referring to it dates from 1270.[6] Elsewhere, there were no fewer than six holy foreskins of the circumcised Jesus in as many churches (at least one surviving), as well as hay from

the manger; various relics of Joseph and Mary, including vials of the latter's breast milk; and especially relics of Jesus, such as a tear he shed at
Lazarus's tomb, one of the vessels in which he changed water to wine, and
especially relics of the crucifixion, including, for example, an entire crown
of thorns, although not the only one, and individual thorns that turned up
here and there.[7]

"True Shrouds," appearing as early as the sixth century, were especially prolific—at least forty-three "by one count in medieval Europe
alone."[8] In 1204, a French crusader named Robert de Clari wrote of a cloth
in Constantinople that he apparently had not seen and that he mistook for
a shroud (*sindon*) when it was likely a facecloth or "napkin" (*sudarium*)
bearing the "features" (i.e., the face) of Jesus. Some Turin shroud devotees
believe this *was* the shroud, although that would still take it back in time
only a century and a half earlier than its known appearance at Lirey.

Some go even further and attempt to equate the Shroud of Turin with
the fourth-century Image of Edessa. Although that was a legendary face-
only cloth, author Ian Wilson rationalized that the shroud had been folded
so that only the face showed![9] There were actually several alleged origi-
nals of this "miraculous" Mandylion, as it was called, two of which presently survive. I have seen both: one is copied from the other, according
to Vatican experts, and they are admittedly painted.[10] In any case, as we
shall see, the Shroud of Turin has a far more recent history, according to a
revealing fourteenth-century document.

THE AFFAIR AT LIREY

By 1357, the Holy Shroud that had surfaced at Lirey in the possession of
Geoffroy de Charny (since killed during the Battle of Poitiers) was placed
on view. Great crowds of pilgrims thronged to see the sensational "relic."
It was exhibited at full length and advertised as the "true Burial Sheet
of Christ." Medallions were struck to commemorate the event. However,
questions were raised; for example, why had New Testament writers
neglected to mention the astonishing imprint of Jesus's body on his burial

cloth? At the urging of "many theologians and other wise persons," an investigation was launched by the local bishop.[11]

We know of this investigation from a lengthy report sent to Pope Clement VII (the first Avignon pope during the Great Western Schism) in 1389, by a later bishop, Pierre d'Arcis. He stated that the shroud was being used as part of a faith-healing scam:

> The case, Holy Father, stands thus. Some time since in this diocese of Troyes the dean of a certain collegiate church, to wit, that of Lirey, falsely and deceitfully, being consumed with the passion of avarice, and not from any motive of devotion but only of gain, procured for his church a certain cloth cunningly painted, upon which by a clever sleight of hand was depicted the twofold image of one man, that is to say, the back and the front, he falsely declaring and pretending that this was the actual shroud in which our Savior Jesus Christ was enfolded in the tomb, and upon which the whole likeness of the Savior had remained thus impressed together with the wounds which He bore. . . . And further to attract the multitude so that money might cunningly be wrung from them, pretended miracles were worked, certain men being hired to represent themselves as healed at the moment of the exhibition of the shroud.[12]

D'Arcis went on, speaking of his predecessor: "Eventually, after diligent inquiry and examination, he discovered the fraud and how the said cloth had been cunningly painted, *the truth being attested by the artist who had painted it*, to wit, that it was a work of human skill and not miraculously wrought or bestowed" (emphasis added). Although action had been taken and the fake shroud hidden away, it had now resurfaced, said d'Arcis, speaking of "the grievous nature of the scandal." As a consequence, Clement ordered that during exhibition of the cloth there must be a loud announcement: "It is not the True Shroud of Our Lord, but a painting or picture made in the semblance or representation of the Shroud."[13]

Although the scandal ended for a time, during the Hundred Years' War Margaret de Charny, Geoffroy's granddaughter, gained custody of the cloth. She claimed to be keeping it safe but in time refused to return it (for which she was eventually excommunicated). Instead she took it on tour in areas of present-day France, Belgium, and Switzerland. When the shroud was further challenged, Margaret could only produce documents officially labeling it a fake, a "representation." In 1453, at Geneva, she sold the cloth to Duke Louis I of Savoy. The Savoys (who later comprised the Italian monarchy) enshrined it in a church at their castle at Chambèry and treated it as a magic talisman. It was, however, damaged in a fire in 1532. Eventually, in a shrewd political move (by a later Duke seeking a more suitable capital), it was transferred to present-day Turin. By then, the scandals at Lirey had been long forgotten.[14]

"PHOTOGRAPHIC" IMAGE

In 1898, Italian photographer Secondo Pia photographed the Shroud of Turin for the first time. He was impressed that the glass-plate negatives showed a more lifelike, quasi-positive image. This began the shroud's modern era, with proponents asking how a mere medieval forger could have produced a perfect "photographic" negative before the invention of photography. Actually, the analogy to photographic images is misleading: Because the "positive" image depicts white hair and beard (the opposite of what would have been expected for a Palestinian Jew in his thirties), the image on the cloth is only a quasi-negative. (As we now know, a medieval artistic technique produces just such images. More on this presently.)[15]

Various hypotheses for the image formation were proposed. That it might have been produced by simple contact with bloody sweat or burial ointments was disproved by the lack of wraparound distortions. The notion of "vaporography"—body vapors supposedly interacting with spices on the cloth to produce a vapor "photo"—was doomed by experiments (mine included) that showed only a resulting blur. Still others postulated a "scorch" from a miraculous burst of radiant energy at the moment of Jesus's resurrection. Yet there was no known radiation capable of achieving the shallow penetration of the images—short of invoking a miracle—and actual scorches on the cloth from the fire of 1532 yielded strong fluorescence, unlike the shroud images, which do not fluoresce at all.[16]

One misguided skeptic, Nicholas Allen, took the word *photographic* quite literally and proposed that the shroud was the world's first photo-graph! His elaborate concept involved a room-sized camera obscura, newly outfitted with crystal lenses, and "film" in the form of a linen sheet treated with a silver salt (none of which, however, is found on the shroud). With this affront to Occam's razor (the principle that the hypothesis with the fewest assumptions is to be preferred), Allen made skeptics look ridic-ulous and desperate.[17] In fact, the artistic hypothesis has overwhelming evidence in its favor.

PATHOLOGY VS. ICONOGRAPHY

The image on the Shroud of Turin yields copious anatomical evidence that betrays its artistic origin. Not only has no burial cloth known to history ever had such imprints, but as the noted pathologist Dr. Michael Baden observes of the overall shroud image, "Human beings don't produce this kind of pattern." He finds the "blood" flows especially suspicious, pointing out their neat artistic appearance, with some on the head seeming to be almost levitated on the outside of the locks. The hair should instead have been matted and caked with blood. "To me, this makes the image less real," Baden stated. "I'd expect to see a pool of blood. Whatever did this doesn't speak for severe scalp lacerations."[18]

There are many additional image flaws that point to artistry. For example, the physique is unnaturally elongated—like figures in Gothic art! The hair erroneously hangs down as for a standing figure, rather than being splayed as for a reclining one. Also, the imprint of a bloody foot is incom-patible with the outstretched leg to which it belongs. Although shroud pro-ponents claim nail wounds are realistically placed in the wrists, rather than the hands (where they would supposedly fail to support the body's weight), those observations are subjective. Besides, only one "wound" shows, and it is an exit wound that actually suggests the imagined nail entered the base of the palm—following the gospel narratives and artistic tradition.[19]

Many other details are suggestive. For instance, most proponents accept that the position of the feet implies they were nailed together rather than separately (the left foot pointing inward). However, that placement is simply a European artistic concept that had become conventional by the fourteenth century, the time of the forger's confession. Also, the shroud depicts the lance wound in Jesus's right side, where artists invariably placed it, although only the gospel of John (19:34) even mentions this wound and does not specify which side was pierced.[20]

Iconographic considerations provide further evidence against authenticity. As St. Augustine lamented in the early fifth century, there is no way to know exactly what Jesus looked like. There is not a hint in the New Testament. Indeed, the earliest portraits (dating from the third century) depict a beardless, Apollo-like youth. However, this concept would soon be paralleled by a more Semitic representation, with long flowing hair and beard and a prominent nose. It was this later tradition that became conventional in Christian art. It is therefore most suspicious that a "shroud"—its whereabouts unrecorded for 1300 years—should suddenly appear, bearing an image of Jesus looking exactly like artists had come to imagine him.

Moreover, the shroud seems the culmination of a lengthy tradition of "not-made-with-hands" portraits (the Mandylion image tradition discussed earlier). From the sixth century came such images, reputedly imprinted by the "bloody sweat" of the living Christ, and by the twelfth century there were accounts of him having pressed "the length of his whole body" upon a cloth. Already (by the eleventh century) artists had begun to represent (incorrectly) a single double-length shroud cloth (although non-imaged) in paintings of the Lamentation (a gathering of Jesus's followers grieving over his body) and Deposition (the placing of his body in the tomb), and by the thirteenth century we find ceremonial shrouds bearing full-length images of Jesus's body in death. In these, as with the shroud, the hands are folded over the loins—a manner incorrect for Jewish practice but an artistic convention dating from the eleventh century (and no doubt a concession to medieval prudery).

Thus, from an iconographic point of view, these various traditions come together in the Shroud of Turin and suggest it is the work of an artist of the thirteenth century or later. The shroud's provenance indicates, of course, a mid-fourteenth-century date, as do the condition and weave of the cloth. (The linen shrouds from Jesus's time were always plain weave, whereas the shroud is woven in a later herringbone twill.)[21]

TESTING THE STAINS

The Shroud of Turin has been subjected to numerous tests that—despite what most people have heard—have yielded powerful evidence that it was produced by an artist working at the middle of the fourteenth century. In 1969, the Archbishop of Turin appointed a secret commission to examine the shroud. That fact was leaked, then unfortunately disavowed, but "at last the Turin authorities were forced to admit what they had previously denied."[22] More detailed studies—again clandestine—began in 1973. The subsequent report, devastating to authenticity, was largely suppressed, whereas, a *rebuttal* was freely offered.

The commission's internationally known forensic serologists made heroic efforts to validate the blood. This had remained bright red, which itself is suspicious, since real blood eventually blackens with age.[23] In fact, the red stains failed all the sophisticated microscopical, chemical, biological, and instrumental analyses. The preliminary tests for peroxidase (a blood enzyme) and traces of hemoglobin and hemoglobin derivatives were negative, as were attempts to detect corpuscles or any other blood components. Microscopic investigation revealed that making up the "blood" images, in part, were reddish granules that failed to dissolve in reagents that dissolve blood.

Further sophisticated tests were likewise unsuccessful in confirming blood, even though these—including microspectroscopic analysis and thin-layer chromatography—permit examination of extremely small quantities of blood. Neutron activation analysis was also conducted but also with negative results. One scientist microscopically detected traces of a red substance, which he concluded had been added subsequently and deliberately—traces of apparent paint. An art expert on the commission concluded that the image had been produced by an artistic imprinting technique, while the member who had detected the apparent paint suggested the technique involved a model or molds.[24]

Additional examinations came in 1978, conducted by the Shroud of Turin Research Project (STURP). Although the members are invariably billed as impartial scientists, chosen for their expertise, in fact they lacked experience with art forgery, most were religious and even shroud enthusiasts, and both of their leaders and some members served on the Executive Council of the Holy Shroud Guild, a Catholic organization devoted to the "cause" of the supposed relic. Having such people examine the shroud is akin to asking the Flat Earth Society to investigate the shape of the planet. The unfortunate consequences of having such a biased group were predictable.

STURP took sticky-tape liftings from the shroud's image and off-image areas, turning them over to one of the world's foremost microanalysts, Walter McCrone. He soon identified the "blood" as tempera paint containing red ocher and vermilion, along with traces of rose madder—pigments used by medieval artists to depict blood. He also discovered that on the image—but not on the background—were significant amounts of the red ocher pigment. For his efforts, McCrone was held to a secrecy agreement, while statements made to the media indicated that no evidence of artistry had been found. STURP representatives paid a surprise visit to McCrone's famous laboratory to confiscate his samples, then gave them to two late additions to STURP, John Heller and Alan Adler, neither of whom was a forensic serologist or pigment expert. The pair soon proclaimed they had "identified the presence of blood." Subsequently, however, at a conference of the International Association for Identification, forensic analyst John F. Fischer explained how results like theirs—using their unacceptable additive approach (red color + iron + protein, etc.)—could be obtained from tempera paint![25]

Other dubious claims surfaced, with proponents typically seeming to have started with the desired answer and worked backward to the evidence. For example, in his book *The DNA of God*, a Texas researcher named Leoncio Garza-Valdez, a member of a new shroud project, claimed there was evidence of human DNA in a shroud "blood" sample. Actually, the scientist at the DNA lab, Victor Tryon, said he could not say how old the DNA was or confirm that it came from blood. "Everyone who has ever touched the shroud or cried over the shroud," he stated "has left a potential DNA signal there." Tryon resigned from the project due to what he disparaged as "zealotry in science."[26]

THE RADIOCARBON DATING

After the death in 1983 of the Shroud's titular owner, Italy's exiled King Umberto II, the Vatican inherited the cloth (which had long been under the custodianship of the Archbishop of Turin). After extensive planning and negotiations, the Vatican gave approval for the cloth at last to be radiocarbon dated (based on the decay of a radioactive isotope of carbon, carbon-14). A new method—accelerator mass spectrometry—had been invented that could use very small samples, thus destroying much less of the cloth than would previously have been required. Walter McCrone, who had discovered that the image had been artistically rendered in tempera paint, had predicted the cloth would ultimately have to be radiocarbon tested. "What I think is going to happen is that the Shroud will be dated," he said, adding with considerable wit, "and that the date will be August the 14th, 1356, plus or minus ten years. Mostly minus."[27] (In other words, McCrone was predicting it would date to the time of the forger's reported confession.)

Three laboratories capable of applying the new dating method—at Oxford, Zurich, and the University of Arizona—were commissioned. On April 21, 1988, a narrow strip measuring 0.4 inches by 2.8 inches was snipped from the main body of the shroud (avoiding a sewed-on "side strip" of the same herringbone weave, apparently contemporaneously added, perhaps to center the image[28]). The cut swatch was then subdivided into three smaller pieces, each roughly the size of a postage stamp. Also sent to each lab were several control swatches of known dates (including a Cleopatra mummy cloth). The samples were all submitted "blind"; that is, the swatches were unidentified so as to help prevent bias.

The obtained results were in very close agreement: The age span of the Shroud of Turin was determined to be 1260–1390 CE, and the results gained added credibility by correct dates being obtained from the controls.[29] In other words, the results confirmed that the shroud dated from about the mid-fourteenth century when an artist admitted it was his handiwork—just as Bishop Pierre d'Arcis reported to Pope Clement VII in 1389. And that date was further supported by the shroud's lack of provenance before that time, its medieval iconography, the weave and condition of the cloth, and so on.

Committed shroud proponents were devastated, but before long they determined to attack the scientific findings rather than accept them. They used a method they had always employed, what I call "shroud science": that is, beginning with the desired answer and then, with "confirmation bias," only seeking evidence that appeared to favor authenticity.[30] There were numerous attempts to discredit, revise, or replace the damning carbon-14 dating.

Some unscientifically invoked the supernatural, suggesting that an imagined burst of radiant energy at the moment of Christ's resurrection had altered the carbon ratio. More down to earth was the notion of Russian scientist Dmitrii Kuznetsov, who claimed to have established that the heat from a fire (like that of 1532) could affect the radiocarbon date. However, his work could not be replicated, and his physics calculations were found to have been plagiarized—complete with an error![31] Another researcher, the previously mentioned Garza-Valdez, claimed to have obtained a swatch of the "miraculous cloth" and found thereon a microbial coating, contamination that could have altered the carbon date. In fact, the shroud samples selected for testing were thoroughly cleansed before radiocarbon dating; moreover, for the shroud's date to have been altered by thirteen centuries (from Jesus's first-century death to the radiocarbon date of 1325, plus or minus 65 years), there would have had to be twice as much contamination, by weight, as the cloth itself.[32]

Some shroud researchers attempted to place the cloth in Jerusalem to counter the evidence it was made in France, and to give it a seeming history and provenance. Swiss criminologist Max Frei alleged that he had found certain pollen grains on the cloth that "could only have originated from plants that grew exclusively in Palestine at the time of Christ," as well as pollens characteristic of Istanbul (formerly Constantinople) and the area of ancient Edessa—seeming to confirm Ian Wilson's "theory" that the shroud was the earlier Mandylion. Unfortunately, Frei had credibility problems. Before his death in 1983, he pronounced authentic the forged "Hitler diaries." Worse, STURP's tape-lifted samples showed few pollen grains, and Walter McCrone subsequently found that so did Frei's—except an area of one tape that appeared to have been "contaminated," possibly deliberately.[33] Another claim followed from attempts to link traces of limestone on an alleged shroud thread to the Jerusalem area. Ian Wilson describes the results as "a strong possibility"—but only when "combined with the other scientific evidence assembled thus far."[34] But that evidence, as we are seeing, is unrelentingly *against* authenticity. (The Archbishop of Turin now refuses to acknowledge the authenticity of any alleged shroud samples that have not been certified by them or to recognize any results obtained from such samples.)

And some so-called shroud image analysis is so outrageously pseudo-scientific as to be laughable. Consider the efforts of a retired geriatric psychiatrist Alan Whanger and his wife Mary, assisted by an Israeli botanist who—looking at greatly enhanced smudgy areas in photos of the shroud—claimed to see images of "the flowers of Jerusalem." The three went on to perceive, Rorschach style, here and there in image and off-image areas, the following: Roman coins over the eyes, small Jewish prayer boxes (called phylacteries), an amulet, and various crucifixion-associated items (compare John 19), such as a large nail, a hammer, a sponge on a reed, a Roman thrusting spear, pliers, two scourges, two brush brooms, two small nails, a loose coil of rope, a cloak with a belt, a tunic, and a pair of sandals. There were also other far-fetched imaginings, including Roman dice.

The trio and others also reported finding ancient Latin and Greek words, such as "Jesus" and "Nazareth." However, even shroud proponent Ian Wilson felt compelled to state "while there can be absolutely no doubting the sincerity of those who make these claims, the great danger of such arguments is the researchers may 'see' what their minds trick them into thinking is there."[35]

Still another claim also involved questioned samples (ostensibly those taken from Walter McCrone), but there were even more problems. STURP's Ray Rogers claimed the carbon-14 dating tests were invalid because they were taken from an alleged medieval "invisible mending" or "rewoven area"—"as unlikely as it seems," Rogers admitted. He himself attempted to date the shroud by the amount of the linen's lignin decomposition. However, the alleged patched area exists only hypothetically and is based on the circular reasoning of shroud defenders. Moreover, a team of Italian chemists presented a scientific paper demonstrating that Rogers's mass-spectrometry pyrolysis spectra did not support his assertions; they called the hypothesis of the carbon-14 sample having been taken from a patched area "unsupported" and, indeed, "pseudoscientific."[36]

COMPANION CLOTH?

Can one fake be used to support the authenticity of another? Yes, it can, and we turn to a notorious example of that approach. Shroud proponents are now ballyhooing another cloth, a supposed companion burial linen, stained with "blood" and measuring 84 × 3 × 53 cm, known as the Oviedo Cloth. It is the supposed missing *sudarium*, or "napkin," that covered Jesus's face in the tomb. As we might suspect, just as there were numerous "true shrouds," there were many allegedly genuine *sudaria*, but all such cloths must be questioned in light of the fact that there is not a hint in the New Testament that the burial linens of Jesus were actually preserved. Of course, as fakes proliferated, so did apocryphal texts claiming otherwise.[37]

There is one obvious problem in attempting to link the image-bearing shroud with the reputed Oviedo sudarium: the latter lacks any image. If such a cloth had indeed covered Jesus's face, "this would have prevented the facial image from being formed on the shroud, and it would presumably have caused it to be formed on the sudarium."[38] Enter shroud rationalizers, who are ever-capable of turning evidence upside down. They postulate that the sudarium was used only temporarily, during the period between crucifixion and burial, having been put aside before the body was enclosed in the shroud. Yet, however clever this rationalization, John's gospel states Jesus was buried in the Jewish manner, with a kerchief covering the face, as is described in the Mishnah. Also, with regard to the burial of Lazarus (John 11:44), who was "bound hand and foot with graveclothes," we are told that "his face was bound about with a napkin."

A major reason for interest in the Oviedo Cloth among Shroud of Turin proponents is in hopes of countering the radiocarbon evidence that has demolished claims of the shroud's authenticity. Advocates hope to link the two cloths because, allegedly "the history of the sudarium is undisputed," and it "was a revered relic preserved from the days of the Crucifixion."[39] We will return to this boast, but writer Mark Guscin says of the alleged relic, "As a historical document, it confirms many of the details contained in the gospels." He adds:

> More importantly, it shows the fourteenth century date for the Shroud obtained by the carbon dating must be mistaken. All the tests carried out on the sudarium show that it must have covered the same face as the Shroud did, and as the sudarium has been in Oviedo since 1075, the Shroud cannot possibly date from the fourteenth century. This, perhaps, is the most valuable testimony of the sudarium. All the arguments in its favour are purely scientific, not depending in any way on faith. The investigations have had a cold, twentieth century approach, and the results point to its being genuine.[40]

But do the Oviedo Cloth's supporters really rely on science, or is theirs a further application of "confirmation bias" like that of shroud advocates? The evidence allegedly supporting the cloth of Oviedo comes, in fact, from some of the same dubious and discredited sources that were involved with the Shroud of Turin.

For example, one claimant alleges *both* cloths contain human blood of type AB, but this source pretends to do what better evidence shows was not done, since, as we have seen, internationally known forensic serologists found the shroud "blood" failed every test, and it was subsequently identified as tempera paint. The claimant was Pierlugi Baima Bollone, a professor of legal medicine, and Bollone's claims are baloney. Ian Wilson, one of the shroud's staunchest defenders, merely remarks in passing that

Bollone "claimed to" have made such a determination.[41] Pro-shroud scientist Ray Rogers went even further, saying, "The things you hear about
typing are nonsense."[42]

As to the previous boast, that the sudarium's history is "undisputed," it
cannot—assertions aside—be established as dating earlier than the eighth
century, and the earliest supposed documentary evidence is from the eleventh. As Guscin observes, "The key date in the history of the sudarium
is 14 March 1075," at which time an oak chest in which it was kept, was
reportedly opened by King Alfonso VI and others, including the famed
knight El Cid.[43] Unfortunately the original document is lost and only a
thirteenth-century "copy" remains in the cathedral archives.

If reports that the Oviedo Cloth has been radiocarbon tested are true,[44]
then the supposed relic is indeed a fake. The claim is that two laboratories
have dated it to the seventh and eighth centuries, respectively, and that
is devastatingly consistent with the historical record. (For more, see my
Relics of the Christ.[45])

FORGING A SHROUD

If it is true, as the preponderance of evidence demonstrates, that the Shroud of
Turin was never a burial cloth but instead the work of a medieval artisan, then
how did he produce the quasi-negative images that have seemed so believable
to so many? That is, if it is not simple contact imprinting, a "vaporograph,"
or a miraculous scorch—nor yet the world's first photograph or an "ordinary
painting"—how could such an image be made in the Middle Ages?

Well, we begin by recognizing that it is the product of an artist of the
1350s, working in the diocese of Troyes, France. Commissioned to make a
realistic shroud, he could still be expected to have been influenced by the
elongated forms of French Gothic art and the conventions of Christian iconography. He lacked today's forensic knowledge, so he made a number of
mistakes, yet he was playing to his audience and only had to make things
look right to them.

Of course, his artistic sense led him to use a single displayable cloth (rather than multiple linens having imprinted body parts here and there). Although we know he should have used myrrh and aloes for the body image and real blood for the wounds, he had not the slightest inkling to do this, since forensic tests were not yet imagined. He thus used, I think, powdered iron-oxide pigment for the main image (thus accounting for the coloring not soaking into the threads) and a suitable red tempera paint for the "blood." He could have used an actual model for the body in the manner of taking a rubbing of a gravestone, except that, for the face, he would have to substitute a bas-relief to minimize distortions. Again his artistic sensibility asserted itself: The results had to look real but, well, better than real; hence, he added the neat "picturelike" trickles of "blood" also. Needless to say, his audience was not made up of impartial forensic experts.

Such a medieval rubbing image would produce shroudlike images *automatically*, thus giving the appearance of an imprint of a body that had been covered with spices. The raised portions of the body/relief would print, while the recesses would be left blank, leaving—voila!—the realistic quasi-negative image that is found on the Shroud of Turin. Some touching up, the addition of brush-painted "blood" the imprinting of flagellation markings on the back, and so on, and the clever (if flawed) work was complete. (That it was produced in an artist's studio is further indicated by McCrone's discovery on its fibers of such additional contaminating pigments as ultramarine, azurite, orpiment, and wood charcoal.[46])

I made experimental face images by a dry-powder rubbing technique as early as 1978. (See the photo section of my book *Inquest on the Shroud of Turin*.[47]) (McCrone had commissioned an artist to make a dilute-tempera *copy* of the shroud face, but, of course, the coloring soaked through to the back of the cloth. McCrone stated, "I'm sorry I got into this part of the controversy, and I wish that I had stuck to my microscope."[48]) My images were, to the impartial viewer, strikingly shroudlike. (The Associated Press once published one of mine, at Easter time, *as* the Shroud of Turin!) However, lacking over six hundred years of handling, the image was still bold and its edges still comparatively sharp—and shroud zealots used these qualities as a pretext to dismiss my work. They even used my technique most unsatisfactorily to make an image themselves and condemned the laughable results.[49] (Adding insult to injury, one source even called me an "amateur detective," when, in fact, I was a twice-promoted investigator and undercover operative for the world's oldest and largest private detective agency.[50])

Forensic anthropologist Emily Craig used a method that combined features of McCrone's and my techniques (she consulted with me), applying dry pigments freehand onto a suitable surface, then transferring the image to cloth by rubbing. As shown on a CBS documentary, the resulting image also exhibited three-dimensional characteristics like those on the shroud.[51]

In 2009, my friend and colleague Luigi Garlaschelli, at the University of Pavia, produced a full-size replica of the shroud, utilizing hypotheses I advanced in my *Inquest on the Shroud of Turin*.[52] Using the procedure described earlier, he laid hand-woven linen over a volunteer, with a bas-relief substituted for the face to avoid critical wraparound distortions. He employed a version of my rubbing technique, with my added hypothesis of an acidic pigment (i.e., iron oxide may be calcined from iron sulfate, a medieval technique, remaining traces of which would be acidic). The Turin shroud image has sparse red ocher (iron-oxide) pigment confined to the tops of the threads, and an attendant yellowish stain of apparent cellulose degradation.

Garlaschelli artificially aged the result and washed off the pigment. (A powdered pigment could also be expected to slough off over the more than six centuries of its existence. Early paintings of the shroud show a much bolder image.) As Garlaschelli observes, his replica shroud image possesses "all the characteristics of the Shroud of Turin." He adds, "In particular, the image is a pseudo-negative, is fuzzy with half-tones [i.e., has tonal gradations and edge-blurring properties], resides on the topmost fibers of the cloth, has some 3-D embedded properties [as is emphasized with the shroud], and does not fluoresce."[53]

In the fall of 2009, Garlaschelli presented his results at Italy's largest science fair, held in Genoa, where I was lecturing also. His illustrated presentation was dedicated to me, he said, adding (too generously), that I was "the brain" and he "only the hands." In fact, I am both humbled and proud that my years of study and experimentation had helped lead to this important recreation—demonstrating shroud science trumped by real science.

THE SHROUD ARTIST

Who, then was the artist who "cunningly painted" (as the bishop said) the image-bearing cloth now known as the Shroud of Turin? Note that he did not merely paint it, like any masterpiece, but, as we understand, did something cleverly different to make a real shroud. (Shroud critic Steven D. Schafersman says of my rubbing images that they are "identical, for all practical purposes, to those on the shroud," adding that the technique "is embarrassingly simple."[54])

Let us put aside the silliness of those who suggest the artist—whose skill they exaggerate—was actually Leonardo da Vinci. A bizarre twist is given to the shroud's touted photo-negative properties in Lynn Pick-nett and Clive Prince's 1994 *Turin Shroud: In Whose Image? The Truth Behind the Centuries-Long Conspiracy of Silence*. The conspiracy-minded duo not only endorse the notion that the shroud image was the world's first photograph, but argue that photography was invented for the purpose by Leonardo da Vinci himself. They are not joking, even though Leonardo (1452–1519) was not even born until a century *after* the shroud first appeared! There's more: maybe, they think conspiratorially, the da Vinci shroud was switched for the earlier Lirey shroud about 1492![55]

Returning to reality, the shroud is not a photograph, and the question of the artist's identity remains. Now, it is occasionally useful, in the field of art history and criticism, to assign a name to the unknown artist of a particular masterwork. The traditional way of doing this is to designate him "Master," followed by an appropriate descriptor—such as a place (for instance, Master of Flémalle, or Master Honoré of Paris), or a work of art (Master of the Castello Nativity, or Master of the Altar of St. Bartholomew).[56] We can follow this tradition.

First, though, is well to recognize that the Shroud of Turin has had many appellations (some more important than others). It was originally (about 1355) the Shroud of Lirey (after the village in north-central France). It was later the Shroud of Chambèry (from 1453). And when Protestant reformer John Calvin published his *Treatise on Relics* (1543), he mentioned disparagingly an alleged burial cloth at Nice. It bore "the full-length likeness of a human body on it,"[57] and we recognize that the cloth now at Turin was kept in Nice at the time Calvin was writing his treatise.[58] This then became the Shroud of Turin when it was finally transferred to that city in 1578. (Earlier "shrouds" had similarly been named for their homes—the Shroud of Cadouin, the Shroud of Besançon, the Shroud of Compiégne, as examples).[59]

Following the tradition of providing appellations to unknown artists, we may now name the creator of what has come to be known as the Shroud of Turin. It is, of course, really the Shroud of Lirey-Chambèry-Turin (to simplify somewhat), but it seems appropriate to use the place name originally connected with the artist historically—that is, Lirey. Therefore, I have given him the title, "Master of the 'Shroud' of Lirey,"[60] which seems entirely suitable, in keeping with the evidence from history and science.

POSTSCRIPT ON THE POSTMORTEM

Despite all the evidence that the "shroud" is no such thing and that it is indeed a medieval forgery, zealots continue to proclaim its authenticity. It is almost as if there were *two* shrouds—one authenticated by believers, the other disproved by art and forensic experts.

The believers—who ultimately refused to accept the results of the famous McCrone laboratory regarding the pigments and paint, and the results of three more prestigious laboratories concerning the radiocarbon dating—continue to publish writings full of denial, pseudoscience, and wishful thinking, though sometimes with moments of interest. Before his death in 2005, STURP's Ray Rogers (with whom I sparred in the pages of *Skeptical Inquirer*[61]) dismissed some of the astonishing nonsense of certain shroud claimants (the burst-of-radiant-energy "theory" for instance), disparaging what he termed "lunatic fringes" and "religious zealots." He had come to believe that the shroud image was the result of "decomposition products of a rotting body," adding that "no miracles or painters are required." Unfortunately, the lack of wraparound distortions and the presence of pigments and paint, together with much other evidence, rules out the "rotting body" scenario.

Shroud proponents lack not only convincing evidence in favor of the shroud but also any viable hypothesis for the image formation. Their response, of course, is not to admit defeat but rather to, first, attempt to switch the burden of proof, and, second, to attempt to parlay the "mystery" of the image into a semblance of the miraculous. Neither attempt is intellectually honest; nor is either successful (except perhaps among those who are misinformed by shroud propaganda). The attempt to argue that the "unexplained" shroud image suggests a miracle is nothing less than a logical fallacy called "an argument from ignorance" (i.e., from a lack of knowledge): one cannot say, we don't know, then conclude that we do know. We must also remember that the burden of proof—in science, just as it is in law—is upon the advocate of the idea. It is never incumbent on anyone else to attempt to prove a negative. And we must make every attempt to avoid confirmation bias, the tendency to seek only the evidence we wish for.

Yet so it is, by rationalizations and questionable evidence, that the shroud's defenders continue, offering one explanation for the contrary gospel evidence (certain passages require clarification), another for the lack of provenance (the cloth must have been hidden away), still another for the forger's confession (the reporting bishop misstated the case), yet another for the paint pigments (maybe an artist who copied the shroud ritualistically pressed it to the image), and so on. By such an approach, people deceive first themselves, then others.

Instead, we must follow the scientific method, allowing the best evidence to lead us to the warranted conclusion—in this case, that the shroud is the handiwork of a medieval artisan. The various pieces of the puzzle effectively interlock and corroborate each other. For example, the artist's confession is supported by the prior lack of historical record, as well as the revealingly red and picturelike "blood" that in turn has been identified as tempera paint. And the radiocarbon date is consistent with the time the artist was revealed.

Having spent some forty years investigating the shroud, I often recall the words of Canon Ulysse Chevalier, the Catholic historian who brought to light the documentary evidence of the shroud's medieval origin and the scandal at Lirey. As he lamented, "The history of the shroud constitutes a protracted violation of the two virtues so often commended by our holy books: justice and truth."[62]

However, when the Vatican actually acquired the Shroud of Turin in 1983, official practices relating to the cloth changed—for the better. Gone were many of the abuses and excesses of the past. Soon, the radiocarbon testing was carried out—not by shroud proponents but by the best experts with strict protocols. The Vatican appears to have accepted the scientific results. And the new pope, Francis I, revealingly referred to the cloth (at Easter 2013) as an "icon" (i.e., a work of art) rather than "relic" (which it would be, in Catholic parlance, if it had actually wrapped the corpse of Jesus).[63] Does he know the difference between *icon* and *relic*? Is the pope Catholic?

ABOUT THE CONTRIBUTORS

Dr. Aaron Adair is an assistant professor at Merrimack College, North Andover, Massachusetts, and the author of *The Star of Bethlehem: A Skeptical View* (2013).

Dr. Rebecca Bradley earned a PhD in archaeology from Cambridge University and worked as an archaeologist in Egypt and the Sudan. She writes for Skeptic Ink, http://www.skepticink.com/lateraltruth/.

Dr. Robert R. Cargill is assistant professor of classics and religious studies at the University of Iowa, having earned a PhD in Near Eastern languages and cultures from UCLA. He is a biblical studies scholar, classicist, archaeologist, author, and digital humanist. His research includes study in the Qumran and the Dead Sea Scrolls, literary criticism of the Bible and the Pseudepigrapha, and the Ancient Near East. He is the author of *The Cities that Built the Bible* (2016).

Dr. David Eller earned a doctorate degree from Boston University in anthropology and taught anthropology in Denver, Colorado. He has published three books on atheism, *Natural Atheism* (2004), *Atheism Advanced* (2008), and *Cruel Creeds, Virtuous Violence: Religious Violence across Culture and History* (2010). He has also written two textbooks, titled *Violence and Culture: A Cross-Cultural and Interdisciplinary Approach* (2005) and *Introducing Anthropology of Religion*, 2nd edition (2015).

Dr. Abby Hafer has a doctorate in zoology from Oxford University and teaches Human Anatomy and Physiology at Curry College in Milton, Massachusetts. She's a popular public speaker and the author of *The Not-So-Intelligent Designer—Why Evolution Explains the Human Body and Intelligent Design Does Not* (2015).

Phil Halper is the producer of the popular YouTube series *Before the Big Bang*, which features interviews with some of the world's leading cosmologists, including Gabriele Veneziano (the father of string theory), Abhay Ashtekar (founder of loop quantum gravity), and Sir Roger Penrose (coauthor of the classic Penrose-Hawking singularity theorems). He earned degrees from Manchester University and Southampton University, and a diploma in astronomy from University College London.

Guy P. Harrison has degrees in history and anthropology and has written several critically acclaimed books on skepticism and belief, including *50 Reasons People Give for Believing in a God* (2008), *50 Simple Questions for Every Christian* (2013), *50 Popular Beliefs That People Think Are True* (2013), *Think: Why You Should Question Everything* (2013), and *Good Thinking: What You Need to Know to Be Smarter, Safer, Wealthier, and Wiser* (2015).

John W. Loftus earned MA, MDiv, and ThM degrees in philosophy, theology, and the philosophy of religion, the last of which was under William Lane Craig. John also studied in a PhD program at Marquette University for a year and a half, in the area of theology and ethics. He is the author of *Why I Became an Atheist: A Former Preacher Rejects Christianity* (Revised Edition, 2012), *The Outsider Test for Faith: How to Know Which Religion Is True* (2013), *How to Defend the Christian Faith: Advice from an Atheist* (2015), and *Unapologetic: Why Philosophy of Religion Must End* (2016). He edited *The Christian Delusion: Why Faith Fails* (2010), *The End of Christianity* (2011), and *Christianity Is Not Great* (2014). He has also coauthored, with Dr. Randal Rauser, *God or Godless?: One Atheist. One Christian. Twenty Controversial Questions* (2013).

Dr. Julien Musolino is a Franco-American cognitive scientist and an associate professor at Rutgers University, where he directs the Psycholinguistics Laboratory and holds a dual appointment in the Department of Psychology and the internationally renowned Center for Cognitive Science. He is the author of over thirty scientific articles, and his research has been funded by the National Institutes of Health and the National Science Foundation. He is the author of *The Soul Fallacy: What Science Shows We Gain from Letting Go of Our Soul Beliefs* (2015).

Dr. Ali Nayeri received his PhD in theoretical physics from IUCAA in 1999. He was appointed as a postdoctoral associate at the Department of Physics at MIT for two years and as a postdoctoral fellow at MIT Center for Theoretical Physics for an additional year. Since then, he has been a research affiliate at MIT. He is currently with the Harvard Physics Department in Cambridge, Massachusetts. His fields of research include early universe and inflation, brane and string cosmology, semi-classical theory of gravity, and alternative cosmologies.

Sharon Nichols is a retired associate professor of geography at the College of DuPage, Glen Ellyn, Illinois. She is also former president of the Illinois Geographical Society and a National Council for Geographic Education (NCGE) Outstanding Faculty, Distinguished Alumni, South Dakota State University. She was awarded the American Heritage Award by Americans United for her part in an ACLU lawsuit against a Ten Commandments Monument in Haskell County, Oklahoma.

Joe Nickell has been called "the modern Sherlock Holmes." Since 1995 he has been the world's only full-time, professional, science-based paranormal investigator. His careful, often innovative investigations have won him international respect in a field charged with controversy. He has written numerous books, including *Relics of the Christ* (2007) and *The Science of Miracles: Investigating the Incredible* (2013).

Jonathan Pearce earned a degree from the University of Leeds, a PGCE from the University of St Mary's, Twickenham, and a Masters in Philosophy from the University of Wales, Trinity St David. Pearce has written *Free Will? An Investigation into Whether We Have Free Will or Whether He Was Always Going to Write This Book* (2010), *The Little Book of Unholy Questions* (2011), and *The Nativity: A Critical Examination*

(2012). Working as a publisher and teacher, he lives in Hampshire, UK, with his partner and twin boys.

Dr. Robert M. Price is a member of the Jesus Seminar and author of numerous books, including *Deconstructing Jesus* (2000), *The Incredible Shrinking Son of Man: How Reliable Is the Gospel Tradition?* (2003), *The Paperback Apocalypse: How the Christian Church Was Left Behind* (2007), *Inerrant the Wind: The Evangelical Crisis of Biblical Authority* (2009), and (coauthored with Edwin A. Suominen) *Evolving out of Eden: Christian Responses to Evolution* (2013).

René Salm is the author of *The Myth of Nazareth: The Invented Town of Jesus* (2008) and *NazarethGate: Quack Archaeology, Holy Hoaxes, and the Invented Town of Jesus* (2015). He maintains several websites, including Mythicist Papers (www.mythicistpapers.com) and www.nazarethmyth.info.

Dr. Victor J. Stenger (1935–2014) was an adjunct professor of philosophy at the University of Colorado and emeritus professor of physics at the University of Hawaii. He was a prolific author with twelve critically acclaimed books that interface between physics and cosmology and philosophy, religion, and pseudoscience. His 2007 book *God: The Failed Hypothesis: How Science Shows that God Does Not Exist* was a *New York Times* bestseller. His last book was *God and the Multiverse* (2014).

Edwin A. Suominen is a retired engineer (BSEE, University of Washington, 1995), patent agent, and inventor with over a dozen patents. After forty years as a Christian fundamentalist, he learned about evolution through some engineering work and underwent an intellectual awakening that resulted in the book *Evolving out of Eden: Christian Responses to Evolution* (2013), coauthored with Dr. Robert M. Price.

Dr. Valerie Tarico earned a doctorate degree in counseling psychology from the University of Iowa and completed postdoctoral studies at the University of Washington. She is the author of the book *Trusting Doubt: A Former Evangelical Looks at Old Beliefs in a New Light* (2010), as well as chapters in *The Christian Delusion: Why Faith Fails* (2010), *The End of Christianity* (2011), and *Christianity Is Not Great* (2014), edited by John W. Loftus. She founded WisdomCommons.org and writes regularly for online news and opinion sites, including AlterNet and the *Huffington Post*. Her articles can be found at AwayPoint.Wordpress.com.

Frank R. Zindler is a retired professor of biology and geology at SUNY–Johnstown after thirty years of teaching. He has worked as a linguist, analyst, and editor of scientific information for a learned society in Ohio. A member of the Jesus Project and a long-time atheist debater against creationist pseudoscience and biblical claims, he has served as editor of American Atheist Press since the murder of the Murray-O'Hair family in 1995.

NOTES

INTRODUCTION

1. Rudolf Bultmann, quoted in *Kerygma & Myth: A Theological Debate*, ed. Hans Werner Bartsch (New York: Harper & Row, 1961), pp. 1–7.

2. Uta Ranke-Heinemann, *Putting Away Childish Things*, trans. Peter Heinegg (San Francisco: HarperSanFrancisco, 1994).

3. Ibid., p. 14.

4. *How to Defend the Christian Faith* (Pitchstone Publishing, pp. 132–135. Much of what I said can be found as a post on my blog titled, *Does Methodological Naturalism Presuppose Its Own Conclusion?* (http://debunkingchristianity.blogspot.com/2014/04/does-methodological -naturalism.html).

5. See my book *The Outsider Test for Faith: How to Know Which Religion Is True* (Amherst, NY: Prometheus Books, 2013).

6. Jerry Coyne, *Faith vs. Fact: Why Science and Religion Are Incompatible* (New York: Viking, 2015), pp. 85–86.

7. Michael Shermer, *The Believing Brain* (New York: Times Books / Henry Holt and Company, 2011), p. 259.

8. Stephen Jay Gould, *Rocks of Ages: Science and Religion in the Fullness of Life* (New York: Ballantine Books, 2002), p. 4.

9. James B. Campbell, Jason W. Busse, and H. Stephen Injeyan, "Chiropractors and Vaccination: A Historical Perspective," *Pediatrics* 105, no. 4 (April 2000).

10. For more, see: https://vimeo.com/44397162, or use the Internet to look up "Sean B. Carroll at Science Writing in the Age of Denial, April 23, 2012."

11. Jerry Coyne, "Templeton Funds Climate-Change Denialist Groups," December 26, 2013, *Why Evolution Is True*, https://whyevolutionistrue.wordpress.com/2013/12/26/ templeton-funds-climate-change-denialist-groups/.

CHAPTER 1: HOW TO THINK LIKE A SCIENTIST

1. I am aware, of course, that many Christians believe science does support Christianity's most important claims. I encourage those who think this to research specific topics from credible scientific sources. For example, many believers mistakenly think that archaeological discoveries have confirmed key stories and events in the Bible. To date, however, archaeologists have only found evidence of some physical places and human beings named in the Bible. No artifact or

site has ever confirmed the occurrence of any supernatural event, such as the Great Flood or the resurrection of Jesus, or the existence of any supernatural beings, like angels, demons, or God himself. To date, there is no known verifiable scientific evidence that confirms any Christian supernatural claims.

2. Some Protestant versions of Christianity promote a symbolic ritual consumption of Jesus's flesh and blood. Catholics, however, are directed by the Vatican to believe that the wafers and wine they ingest during Mass or Holy Eucharist actually become human flesh and blood during ingestion. They call this transubstantiation.

3. Guy P. Harrison, *Good Thinking: What You Need to Know to Be Smarter, Safer, Wealthier, and Wiser* (Amherst, NY: Prometheus Books, 2015).

4. David W. Moore, "Three in Four Americans Believe in Paranormal," Gallup News Service, June 16, 2005, http://www.gallup.com/poll/16915/three-four-americans-believe -paranormal.aspx (accessed November 30, 2015).

5. Linda Lyons, "Paranormal Beliefs Come (Super) Naturally to Some," Gallup News Service, November 1, 2005, http://www.gallup.com/poll/19558/ paranormal-beliefs-come -supernaturally-some.aspx (accessed November 30, 2015).

6. My book, *Good Thinking* includes a tour of the human brain. I also recommend the following books as excellent sources of general information about brain anatomy and function:

Rita Carter, *The Human Brain Book* (New York: DK Adult, 2009).

Editors of *Scientific American* Magazine, *The Scientific American Book of the Brain* (New York: Lyon, 1999).

Gary Marcus, *Kluge: The Haphazard Construction of the Human Mind* (New York: Houghton Mifflin, 2008).

Michael S. Sweeney, *Brain: The Complete Mind: How It Develops, How It Works, and How to Keep It Sharp* (Washington, DC: National Geographic, 2009).

CHAPTER 2: A MIND IS A TERRIBLE THING

1. Scott Atran and Joseph Henrich, "The Evolution of Religion: How Cognitive By-Products, Adaptive Learning Heuristics, Ritual Displays, and Group Competition Generate Deep Commitments to Prosocial Religions," *Biological Theory* 5, no. 1 (2010): 18.

2. Pascal Boyer, *Religion Explained: The Evolutionary Origins of Religious Thought* (New York: Basic Books, 2001), p. 4.

3. International Association for the Cognitive Science of Religion, www.iacsr.com.

4. Clifford Geertz, *The Interpretation of Cultures* (New York: Basic Books, 1973).

5. Stewart Guthrie, "A Cognitive Theory of Religion," *Current Anthropology* 21, no. 2 (1980): 187. See also Stewart Guthrie, *Faces in the Clouds: A New Theory of Religion* (New York and Oxford: Oxford University Press, 1993).

6. A. Irving Hallowell, "Ojibwa Ontology, Behavior, and World View," in Paul Radin, ed. *Contributions to Anthropology: Selected Papers of A. Irving Hallowell* (Chicago: University of Chicago Press, 1976), p. 362.

7. Ibid., pp. 366–67.

8. Guthrie, *Faces in the Clouds*, p. 4.

9. Justin Barrett, *Why Would Anyone Believe in God?* (Lanham, MD: AltaMira, 2004), p. 31.

10. Scott Atran, *In Gods We Trust: The Evolutionary Landscape of Religion* (Oxford: Oxford University Press, 2002), pp. 57–58.

11. Boyer, *Religion Explained*, p. 42.

12. Atran, *In Gods We Trust*, p. 4.

13. Ibid., p. ix.

14. Boyer, *Religion Explained*, p. 311.

15. Ibid., p. 321.

16. Ibid., p. 311.

17. William James, *The Varieties of Religious Experience: A Study in Human Nature* (1902; New York: New American Library, 1958), p. 40.

18. Ibid., p. 37.

19. Aku Visala, *Naturalism, Theism and the Cognitive Study of Religion: Religion Explained?* (Surrey, UK: Ashgate, 2011), p. 9.

20. Lee Kirkpatrick, *Attachment, Evolution, and the Psychology of Religion* (New York: Guilford, 2005), pp. 253–54; emphasis added.

21. Atran and Henrich, "Evolution of Religion," p. 19.

22. Visala, *Naturalism, Theism and the Cognitive Study of Religion*, p. 9.

23. Gary Marcus, *Kluge: The Haphazard Construction of the Human Mind* (New York: Houghton Mifflin, 2008), p. 144.

24. Ibid., p. 51.

25. Malcolm Gladwell, *Blink: The Power of Thinking without Thinking* (New York: Little, Brown and Company, 2005).

26. Marcus, *Kluge*, p. 51.

27. Ibid., p. 52.

28. Amos Tversky and Daniel Kahneman, "Judgment under Uncertainty: Heuristics and Biases," *Science* 185, no. 4157 (1974): 1124–31.

29. Kendra Cherry, "What Is a Heuristic?" http://psychology.about.com/od/hindex/g/heuristic.htm (accessed November 12, 2015).

30. Tversky and Kahneman, "Judgment under Uncertainty," p. 1126.

31. Ibid., p. 1127.

32. Marcus, *Kluge*, p. 50.

33. Tversky and Kahneman, "Judgment under Uncertainty," p. 1131.

34. Eric Fernandez, "A Visual Guide to Cognitive Biases," https://www.scribd.com/doc/30548590/Cognitive-Biases-A-Visual-Study-Guide (accessed July 20, 2015).

35. Marcus, *Kluge*, p. 19.

36. Ibid., p. 27.

37. Robert A. Burton, *On Being Certain: Believing You Are Right Even When You're Not* (New York: St. Martin's, 2008), p. 98.

38. Justin Kruger and David Dunning, "Unskilled and Unaware: How Difficulties

in Recognizing One's Own Incompetence Lead to Inflated Self-Assessments," *Journal of Personality and Social Psychology* 77, no. 6 (1999): 1121–34.

39. Thomas Kida, *Don't Believe Everything You Think: The 6 Basic Mistakes We Make in Thinking* (Amherst, NY: Prometheus Books, 2006).

40. Richard Swinburne, *The Coherence of Theism* (Oxford: Clarendon, 1977), p. 2.

41. See also Jaco Gericke, "Can God Exist if Yahweh Doesn't," in John W. Loftus, ed. *The End of Christianity* (Amherst, NY: Prometheus Books, 2011).

42. Quoted in Ed L. Miller, *Classical Statements of Faith and Reason* (New York: Random House, 1970), p. 5.

43. Dan Ariely, *Predictably Irrational: The Hidden Forces that Shape Our Decisions* (New York: HarperCollins, 2008).

44. Human Dimension Capabilities Development Task Force, "Cognitive Biases and Decision Making: A Literature Review and Discussion of Implications for the US Army," http://usacac.army.mil/sites/default/files/publications/HDCDTF_WhitePaper_Cognitive%20Biases%20and%20Decision%20Making_Final_2015_01_09_0.pdf (accessed November 13, 2015), p. 2.

45. Carl Sagan, *The Demon-Haunted World: Science as a Candle in the Dark* (London: Headline Books, 1996), pp. 189–206.

46. John W. Loftus, *The Outsider Test for Faith: How to Know Which Religion Is True* (Amherst, NY: Prometheus Books, 2013).

47. Marcus, *Kluge*, pp. 165–72.

CHAPTER 3: WHAT SCIENCE TELLS US ABOUT RELIGION

1. See Bill McKibben's organization 350.org.

2. Bill McKibben, *Eaarth: Making a Life on a Tough, New Planet* (New York: Times Books, 2010); Bill McKibben, *The End of Nature* (New York: Random House, 1989).

3. The Black Robe Regiment has a webpage devoted to espousing their cause for a sanitized version of American History that promotes Christian patriotism over history. They especially do not want our history books to say anything critical of the United States and claim it is the AP America History text that is historical revisionist. See: Sarah Gutekunst, "Oklahoma's Revision of History Is the Opposite of Patriotic," *Tartan*, February 22, 2015, http://thetartan.org/2015/2/23/forum/aphistory.

4. Cynthia Dunbar was the Tenth District Representative on the Texas Board of Education from 2001 to 2007 and wrote *One Nation Under God: How the Left Is Trying to Erase What Made Us Great* (Oviedo, FL: Higher Life Development Services, 2008).

5. Eric Owens, "Florida State Rep. Says High School World History Textbook Is a Big Islam Lovefest," *Daily Caller*, July 30, 2013, http://dailycaller.com/2013/07/30/florida-state-rep-says-high-school-world-history-textbook-is-a-big-islam-lovefest/.

6. Peter Gleick, "Why Anti-Science Ideology Is Bad for America," *Forbes*, August 31, 2011, http://www.forbes.com/sites/petergleick/2011/08/31/why-anti-science-ideology-is-bad-for-america.

7. Brian Tashman, "James Inhofe Says the Bible Refutes Climate Change," *Right Wing Watch*, March 8, 2012, http://www.rightwingwatch.org/content/james-inhofe-says-bible-refutes-climate-change.

8. Tristram Korten, "In Florida, Officials Ban Term 'Climate Change,'" *Miami Herald*, March 8, 2015, http://www.miamiherald.com/news/state/florida/article12983720.html.

9. Naomi Oreskes and Erik M. Conway, *Merchants of Doubt: How a Handful of Scientists Obscured the Truth on Issues from Tobacco Smoke to Global Warming* (New York: Bloomsbury, 2010), p. 248.

10. Chris Mooney, "What Science Denial Is about Much More Than Corporate Interests," *Washington Post*, March 13, 2015, https://www.washingtonpost.com/news/energy-environment/wp/2015/03/13/the-economics-of-dubious-science-why-its-all-about-supply-and-demand/.

11. Susan L. Smalley, "Willful Ignorance: Penn State and 'Don't Ask, Don't Tell,'" *Look Around and Look Within: The Science and Art of Human Behavior* (blog), *Psychology Today*, November 29, 2011, https://www.psychologytoday.com/blog/look-around-and-look-within/201111/willful-ignorance-penn-state-and-dont-ask-dont-tell.

12. Lee McIntyre, "The Price of Denialism," *New York Times*, November 7, 2015. This article defines the costs to knowledge, progress, a free society, and to the world that denialism presents.

13. Michael Specter, *Denialism: How Irrational Thinking Hinders Scientific Progress, Harms the Planet, and Threatens Our Lives* (New York: Penguin, 2009).

14. Robert De Filippis, "Willful Ignorance: An American Ideal?" *Op Ed News*, September 23, 2014, http://www.opednews.com/articles/Willful-Ignorance-An-Amer-by-Robert-De-Filippis-Anti-intellectualism_Awareness_Beliefs_Change-140923-201.html.

15. Mark Hoofnagle, "Hello Scienceblogs," *Denialism Blog: Don't Mistake Denialism for Debate*, April 30, 2007, scienceblogs.com/denialism/2007/04/30/hello-to-scienceblogs/.

16. Smalley, "Willful Ignorance."

17. T. S. Eliot, "Little Gidding," *Four Quartets*: "We shall not cease from exploration and the end of all our exploring will be to arrive where we started and know the place for the first time."

18. Adam Frank, "Science Denialism Has Consequences," NPR, February 4, 2015, http://www.npr.org/sections/13.7/2015/02/03/383501038/science-denialism-has-consequences; Jeffrey Baumgartner, "Anti-Science Threatens America's Innovation," *Jeffrey Baumgartner* (blog), http://www.creativejeffrey.com/creative/antiscience.php?topic=creative.

19. Human Origins Initiative, "Bigger Brains: Complex Brains for a Complex World," *What Does It Mean to Be Human?* Smithsonian National Museum of Natural History, http://humanorigins.si.edu/human-characteristics/brains.

20. Ibid.

21. Robin Dunbar, "The Social Brain: Mind, Language, and Society in Evolutionary Perspective," *Annual Review of Anthropology* 32 (2003): 163–81.

22. B. K. Bryant, "An Index of Empathy for Children and Adolescents," *Child Development* 53 (1982): 413–15.

23. Simon Baron-Cohen, "Theory of Mind in Normal Development and Autism," *Prisme* 34 (2001): 174–83. Baron-Cohen has studied autism and published many papers espousing ToM as a brain module lacking in autistic spectrum individuals.

24. Symboling is considered a prerequisite for language.

25. Dennis O'Neil, "What Is Language?" *Language and Culture: An Introduction to Human Communications*, August 31, 2006, http://anthro.palomar.edu/language/language_2.htm.

26. Culture refers to the entire way of life of a people.

27. Justin L. Barrett, *Why Would Anyone Believe in God?* Cognitive Science of Religion (Walnut Creek, CA: Altamira, 2004).

28. Stewart Guthrie, "A Cognitive Theory of Religion," *Current Anthropology* 21, no. 2 (1980): 184.

29. Ibid., p. 191.

30. Pascal Boyer, *Religion Explained: The Evolutionary Origins of Religious Thought* (New York: Basic Books, 2001), p. 2.

31. Ibid., p. 4.

32. Lesley Newsom and Peter Richardson, "Religion: The Dynamics of Cultural Adaptations," in *Evolution, Religion, and Cognitive Science: Critical and Constructive Essays*, ed. Fraser Watts and Leon P. Turner (Oxford, UK: Oxford University Press, 2014), pp. 192–218.

33. The term "culture" was introduced in the United States by anthropologist Alfred. L. Kroeber, who brought it to the Berkeley School of Geography where geographer Carl O. Sauer incorporated it into Cultural Geography.

34. Brian Fagan, *People of the Earth: Introduction to World Prehistory*, 3rd edition (Boston: Little, Brown, 1980).

35. Clyde Kluckhohn, Louis R. Gottschalk, and Robert Angell, "The Personal Document in Anthropological Science," *The Use of Personal Documents in History, Anthropology, and Sociology*, Bulletin 53 (New York: Social Science Research Council, 1945), pp. 79–174.

36. Clyde Kluckhohn and William H. Kelly, "The Concept of Culture," *The Science of Man in the World Crisis*, ed. Ralph Linton (New York: Columbia University Press, 1945).

37. Edward B. Tylor, *Primitive Culture: Researches into the Development of Mythology, Philosophy, Religion, Art, and Custom*, vol. 1 (London: John Murray, 1871), p. 1.

38. Alfred L. Kroeber and Clyde Kluckhohn, *Culture: A Critical Review of Concepts and Definitions* (New York: Vintage Books, 1952), p. 84.

39. This means that all culture traits are, in fact, products of the culture. The products don't have to have arisen from within the culture; they could be adoptions or adaptations from other cultures.

40. Sharon Nichols, "The Definitions of Culture," paper, Geography Dept., South Dakota State University, 1981.

41. Cetaceans (whales and dolphins) have social structures, lifestyles, ways of communicating, and hunting techniques that are taught to their young. (Kate Douglas, "Six 'Uniquely' Human Traits Now Found in Animals," *Scientific American*, May 22, 2008, https://www.newscientist.com/article/dn13860-six-uniquely-human-traits-now-found-in-animals/. The "culture" segment was originally printed in *New Scientist*: Stephanie Pain, "Culture Shock," *New Scientist*, March 24, 2001, https://www.newscientist.com/article/mg16922834-600-culture-shock.)

42. Alexander Heidel, *The Babylonian Genesis* (Chicago: University of Chicago Press, 1942), pp. 82–140.

43. Clifford Geertz, "Religion as a Cultural System," *The Interpretation of Cultures* (London: Fontana, 1993), pp. 87–125.

44.   (1) Cattle are a source of manure, important to Indian agriculture. The cows help human infants survive due to their life-giving milk. Only Hindus forbid eating cattle.

(2) This is a loaded question and is sure to solicit very strong opposing views.

(3) This may be how many or most of us operate on a subconscious level, but it is not a good way to ascertain what is true and real.

(4) This is a biblical injunction and has had disastrous consequences. Great debate question, though!

(5) Overreliance on one type of energy source prevents research money being spent on other types, which may be more economical when the total cradle–to-grave costs are added in; costs to ecosystems, environmental degradation, and global warming are problems to consider.

(6) This is an untrue statement, or at least an unreliable proposition.

(7) We should teach facts, not pseudoscience or revisionist history, otherwise we are not teaching "knowledge."

(8) Heisenberg's Uncertainty Principle says you cannot know the exact position and momentum (of anything moving) simultaneously. Faster than light travel, so far as we know, is not possible. If you change the laws of the universe, we wouldn't have the universe that we have. These disprove the statement that "anything is possible."

(9) Religion fails to explain the natural world.

(10) This is a true statement.

45. Sharon Nichols, "Thinking about Our Thinking Process," (speech, Critical Thinking Workshop, National Council for Geographic Education [NCGE] Annual Meeting, 2000).

46. Adapted from Biology Lecture Notes and class materials developed by Professor Lynn Fancher, biologist at the College of DuPage, Glen Ellyn, Illinois, 1997.

47. Ibid.

48. "Distinguishing Between Fact, Opinion, Belief, and Prejudice," *Writing @ CSU: The Writing Studio*, http://writing.colostate.edu/guides/teaching/co300man/pop12d.cfm, adapted from H. Ramsey Fowler, *The Little, Brown Handbook* (Boston: Little, Brown, 1986).

49. Isaac Asimov, "A Cult of Ignorance," *Newsweek*, January 21, 1980, p. 19.

50. Carl Sagan, *The Demon-Haunted World: Science as a Candle in the Dark* (New York: Random House, 1995).

51. "An extremist is someone who supports an idea, cause, or set of values so adamantly and without compromise that said person will use their [*sic*] ideas to justify anything they do. . . . The truth of the matter is that extremists are by definition almost entirely immersed in their individual and/or shared sense of superiority. Therefore, they typically convince themselves that their particular strain of extremism is not 'extreme.' In the warped mind of the extremist, extremism is the only rational course!" (Don Emmanuel, "extremist," *Urban Dictionary*, January 31, 2008, http://www.urbandictionary.com/define.php?term=extremist. Note: While *Urban Dictionary* is not a scientific, an academic, or even a vetted source, I find this definition serves better than any other.)

52. Michael Muskal, "So Just What Are Religious Freedom Laws Designed to Protect?" *Los Angeles Times*, April 17, 2016, http://www.latimes.com/nation/la-na-religious-freedom-law -20150402-story.html.

53. Susan M. Shaw, "Privilege, Christian Fragility, and Religious Freedom," *Huffington Post*, April 14, 2016, http://www.huffingtonpost.com/susan-m-shaw/christian-privilege -chris_b_9676426.html.

54. Carl Sandburg, "The Hammer," *The Complete Poems of Carl Sandburg: Revised and Expanded Edition* (New York: Harcourt, 1970). Originally published in *Poetry: A Magazine of Verse*, March 1914.

CHAPTER 4: CHRISTIANITY AND COSMOLOGY

1. A comprehensive review of cosmology in the ancient Near East can be found in Edward T. Babinski, "The Cosmology of the Bible," in *The Christian Delusion: Why Faith Fails*, ed. John W. Loftus (Amherst, NY: Prometheus Books, 2010), pp. 109–47.

2. Epicurus, *The Art of Happiness*, trans. George K. Strodach (New York: Penguin Books, 2012).

3. For a translation, see Lucretius, *The Nature of Things*, trans. A. E. Stallings (New York: Penguin, 2007).

4. From NASA's website at http://starchild.gsfc.nasa.gov/docs/StarChild/universe _level2/cosmology.html (accessed August 8, 2013).

5. Image created by Victor J. Stenger.

6. While most ancient cosmologies remained Earth-centered, Aristarchus of Samos anticipated Copernicus by 1,700 years and placed the sun at the center.

7. Helge Kragh, *Conceptions of Cosmos: From Myths to the Accelerating Universe: A History of Cosmology* (Oxford: Oxford University, 2007), p. 34.

8. Stephen Greenblatt, *The Swerve: How the World Became Modern* (New York: W. W. Norton, 2011).

9. Jim al-Khalili, *The House of Wisdom: How Arabic Science Saved Ancient Knowledge and Gave Us the Renaissance* (New York: Penguin, 2011).

10. Kragh, *Conceptions of Cosmos*, pp. 21–23.

11. Bruce Wrightsman, "Andreas Osiander's Contribution to the Copernican Achievement," chapter 7 in *The Copernican Achievement*, ed. Robert S. Westman (Los Angeles: University of California Press, 1975).

12. Dava Sobel, *A More Perfect Heaven: How Copernicus Revolutionized the Cosmos* (New York: Walker, 2011), p. 178.

13. Galileo Galilei, *Sidereus Nuncius, Or, the Sidereal Messenger*, trans. Albert Van Helden (1610; Chicago: University of Chicago Press, 1989).

14. Ibid., pp. 65–83.

15. I am following my usual convention of referring to the Judaic-Christian-Islamic supreme divinity as God and all other divinities as god. I will also assume God is personal and male, is in tradition, while god is impersonal and asexual.

16. H. G. Alexander, ed., *The Leibniz-Clarke Correspondence: Together With Extracts From Newton's Principia and Opticks* (Manchester, UK: Manchester University Press, 1956), p. 11.

17. Jacques Laskar, "Stability of the Solar System," *Scholarpedia*, http://www .scholarpedia.org/article/Stability_of_the_solar_system#Laplace-Lagrange_stability_of_the _Solar_System (accessed December 4, 2012).

18. Pierre Simon Laplace, *Exposition Du Système Du Monde.* (Paris: Impr. du Cercle-Social, An IV de la République française, 1796).

19. Georges Lemaître. "Un Univers Homogène De Masse Constante Et De Rayon Croissant Rendant Compte De La Vitesse Radiale Des Nébuleuses Extra-Galactiques (A Homogeneous Universe of Constant Mass and Growing Radius Accounting for the Radial Velocity of Extragalactic Nebulae)." *Annales de la Société Scientifique de Bruxelles* 47 (1927): 49.

20. Georges Lemaître, "Expansion of the Universe, a Homogeneous Universe of Constant Mass and Increasing Radius Accounting for the Radial Velocity of Extra-Galactic Nebulae," *Monthly Notices of the Royal Astronomical Society* 91 (1931): 490–501.

21. Pope Pius XII, "The Proofs for the Existence of God in the Light of Modern Natural Science," address to the Pontifical Academy of Sciences, November 22, 1951, reprinted as "Modern Science and the Existence of God," *The Catholic Mind* 49 (1972): 182–92. Available online at http://www.papalencyclicals.net/Pius12/P12EXIST.HTM (accessed January 27, 2013).

22. Kragh, *Conceptions of Cosmos*, pp. 150–51.

23. Alexander Vilenkin, "Creation of Universes From Nothing," *Physics Letters B* 117, no. 1 (1982): 25–28; André Linde, "Quantum Creation of the Inflationary Universe," *Physics Letters B* 108 (1982): 389–92; David Atkatz, and Heinz Pagels, "Origin of the Universe as a Quantum Tunneling Event," *Physical Review D* 25 (1982): 2065–73; Victor J. Stenger, "A Scenario for the Natural Origin of the Universe," *Philo* 9, no. 2 (2006): 93–102.

24. Ewin MacAskill, "George Bush: 'God Told Me to End the Tyranny in Iraq,'" *Guardian*, Oct. 7, 2005, http://www.theguardian.com/world/2005/oct/07/iraq.usa (accessed November 28, 2013).

25. Erik A. Petigura, Andrew W. Howard, and Geoffrey W. Marcy, "Prevalence of Earth-Size Planets Orbiting Sun-Like Stars," *Proceedings of the National Academy of Sciences* 110, no. 48 (2013): 19175–76; Seth Shostak, "The Numbers Are Astronomical," *Huffington Post*, November 4, 2013, http://www.huffingtonpost.com/seth-shostak/the-numbers-are-astronomi _b_4214484.html (accessed November 6, 2013).

26. Alan H. Guth, *The Inflationary Universe: The Quest for a New Theory of Cosmic Origins* (Reading, MA: Addison-Wesley Publishing, 1997), p. 186.

27. Catholic News Agency, "Believing in Aliens Not Opposed to Christianity, Vatican's Top Astronomer Says." May 13, 2008, http://www.catholicnewsagency.com/news/believing _in_aliens_not_opposed_to_christianity_vaticans_top_astronomer_says/ (accessed August 15, 2013).

CHAPTER 5: BEFORE THE BIG BANG

1. Walter Issacson, *Einstein: His Life and Universe* (New York, Simon & Schuster,2011).

2. Ari Belenkiy, "The Waters I Am Entering No One Yet Has Crossed: Alexander Freidman and the Origins of Modern Cosmology," in *Origins of the Expanding Universe: 1912–1982*, ed. Michael Way and Deirdre Hunter, Astronomical Society of the Pacific Conference Series, vol. 471 (San Francisco: Astronomical Society of the Pacific, 2013).

3. Abbé G. Lemaître, "Un Univers homogène de masse constant et de rayon croissant rendant compte de la vitesse radiale des nébuleuses extra-galactiques," *Annales de la Société de Bruxelles* A47 (1927): 49–59. (Later published in English as: "A Homogeneous Universe of Constant Mass and Increasing Radius Accounting for the Radial Velocity of Extra-Galactic Nebula," *Monthly Notices of the Royal Astronomical Society*, 91, no. 5 [March 13, 1931]: 483–90.)

4. Edwin Hubble, "A Relation Between Distance and Radial Velocity among Extra-Galactic Nebulae," *Proceedings of the National Academy of Sciences* 15, no. 3 (March 15, 1929): 168–73.

5. Roger Penrose, "Gravitational Collapse and Space-Time Singularities," *Physical Review Letters* 14, no. 3 (January 18, 1965): 57.

6. Abbé G. Lemaître, "Contributions to a British Association Discussion on the Evolution of the Universe," *Nature* 128, no. 3234 (October 24, 1931): 704–706.

7. Helge Kragh, *Cosmology and Controversy: The Historical Development of Two Theories of the Universe* (Princeton, NJ: Princeton University Press, 1999).

8. Albert Einstein, *The Meaning of Relativity* (Princeton, NJ: Princeton University Press, 1945) quoted in Stephen W. Hawking and W. Israel, eds., *General Relativity: An Introductory Survey* (Cambridge, UK: Cambridge University Press, 1979), p. 509.

9. H. Bondi and T. Gold, "The Steady-State Theory of the Expanding Universe," *Monthly Notices of the Royal Astronomical Society* 108, no. 3 (June 1, 1948): 252.

10. James M. Rochford, *Evidence Unseen* (Columbus, OH: New Paradigm Publishing, 2013), http://www.evidenceunseen.com/chapter-3-does-the-big-bang-really-threaten-atheism/.

11. Matthew Stanley, *Practical Mystic: Religion, Science, and A. S. Eddington* (Chicago: University of Chicago Press, 2007).

12. Kragh, *Cosmology and Controversy*.

13. Hans Halvorson and Helge Kragh, "Theism and Physical Cosmology," preprint, 2010, http://philsci-archive.pitt.edu/8441/.

14. William Lane Craig, "The Ultimate Question of Origins: God and the Beginning of the Universe," Reasonable Faith, http://www.reasonablefaith.org/the-ultimate-question-of-origins-god-and-the-beginning-of-the-universe#ixzz3taXFCNsv.

15. William Lane Craig and Stephen Law, "Does God Exist?" (debate, Westminster Central Hall, London, October 2011); transcript available at http://www.reasonablefaith.org/does-god-exist-the-craig-law-debate.

16. William Lane Craig, interviewed by Lauren Green, "BICEP2 Project and the Big Bang—William Lane Craig, PhD," YouTube video, 7:54, posted by "firstcauseargument," March 26, 2014, https://www.youtube.com/watch?v=K9V3Sa9vlG8.

17. Brian Bull and Fritz Guy, "The Genesis Account: Six Hebrew Words Make

All the Difference," *Spectrum*, November 12, 2012, http://spectrummagazine.org/article/brian-bull/2012/11/11/genesis-account-six-hebrew-words-make-all-difference.

18. Rabbi Rafael Salber, "The History of Creation Ex-Nihilo within Jewish Thought," *Reshimu*, September 12, 2008, http://www.hashkafacircle.com/journal/R2_RS_exni.pdf.

19. Edward S. Curtis, *The North American Indian: Being a Series of Volumes Picturing and Describing the Indians of the United States and Alaska*, Landmarks in Anthropology (1907; New York: Johnson Reprint Corporation, 1970).

20. S. W. Hawking and R. Penrose, "The Singularities of Gravitational Collapse and Cosmology," *Proceedings of the Royal Society A* 314, no. 1519 (January 27, 1970): 529–48.

21. R. A. Alpher, H. Bethe, and G. Gamow, "The Origin of Chemical Elements," *Physical Review* 73, no. 7 (April 1948): 803–804.

22. Simon Singh, *Big Bang: The Origin of the Universe* (New York: Harper Perennial, 2005).

23. Alan Guth, "Eternal Inflation and Its Implications," *Journal of Physics A: Mathematical and Theoretical* 40, no. 25 (June 6, 2007): 6811.

24. William Lane Craig, "Cosmos and Creator," *Origins and Design* 17, no. 2 (Spring 1996), http://www.arn.org/docs/odesign/od172/cosmos172.htm.

25. Alan Guth, *The Inflationary Universe: The Quest for a New Theory of Cosmic Origins* (New York: Basic Books, 1998).

26. Arvind Borde, Alan H. Guth, and Alexander Vilenkin, "Inflationary Spacetimes Are Incomplete in Past Directions," *Physical Review Letters* 90, no. 15 (April 18, 2003): 151301.

27. Steve Bradt, "3 Questions: Alan Guth on New Insights into the 'Big Bang,'"*MIT News*, March 20, 2014, http://news.mit.edu/2014/3-q-alan-guth-on-new-insights-into-the-big-bang.

28. Guth, "Eternal Inflation and Its Implications."

29. Alex Vilenkin, *Many Worlds in One: The Search for Other Universes* (New York: Hill and Wang, 2007).

30. Borde, Guth, and Vilenkin, "Inflationary Spacetimes Are Incomplete in Past Directions."

31. Vilenkin, *Many Worlds in One*.

32. Leonard Susskind, "Was There a Beginning?" preprint, April 24, 2012, http://arxiv.org/pdf/1204.5385.pdf.

33. Anthony Aguirre, "Eternal Inflation, Past and Future," preprint, December 4, 2007, http://arxiv.org/abs/0712.0571.

34. William Lane Craig and Sean Carroll, "God and Cosmology: The Existence of God in Light of Contemporary Cosmology," (debate, New Orleans Baptist Theological Seminary, New Orleans, LA, March 2014); transcript available at http://www.reasonablefaith.org/god-and-cosmology-the-existence-of-god-in-light-of-contemporary-cosmology.

35. Anthony Aguirre and John Kehayias, "Quantum Instability of the Emergent Universe," *Physical Review D* 88, no. 10 (November 15, 2013): 103504.

36. Aguirre, "Eternal Inflation, Past and Future."

37. Chris Wetterich, "Eternal Universe," *Physical Review* D 90, no. 4 (August 15, 2014): 043520.

38. Yasunori Nomura, "Static Quantum Multiverse," *Physical Review D* 86, no.8 (October 15, 2012): 083505.

39. Sean Carroll, "What If Time Really Exists?" Foundational Questions Institute, November 2008, fqxi.org/data/essay-contest-files/Carroll_fqxitimecontest.pdf.

40. Craig and Carroll, "God and Cosmology."

41. William Lane Craig, "'Honesty, Transparency, Full Disclosure' and the Borde-Guth-Vilenkin Theorem," Q&A with William Lane Craig, September 23, 2013, Reasonable Faith, www.reasonablefaith.org/honesty-transparency-full-disclosure-and-bgv-theorem.

42. Alan Guth, "Did the Universe Have a Beginning? Eternal Inflation: Implications," *Counterbalance*, www.counterbalance.org/cq-guth/etern1-frame.html.

43. Craig, "Ultimate Question of Origins."

44. Guth, "Eternal Inflation and Its Implication."

45. Craig, "Cosmos and Creator."

46. William Lane Craig and Peter Milican, "Does God Exist?" (debate, University of Birmingham, Birmingham, UK, 2014);transcript available at www.reasonablefaith.org/transcript/does-god-exist-craig-vs-millican.

47. Roger A. Freedman, Robert M. Geller, and William J. Kaufmann III, *Universe*, ninth edition (New York: W. H. Freeman, 2010).

48. Lawrence M. Krauss and Michael S. Turner, "The Cosmological Constant Is Back," *General Relativity and Gravitation* 27, no. 11 (November 1995): 1137–44.

49. National Aeronautics and Space Administration (NASA), "WMAP Produces New Results," press release, http://map.gsfc.nasa.gov/news/7yr_release.html. Note, this is an updated version; earlier versions had already found this result: "Inflation was an amazing concept when it was first proposed 25 years ago, and now we can support it with real data" (NASA, "NASA Satellite Glimpses Universe's First Trillionth of a Second," news release, March 16, 2006, http://www.nasa.gov/home/hqnews/2006/mar/HQ_06097_first_trillionth_WMAP.html).

50. Eric Hand, "Cosmology: The Test of Inflation," *Nature* 458, no. 7240 (April 16, 2009): 820–24.

51. Craig and Law, "Does God Exist?"

52. Abhay Ashtekar and David Sloan, "Probability of Inflation in Loop Quantum Cosmology," *General Relativity and Gravitation* 43, no. 12 (December 2011): 3619–55.

53. Roger Penrose, interview, "Before the Big Bang 2: Conformal Cyclic Cosmology Explained," YouTube video, 39:04, posted by skydivephil, January 24, 2014, https://www.youtube.com/watch?v=sM47acQ7pEQ.

54. Guth, *The Inflationary Universe*.

55. João Magueijo, "New Varying Speed of Light Theories," *Reports on Progress in Physics* 66, no. 11 (November 2003): 2025.

56. João Magueijo, *Faster than the Speed of Light: The Story of a Scientific Speculation* (London: Arrow Books, 2004).

57. J. K. Webb et al. "Indications of a Spatial Variation of the Fine Structure Constant," *Physical Review Letters* 107, no. 19 (November 4, 2011): 191101.

58. Itzhak Bars, Paul J. Steinhardt, and Neil Turok, "Cyclic Cosmology, Conformal Symmetry and the Metastability of the Higgs," *Physics Letters B* 726, no. 1–3 (October 7, 2013): 50–55.

59. Saswato R. Das, "How the Higgs Boson Might Spell Doom for the Universe,"

Scientific American, March 26, 2013, http://www.scientificamerican.com/article/how-the -higgs-boson-might-spell-doom-for-the-universe/.

60. Roger Penrose, "Before the Big Bang: An Outrageous New Perspective and Its Implications for Particle Physics," (paper presented at EPAC 2006, Edinburgh, Scotland, June 2006); available at http://epaper.kek.jp/e06/PAPERS/THESPA01.PDF.

61. Penrose, interview, "Before the Big Bang 2."

62. V. G. Gurzadyan and R. Penrose, "On CCC-Predicted Concentric Low-Variance Circles in the CMB Sky," *European Physical Journal Plus* 128, no. 2 (February 2013): 22.

63. I. K. Wehus and H. K. Eriksen, "A Search for Concentric Circles in the 7 Year Wilkinson Microwave Anisotropy Probe Temperature Sky Maps," *The Astrophysical Journal Letters* 733, no. 2 (May 9, 2011): L29.

64. Krzysztof A. Meissner, Paweł Nurowski, and Błażej Ruszczycki, "Structures in the Cosmic Microwave Background," *Proceedings of the Royal Society A* 469, no. 2155 (July 8, 2013).

65. William Lane Craig and Kevin Harris, "Truth Free Will, and Cosmology," Reasonable Faith Podcast, August 30, 2012, www.reasonablefaith.org/truth-free-will-and-cosmology.

66. Penrose, interviews, "Before the Big Bang 2."

67. Fabio Finelli and Robert Brandenberger, "Generation of a Scale-Invariant Spectrum of Adiabatic Fluctuations in Cosmological Models with a Contracting Phase," *Physical Review D* 65, no. 10 (May 15, 2002): 103523.

68. Nikodem J. Popławski, "Cosmology with Torsion: An Alternative to Cosmic Inflation," *Physics Letters B* 694, no. 3 (November 8, 2010): 181–85.

69. Lee Smolin, "The Status of Cosmological Natural Selection," preprint, December 18, 2006, http://arxiv.org/abs/hep-th/0612185.

70. Alexander Vilenkin, "Creation of Universes from Nothing," *Physics Letters B* 117, no. 1–2 (November 4, 1982): 25–28.

71. Edward P. Tryon, "Is the Universe a Vacuum Fluctuation?" *Nature* 246, no. 5433 (December 14, 1973): 396–97.

72. J. B. Hartle and S. W. Hawking, "Wave Function of the Universe," *Physical Review D* 28, no. 12 (December 15, 1983): 2960.

73. James Hartle and Thomas Hertog, "Quantum Transitions between Classical Histories," *Physical Review D* 92, no. 6 (September 15, 2015): 063509.

74. Ahmed Farag Ali and Saurya Das, "Cosmology from Quantum Potential," *Physics Letters B* 741 (February 4, 2015): 276–79.

75. Aron C. Wall, "The Generalized Second Law Implies a Quantum Singularity Theorem," *Classical and Quantum Gravity* 30, no. 16 (August 21, 2013): 165003.

76. Craig and Carroll, "God and Cosmology."

77. Aron Wall, February 25, 2014 (10:44 p.m.), comment on Sean Carroll, "Post Debate Reflections," Sean Carroll (blog), February 24, 2014, www.preposterousuniverse.com/blog/2014/02/24/post-debate-reflections/.

78. Abhay Ashtekar, "Singularity Resolution in Loop Quantum Cosmology: A Brief Overview," *Journal of Physics: Conference Series* 189 (2009): 012003.

79. Abhay Ashtekar, "Loop Quantum Cosmology and the Very Early Universe Interplay between Theory and Observations," Perimeter Institute Recorded Seminar Archive (PIRSA),

1:19:27 (lecture, Perimeter Institute for Theoretical Physics, March 18, 2013), http://pirsa.org/displayFlash.php?id=15010090.

80. Ashtekar and Sloan, "Probability of Inflation in Loop Quantum Cosmology."

81. Ivan Agullo, Abhay Ashtekar, and William Nelson, "The Pre-Inflationary Dynamics of Loop Quantum Cosmology: Confronting Quantum Gravity with Observations," *Classical and Quantum Gravity* 30, no. 8 (April 21, 2013): 085014.

82. Craig, "Ultimate Question of Origins."

83. William Lane Craig and J. P. Moreland, eds., *The Blackwell Companion to Natural Theology* (Chichester, UK: Wiley-Blackwell, 2012).

84. Craig and Milican, "Does God Exist?"

85. Ivan Agullo and Noah A. Morris, "Detailed Analysis of the Predictions of Loop Quantum Cosmology for the Primordial Power Spectra," preprint, September 18, 2015, http://arxiv.org/pdf/1509.05693v1.pdf.

86. Aurélien Barrau, Carlo Rovelli, and Francesca Vidotto, "Fast Radio Bursts and White Hole Signals," *Physical Review D* 90, no. 12 (December 15, 2014): 127503.

87. M. Gasperini and G. Veneziano, "The Pre-Big Bang Scenario in String Cosmology," *Physics Reports* 373, no. 1–2 (January 2003): 1–212.

88. M. Gasperini and G. Veneziano, "String Theory and Pre-Big Bang Cosmology," preprint, March 6, 2007, http://arxiv.org/pdf/hep-th/0703055v1.pdf.

89. Craig and Moreland, *Blackwell Companion to Natural Theology*.

90. Gabriele Veneziano, "The Myth of the Beginning of Time," *Scientific American*, February 1, 2006.

91. Justin Khoury et al. "The Ekpyrotic Universe: Colliding Branes and the Origin of the Hot Big Bang," *Physical Review D* 64, no. 12 (December 15, 2001): 123522.

92. Craig and Moreland, *Blackwell Companion to Natural Theology*.

93. R. Brandenberger and C. Vafa, "Superstrings in the Early Universe," *Nuclear Physics B* 316, no. 2 (April 10, 1989): 391–410.

94. B. A. Bassett et al. "Aspects of String-Gas Cosmology at Finite Temperature," *Physical Review D* 67, no. 12 (June 15, 2003): 123506.

95. Ali Nayeri, Robert H. Brandenberger, and Cumrun Vafa, "Producing a Scale-Invariant Spectrum of Perturbations in a Hagedorn Phase of String Cosmology," *Physical Review Letters* 97, no. 2 (July 14, 2006): 021302.

96. David H. Lyth and David Wands, "Generating the Curvature Perturbation without an Inflaton," *Physics Letters B* 524, no. 1–2 (January 3, 2002): 5–14.

97. Brian A. Powell, "Tensor Tilt from Primordial B Modes," *Monthly Notices of the Royal Astronomical Society* 419, no. 1 (January 1, 2012): 566–72.

98. Gasperini and Veneziano, "String Theory and Pre-Big Bang Cosmology." preprint, March 6, 2007, http://arxiv.org/pdf/hep-th/0703055v1.pdf.

99. Brandenberger and Vafa, "Superstrings in the Early Universe."

100. Petr Hořava, "Quantum Gravity at a Lifshitz Point," *Physical Review D* 79, no. 8 (April 15, 2009): 084008.

101. Robert Brandenberger, "Matter Bounce in Hořava-Lifshitz Cosmology," *Physical Review D* 80, no. 4 (August 15, 2009): 043516.

102. Alan Guth, "The Universe Began in a State of Extraordinarily Low Entropy," Annual Question 2014: What Scientific Idea Is Ready for Retirement? *Edge*, http://edge .org/response-detail/25538.

103. Sean M. Carroll and Jennifer Chen, "Spontaneous Inflation and the Origin of the Arrow of Time," preprint, October 27, 2004, http://arxiv.org/abs/hep-th/0410270.

104. Abhay Ashtekar, interview, "Before the Big Bang 1: Loop Quantum Cosmology Explained," YouTube video, 43:26, posted by skydivephil, August 1, 2013, https://www .youtube.com/watch?v=IFcQuEw0oY8.

105. T. Padmanabhan, "A Dialogue on the Nature of Gravity," preprint, November 22, 2009, http://arxiv.org/abs/0910.0839.

106. Penrose, "Before the Big Bang: An Outrageous New Perspective."

107. Neil Turok, Q&A at "The Mathematics of CCC: Mathematical Physics with Positive Lambda," conference organized by the Clay Mathematics Institute, University of Oxford, September 11–13, 2013.

108. Julian Barbour, Tim Koslowski, and Flavio Mercati, "Identification of a Gravitational Arrow of Time," *Physical Review Letters* 113, no. 18 (October 31, 2014): 181101.

109. Guth, "Universe Began in a State of Extraordinarily Low Entropy."

110. Craig and Carroll, "God and Cosmology."

111. Ibid.

112. Craig and Moreland, *Blackwell Companion to Natural Theology*.

113. Craig, "Cosmos and Creator."

114. Guth, *Inflationary Universe*.

115. John D. Barrow, "Does Infinity Exist?" *Plus*, July 2, 2012, https://plus.maths.org/ content/does-infinity-exist.

116. Craig and Moreland, *Blackwell Companion to Natural Theology*.

117. Paul Benacerraf and Hilary Putnam, eds., *Philosophy of Mathematics: Selected Readings*, second edition (Cambridge, UK: Cambridge University Press, 1984).

118. Planck Collaboration et al., "*Planck* 2013 Results: XVI: Cosmological Parameters," *Astronomy & Astrophysics* 571 (November 2014): A16.

119. Guth, "Eternal Inflation and Its Implications."

120. Penrose, interview, "Before the Big Bang 2."

121. William Lane Craig, "The Triumph of Lorentz," Q&A with William Lane Craig, October 3, 2011, Reasonable Faith, www.reasonablefaith.org/the-triumph-of-lorentz.

122. "The PhilPapers Surveys," PhilPapers, November 2009, http://philpapers.org/surveys/ results.pl.

123. Yuri Balashov and Michel Janssen, "Presentism and Relativity," *British Journal for the Philosophy of Science* 54, no. 2 (June 2003): 327–46.

124. Craig, "Triumph of Lorentz."

125. Eugenie Samuel Reich, "Flaws Found in Faster-than-Light Neutrino Measurement," *Nature News*, February 22, 2012, www.nature.com/news/flaws-found-in-faster-than -light-neutrino-measurement-1.10099.

126. Vilenkin, *Many Worlds in One*.

127. "Doctrinal Statement," Biola University, www.biola.edu/about/doctrinal-statement.

128. Halvorson and Kragh, "Theism and Physical Cosmology."

129. John Barrow, *The Book of Nothing: Vacuums, Voids, and the Latest Ideas about the Origins of the Universe* (New York: Vintage Books, 2002).

130. Sean Carroll, "Guest Post: Don Page on God and Cosmology," Sean Carroll (blog), March 20, 2015, http://www.preposterousuniverse.com/blog/2015/03/20/guest-post-don-page -on-god-and-cosmology/.

131. William Lane Craig and Kevin Harris, "Science, Philosophy, and the Sean Carroll Debate," Reasonable Faith Podcast, August 9, 2015, http://www.reasonablefaith.org/ science-philosophy-and-the-sean-carroll-Debate#ixzz3tp08OwNa.

132. Craig and Moreland, *Blackwell Companion to Natural Theology*.

133. J. Richard Gott III and Li-Xin Li, "Can the Universe Create Itself?" *Physical Review D* 58, no. 2 (July 15, 1998): 023501.

134. William Lane Craig, "Excursus on Natural Theology: Existence of God," section two, part six of *Defenders Podcast: Series 2* (transcript), www.reasonablefaith.org/ defenders-2-podcast/transcript/s4-6.

135. Vilenkin, "Creation of Universes from Nothing."

136. Vilenkin, *Many Worlds in One*.

137. Brad Lemley and Larry Fink, "Guth's Grand Guess," *Discover*, April 1, 2002, http:// discovermagazine.com/2002/apr/cover/.

138. Antonio Damasio, *Self Comes to Mind: Constructing the Conscious Brain* (New York: Vintage Books, 2010).

139. Craig and Law, "Does God Exist?"

CHAPTER 6: INTELLIGENT DESIGN ISN'T SCIENCE

1. The *Wedge Strategy* was originally produced by the Discovery Institute in 1998. It was then leaked to the internet. The entire document's text and a scanned image of the original can be found online at the website of the National Center for Science Education (NCSE): http://ncse.com/ creationism/general/wedge-document (posted October 14, 2008; accessed November 29, 2015).

2. Kerwin Lee Klein, *From History to Theory* (Berkeley, CA: University of California Press, 2001), p. 153.

3. Edwards v. Aguillard, 482 U.S. 578 (1987).

4. National Center for Science Education (NCSE), September 25, 2008, http://ncse.com/ creationism/legal/cdesign-proponentsists (accessed November 29, 2015).

5. Editorial. "'Strengths and Weaknesses'—Will the Texas Board of Education Evolve Backward?" (editorial), *Washington Post*, March 25, 2009; James Gill, "Louisiana's Science Education Act Lacking in Science," *Times Picayune*, June 1, 2011; "'Monkey' Business, Again, in Tennessee" (editorial), *Los Angeles Times*, April 10, 2012; and Hedy Weinberg, "Gov. Bill Haslam Should Veto 'Monkey Bill,'" *Knoxville News Sentinel*, April 5, 2012.

6. Missouri House Bill 1472, 2014, http://www.house.mo.gov/billsummary.aspx?bill=H B1472&year=2014&code=R (accessed November 29, 2015).

7. The National Center for Science Education (NCSE) has archived news about antiscience legislation in past years; see especially information on Ohio House Bill 597 (2014), http://ncse.com/news/ohio (accessed November 29, 2015).

8. NCSE News, Oklahoma; see especially information on Oklahoma House Bill 1674 (2014), http://ncse.com/news/oklahoma (accessed November 29, 2015).

9. NCSE News, South Dakota; see especially information on South Dakota Senate Bill 112 (2014), http://ncse.com/news/south-dakota (accessed November 29, 2015).

10. NCSE News, Virginia; see especially information on Virginia House Bill 207 (2014), http://ncse.com/news/virginia (accessed November 29, 2015).

11. "Continuing Concern Over Antievolutionism in Turkey," NCSE, July 10, 2013, http://ncse.com/news/2013/07/continuing-concern-over-antievolutionism-turkey-0014890 (accessed November 29, 2015); "Creationist Legislation in Brazil," NCSE, November 21, 2014, http://ncse.com/news/2014/11/creationist-legislation-brazil-0016007 (accessed November 29, 2015).

12. Kitzmiller v. Dover Area School District, 400 F. Supp. 2d 707 (M.D. Pa. 2005).

13. Barbara Forrest and Paul R. Gross, *Creationism's Trojan Horse: The Wedge of Intelligent Design* (Oxford, UK: Oxford University Press, 2007).

14. Jerry Jinks, "The Science Processes," Illinois State University, 1997, http://my.ilstu.edu/~jdpeter/THE%20SCIENCE%20PROCESSES.htm (accessed November 29, 2015).

15. "A Timeline of HIV/AIDS," AIDS.gov, 2014, https://www.aids.gov/hiv-aids-basics/hiv-aids-101/aids-timeline/ (accessed November 29, 2015).

16. Pride Chigwedere, George R. Seage III, Sofia Gruskin, Tun-Hou Lee, and M. Essex, "Estimating the Lost Benefits of Antiretroviral Drug Use in South Africa," *Journal of Acquired Immune Deficiency Syndromes* 49, no. 4 (December 1, 2008): 410–15.

17. William Paley, *Natural Theology: Or, Evidence of the Existence and Attributes of the Deity, Collected from the Appearances of Nature* (Philadelphia: John Morgan, 1802).

18. Charles B. Thaxton, Walter L. Bradley, and Roger L. Olsen, *The Mystery of Life's Origin: Reassessing Current Theories* (New York: Philosophical Library, 1984).

19. Percival Davis and Dean H. Kenyon, *Of Pandas and People: The Central Question of Biological Origins* (Richardson, TX: Foundation for Thought and Ethics, 1989).

20. *Understanding Science: How Science Really Works*, University of California at Berkeley, 2014, http://undsci.berkeley.edu/ (accessed November 29, 2015).

21. William A. Dembski, "Is Intelligent Design Testable? A Response to Eugenie Scott," Discovery Institute: Center for Science and Culture, 2001, http://www.discovery.org/a/584 (accessed November 29, 2015).

22. Mary L. McHugh, "The Chi-Square Test of Independence," *Biochemia Medica* 23, no. 2 (June 2013): 143–49.

23. "Creationist Legislation in Brazil."

24. "Continuing Concern Over Antievolutionism in Turkey."

25. Abby Hafer, "No Data Required: Why Intelligent Design Is Not Science," *The American Biology Teacher* 77, no. 7 (September 2015): 507–13.

26. Anna Clark, "The Texas School Board Isn't as Powerful as You Think," *Columbia Journalism Review*, December 1, 2014.

CHAPTER 7: SAYING SAYONARA TO SIN

* The material in this chapter was adapted from Robert M. Price and Edwin A. Suominen, *Evolving out of Eden: Christian Responses to Evolution* (Valley, WA: Tellectual, 2013).

1. Ken Ham, "Beware of Those Who Want the Church to Compromise," Ken Ham: Blog, Answers in Genesis, July 28, 2012, https://answersingenesis.org/blogs/ken-ham/2012/07/28/beware-of-those-who-want-the-church-to-compromise/ (accessed August 2012).

2. Spencer Wells (*Deep Ancestry: The Landmark DNA Quest to Decipher our Distant Past* [Washington, DC: National Geographic Society, 2006], Kindle edition) puts the date at 60,000 years ago, while David Wilcox ("Finding Adam: The Genetics of Human Origins," in *Perspectives on an Evolving Creation*, ed. Keith B. Miller [Grand Rapids, MI: W. B. Eerdmans, 2003], pp. 234–52) would have it a bit more recent, at 50,000 years ago. The anthropologist John Hawks says that most "estimates put it within the last 70,000 years," which he thinks is too young ("Mailbag: Y Chromosome Adam," *John Hawks Weblog*, February 17, 2011, johnhawks.net/weblog/mailbag/y-chromosome-adam-2011.html [accessed July 2012]).

3. The fact that all living men share variants of his Y chromosome is simply the result of genetic drift, where one gene (or chromosome) winds up doing a bit better than its rivals due to nothing more than sampling error, and eventually comes to dominate (Wells, *Deep Ancestry*).

4. Wells, *Deep Ancestry*. Paul Ehrlich (*Human Natures: Genes, Cultures, and the Human Prospect* [Washington: Island, 2000]) notes a distinction between mitochondrial ancestry and "the rest of our genetic endowment," the DNA in the nucleus (p. 99). Women contribute as much to the latter, much more extensive type of DNA, as men—actually more, since the X chromosome is bigger than the Y. But it's not as easy to trace ancestry with the chromosomes other than the Y because their formation includes crossover between each parent's paternal and maternal lines. The Y chromosome and mitochondrial DNA, however, stand alone, changing only by mutation. "All modern human beings do not share just a single female ancestor" (Ehrlich, p. 99), or a male one either. Genetic Adam's rivals contributed lots of DNA to our legacy, even if their Y chromosomes didn't make the cut.

5. Brian Fagan, *Cro-Magnon: How the Ice Age Gave Birth to the First Modern Humans* (New York: Bloomsbury, 2011), Kindle edition.

6. Naama Goren-Inbar et al., "Evidence of Hominin Control of Fire at Gesher Benot Ya'aqov, Israel," *Science*, n.s., 304, no. 5671 (April 30, 2004): 725–27.

7. Jerry Coyne, "How Big Was the Human Population Bottleneck? Another Staple of Theology Refuted" *Why Evolution Is True*, September 18, 2011, https://whyevolutionistrue.wordpress.com/2011/09/18/how-big-was-the-human-population-bottleneck-not-anything-close-to-2/.

No *Homo sapiens* fitting the time or location for the biblical Adam can be imagined, either. "The Genetic case for an African origin for *Homo sapiens* seems overwhelming. The archaeologists have also stepped forward with new fossil discoveries, including a robust 195,000-year-old modern human from Omo Kibish, in Ethiopia, and three 160,000-year-old *Homo sapiens* skulls from Herto, also in Ethiopia. Few anthropologists now doubt that Africa was the cradle of *Homo sapiens* and home to the remotest ancestors of the first modern Europeans—the Cro-Magnons. The seemingly outrageous chronology of two decades ago is now accepted as historical reality" (Fagan, *Cro-Magnon*).

8. H. Wade Seaford, quoted in John W. Loftus, *Why I Became an Atheist: A Former Preacher Rejects Christianity* (Amherst, NY: Prometheus Books, 2008), Kindle edition.

9. Pelagius (ca. 360–418) was a dour monk who lived in Britain. He debated Augustine of Hippo, a bishop in North Africa, over the topic of Original Sin and the nature of God's saving grace. His views were condemned as heresy as, in effect, teaching a doctrine of self-salvation.

10. Augustine, "A Treatise on the Merits and Forgiveness of Sins, and on the Baptism of Infants," book 3, chap. 19, in *The Collected Works of 46 Books by St. Augustine*, ed. Philip Schaff (1887; reprint ed., Amazon Digital Services), Kindle edition.

11. Albert Mohler Jr., "False Start? The Controversy Over Adam and Eve Heats Up," Albert Mohler (blog), August 22, 2011, www.albertmohler.com/2011/08/22/false-start-the -controversy-over-adam-and-eve-heats-up/ (accessed May 2012).

12. Martin Luther, *Lectures on Genesis*, vol. 1, trans. George V. Schick (Saint Louis, MO: Concordia Publishing House, 1958).

13. Peter Enns, *The Evolution of Adam* (Grand Rapids, MI: Brazos, 2012).

14. Ansfridus Hulsbosch, *God in Creation and Evolution* (Lanham, MD: Sheed and Ward, 1966).

15. Moral Rearmament began in 1938 as an interdenominational revival movement based on small group meetings in which people freely and tearfully confessed their sins in everyone's hearing. It is also known as "Buchmanism" after the minister who started it, Frank N. D. Buchman. It is also called the Oxford Group, the original name given the program in the 1920s as a part of Buchman's ministry. It morphed into "Moral Rearmament" in the hope of reinforcing the integrity and moral courage of Europeans and Americans about to enter World War Two.

16. Jerry Korsmeyer, *Evolution and Eden: Balancing Original Sin and Contemporary Science* (Mahwah, NJ: Paulist, 1998), p. 57.

17. C. John Collins, *Did Adam and Eve Really Exist?* (Wheaton, IL: Crossway, 2011), pp. 25–26.

18. Ibid., p. 61.

19. Denis O. Lamoureux, *Evolutionary Creation: A Christian Approach to Evolution* (Eugene, OR: Wipf & Stock, 2008), pp, 291–92.

20. Robin Collins, "Evolution and Original Sin," in *Perspectives on an Evolving Creation*, ed. Keith B. Miller (Grand Rapids, MI: W.B. Eerdmans, 2003), p. 495.

21. Rudolf Otto, *The Idea of the Holy: An Inquiry into the Non-rational Factor in the Idea of the Divine and its Relation to the Rational*, trans. John W. Harvey (London: Oxford University Press, 1924).

22. Karl Jaspers, *The Origin and Goal of History*, trans. Michael Bullock (New Haven, CT: Yale University Press, 1953).

23. "Satan" was at first a title, not a proper name, and it belonged to one of the sons of God (Job 1:6) who served as God's special assistant to monitor and test the mettle of God's favorites, like Job (Job 1:9–11), David (1 Chron. 21:1), and Joshua, the favored candidate for the high priesthood (Zech. 3:1). Isaiah 14 refers to the humiliation of the minor deity Halal (the planet Venus), while Ezekiel describes the expulsion of Adam from the cherub-guarded Eden; neither has to do with a fallen angel named Lucifer or Satan. Nor was the Serpent in Eden identified with Satan until much, much later, in apocryphal books like *The Life of Adam and Eve* and *The Secrets of Enoch*.

24. Sam K. Williams, *Jesus' Death as Saving Event: The Background and Origin of a Concept*, Harvard Dissertations in Religion 2 (Missoula, MT: Scholars, 1975).

25. Korsmeyer, *Evolution and Eden*, p. 22.

26. Tatha Wiley, *Original Sin: Origins, Developments, Contemporary Meanings* (Mahwah, NJ: Paulist, 2002), Kindle edition.

27. Luther, *Lectures on Genesis*, ch. 3, v. 16.

Eve seems to be forgotten in this discussion, though Luther certainly doesn't leave her out. He writes about the "curse" of childbearing in a way that glorifies it as an honor and blessing of sorts, while still acknowledging its discomforts and danger. Despite being somewhat ahead of his time in the way he treated his wife, Katherine, Luther shows no shortage of sexism in his ultimate conclusion about Eve's fate after the Fall: If she "had persisted in the truth, she would not only not have been subjected to the rule of her husband, but she herself would also have been a partner in the rule which is now entirely the concern of males." Women, he claims, "are generally disinclined to put up with this burden, and they naturally seek to gain what they have lost through sin. If they are unable to do more, they at least indicate their impatience by grumbling. However, they cannot perform the functions of men, teach, rule, etc." Only in "procreation and in feeding and nurturing their offspring they are masters."

28. Ibid.

29. Martin Luther, *The Bondage of the Will*, trans. Henry Cole, 1823, §152.

30. Wiley, *Original Sin*.

31. Richard Dawkins, *The Selfish Gene: 30th Anniversary Edition with a New Introduction by the Author* (New York: Oxford University Press, 2006), p. 45.

32. Ibid., p. 196 (emphasis ours).

33. Daniel J. Fairbanks, *Relics of Eden: The Powerful Evidence of Evolution in Human DNA* (Amherst, NY: Prometheus Books, 2007), p. 111.

34. Russell Kolts, *The Compassionate Mind Approach to Managing Your Anger* (London: Robinson, 2011), pp. 28–29.

35. Ibid.

36. Ibid.

37. Collins, "Evolution and Original Sin," p. 495.

38. Ibid.

39. Daryl P. Domning, "Evolution, Evil and Original Sin," *America: The National Catholic Weekly*, November 12, 2001, www.americamagazine.org/issue/350/article/evolution-evil-and-original-sin (accessed June 2012).

40. Ibid.

41. Ibid.

CHAPTER 8: THE SOUL FALLACY

1. Mark Baker and Stewart Goetz, eds. *The Soul Hypothesis: Investigations into the Existence of the Soul* (New York: Continuum, 2011), p. 1.

2. Francis Crick, *The Astonishing Hypothesis: The Scientific Search for the Soul* (New York: Charles Scribner's Sons, 1994).

3. Joshua D. Greene, "Social Neuroscience and the Soul's Last Stand," in *Social Neuroscience: Toward Understanding the Underpinning of the Social Mind*, ed. A. Todorov, S. Fiske, and D. Prentice (New York: Oxford University Press, 2011).

4. Owen Flanagan, *The Problem of the Soul: Two Visions of the Mind and How to Reconcile Them* (New York: Basic Books, 2002).

5. Jan N. Bremmer, *The Early Greek Concept of the Soul* (Princeton: Princeton University Press, 1983). See also work by the religious historian Ernst Arbman (1891–1917), which reveals that beliefs about free souls and body souls were widespread among native North Americans and the peoples of Northern Asia, Northern Europe, and India.

6. Arthur C. Clarke, "Hazards of Prophecy: The Failure of Imagination," chapter 2 in *Profiles of the Future: An Enquiry into the Limits of the Possible* (New York: Harper & Row, 1962).

7. Jerome W. Elbert, *Are Souls Real?* (Amherst, NY: Prometheus Books, 2000), p. 36.

8. René Descartes, *The World and Other Writings*, ed. and trans. S. Gaukroger (Cambridge: Cambridge University Press, 1998), p. 169.

9. Larry Shannon-Missal, "Americans' Belief in God, Miracles and Heaven Declines," Harris Poll, December 16, 2013, http://www.theharrispoll.com/health-and-life/Americans_Belief _in_God__Miracles_and_Heaven_Declines.html (accessed April 13, 2016).

10. "Religious Beliefs and Practices," chapter 1 in *U.S. Religious Landscape Survey: Religious Beliefs and Practices*, June 1, 2008, http://www.pewforum.org/2008/06/01/chapter-1 -religious-beliefs-and-practices/ (accessed May 22, 2014).

11. Albert L. Winseman, "Eternal Destinations: Americans Believe in Heaven, Hell" Gallup, May 25, 2004, http://www.gallup.com/poll/11770/Eternal-Destinations-Americans -Believe-Heaven-Hell.aspx (accessed May 22, 2014).

12. "Americans Describe Their Views about Life after Death," Barna Group, October 21, 2003, https://www.barna.org/component/content/article/5-barna-update/45-barna-update-sp-657/128 -americans-describe-their-views-about-life-after-death (accessed April 13, 2016).

13. Pippa Norris and Ronald Inglehart, *Sacred and Secular: Religion and Politics Worldwide* (New York: Cambridge University Press, 2004).

14. Bertrand Russell, *Religion and Science* (Oxford: Oxford University Press, 1997), p. 137.

15. Steven Pinker, "The Untenability of Faithism," *Current Biology* 25, no. 15 (August 3, 2015): R639.

16. Baker and Goetz, *Soul Hypothesis*, p. 101.

17. Paul M. Churchland, *Matter and Consciousness* (Cambridge, MA: MIT Press, 2013), pp. 24–25.

18. Julien Musolino, *The Soul Fallacy: What Science Shows We Gain from Letting Go of Our Soul Beliefs* (Amherst, NY: Prometheus Books, 2015), pp. 111–12.

19. Dean Mobbs and Caroline Watts, "There Is Nothing Paranormal about Near-Death-Experiences: How Neuroscience Can Explain Seeing Bright Lights, Meeting the Dead, or Being Convinced You Are One of Them," *Trends in Cognitive Sciences* 15, no. 10 (2011): 447–49.

20. Ibid.

21. Jean Bricmont, "Qu'est-ce que le materialisme scientifique?" *Dogma*, http://www .dogma.lu/txt/JB-MatSc.htm (accessed May 27, 2014).

22. Sean Carroll, "Physics and the Immortality of the Soul," *Scientific American*, guest blog, May 23, 2011, http://blogs.scientificamerican.com/guest-blog/physics-and-the-immortality-of -the-soul/ (accessed December 09, 2015).

23. Paul M. Churchland, *Matter and Consciousness* (Cambridge, MA: MIT Press, 2013), p. 35.

24. Michel Desmurget et al., "Movement Intention after Parietal Cortex Stimulation in Humans," *Science* 324, no. 5928 (2009): 811–13.

25. J. V. Haxby, "Distributed and Overlapping Representations of Faces and Objects in Ventral Temporal Cortex," *Science* 293 (2009): 2425–30.

26. For a critical review of D'Souza's claims, see Victor J. Stenger, "Life After Death: Examining the Evidence," in *The End of Christianity*, ed. John W. Loftus (Amherst, NY: Prometheus Books, 2011).

27. Dinesh D'Souza, *Life After Death: The Evidence* (Washington, DC: Regnery, 2009), p. 139.

28. Robert Putman and David Campbell, *American Grace: How Religion Divides and Unites Us* (New York: Simon & Schuster, 2012). See also Paul Bloom, "Religion, Morality, and Evolution," *Annual Review of Psychology* 63 (2012): 179–99.

29. Phil Zuckerman, *Society Without God: What the Least Religious Nations Can Tell us about Contentment* (New York: New York University Press, 2008).

30. Michael Shermer, blurb on back cover of *The Soul Fallacy*. See also John W. Loftus, "Thou Shalt Not Suffer a Witch to Live: The Wicked Christian Witch Hunts," in *Christianity Is Not Great*, ed. John W. Loftus (Amherst, NY: Prometheus Books, 2014).

31. Lydia Saad, "Public Opinion about Abortion—An In-Depth Review," Gallup, January 22, 2002, http://www.gallup.com/poll/9904/Public-Opinion-About-Abortion-InDepth-Review. aspx (accessed June 1, 2014).

32. "God Alone Has the Right to Initiate and Terminate Life: Answers," *Christian Life Resources*, http://www.christianliferesources.com/article/god-alone-has-the-right-to-initiate-and -terminate-life-answers-715 (accessed June 01, 2014). See also Ronald A. Lindsay, "The Christian Abuse of the Sanctity of Life," in *Christianity Is Not Great*, ed. John W. Loftus (Amherst, NY: Prometheus Books, 2014).

33. Michael Tonry, *Thinking about Crime: Sense and Sensibility in American Penal Culture* (Oxford: Oxford University Press, 2004), Kindle edition.

34. Joshua Greene and Jonathan Cohen, "For the Law, Neuroscience Changes Nothing and Everything," *Philosophical Transactions of the Royal Society of London B* 359, no. 1451 (November 29, 2004): 1775.

35. Ibid.

CHAPTER 9: FREE WILL

1. This is actually a hotly debated area of experimental research, with some researchers claiming there is a folk intuition of compatibilism, but perhaps not the philosophical type to be explained later. Others complain that, in many respects, the jury is still out. See Eddy Nahmias et al., "Surveying Freedom: Folk Intuitions about Free Will and Moral Responsibility," *Philosophical Psychology* 18, no. 5 (2005): 561–84; and Oisín Deery, Taylor Davis, and Jasmine Carey, "The Free-Will Intuitions Scale and the Question of Natural Compatibilism," *Philosophical Psychology* 28, no. 6 (2015): 776–801.

2. Saul Smilansky, "Free Will, Fundamental Dualism, and the Centrality of Illusion," chapter 22, in *The Oxford Handbook of Free Will*, ed. Robert Kane (Oxford: Oxford University Press, 2003), pp. 489–90.

3. The Kalam Cosmological Argument is usually stated as follows: Everything that begins to exist has a cause for its existence. The universe began to exist. Therefore, the universe has a cause for its existence.

God is argued as being the only entity that can be an uncaused causer (here, for the universe). However, theists seem to argue that humans can be prime movers in causal chains involving human free choice. Both positions are mutually exclusive, then.

4. David Eagleman, *Incognito: The Secret Lives of the Brain* (New York: Vintage Books, 2012).

5. Adrian Raine, *The Anatomy of Violence: The Biological Roots of Crime* (New York: Vintage Books, 2014).

6. Norenzayan A, Gervais WM, Trzesniewski KH, 2012. Mentalizing Deficits Constrain Belief in a Personal God. *PLoS ONE* 7(5): e36880. doi:10.1371/journal.pone.0036880.

7. Nicholas G. Shakeshaft et al., "Strong Genetic Influence on a UK Nationwide Test of Educational Achievement at the End of Compulsory Education at Age 16," *PLoS ONE* 8, no. 12 (December 2013): e80341.

8. Dean H. Hamer, *The God Gene: How Faith Is Hardwired Into Our Genes* (New York: Anchor Books, 2005).

9. Ara Norenzayan, Will M. Gervais, and Kali H. Trzesniewski, "Mentalizing Deficits Constrain Belief in a Personal God," *PLoS ONE* 7, no. 5 (May 2012): e36880.

10. Peter K. Hatemi, John R. Alford, John R. Hibbing, Nicholas G. Martin, and Lindon J. Eaves, "Is There a "Party" in Your Genes?" *Political Research Quarterly* 62, no. 3 (September 2009): 584–600.

11. See, for example, Jacques Balthazart, *The Biology of Homosexuality*, Oxford Series in Behavioral Neuroendocrinology (New York: Oxford University Press, 2011).

12. Steven Pinker, *The Blank Slate: The Modern Denial of Human Nature* (New York: Viking, 2002), p. 47.

13. Ibid., p. 51.

14. Benjamin Libet, Curtis A. Gleason, Elwood W. Wright, and Dennis K. Pearl, "Time of Conscious Intention to Act in Relation to Onset of Cerebral Activity (Readiness-Potential): The Unconscious Initiation of a Freely Voluntary Act," *Brain* 106, no. 3 (September 1983): 632–42; and Benjamin Libet, "Unconscious Cerebral Initiative and the Role of Conscious Will in Voluntary Action," *Behavioral and Brain Sciences* 8, no. 4 (December 1985): 529–66.

15. Chun Siong Soon, Marcel Brass, Hans-Jochen Heinze, and John-Dylan Haynes, "Unconscious Determinants of Free Decisions in the Brain," *Nature Neuroscience* 11, no. 5 (May 2008): 543–45.

16. J. M. Pierre, "The Neuroscience of Free Will: Implications for Psychiatry," *Psychological Medicine* 44, no. 12 (September 2014): 2467.

17. Chun Siong Soon, Anna Hanxi He, Stefan Bode, and John-Dylan Haynes, "Predicting Free Choices for Abstract Intentions," *Proceedings of the National Academy of Sciences of the United States of America* 110, no. 15 (April 9, 2013): 6217–22.

18. Elisa Filevich, Simone Kühn, and Patrick Haggard, "There Is No Free Won't: Antecedent Brain Activity Predicts Decisions to Inhibit," *PLoS ONE* 8, no. 2 (February 2013): e53053.

19. Liane Young et al., "Disruption of the Right Temporoparietal Junction with Transcranial Magnetic Stimulation Reduces the Role of Beliefs in Moral Judgments," *Proceedings of the National Academy of Sciences of the United States of America* 107, no. 15 (April 13, 2010): 6753–58.

20. Charles Whitman, Whitman Letter, "The Whitman Archive," *Austin American-Statesman*, July 31, 1966.

21. John A. Bargh, Mark Chen, and Lara Burrows, "Automaticity of Social Behavior: Direct Effects of Trait Construct and Stereotype-Activation on Action," *Journal of Personality and Social Psychology* 71, no. 2 (August 1996): 230–44.

22. Ap Dijksterhuis and Ad van Knippenberg, "The Relation between Perception and Behavior, or How to Win a Game of Trivial Pursuit," *Journal of Personality and Social Psychology* 74, no. 4 (April 1998): 865–77.

23. Pierre, "Neuroscience of Free Will," p. 2468.

24. Alison George, "The Yuck Factor: The Surprising Power of Disgust," *New Scientist*, July 11, 2012, http://www.newscientist.com/article/mg21528731.800-the-yuck-factor-the-surprising-power-of-disgust.html (accessed May 20, 2015); and David A. Pizarro, "Research: Emotion and Judgment," David A. Pizarro (website), http://www.peezer.net/research/ (accessed May 22, 2015).

25. Yoel Inbar, David A. Pizarro, and Paul Bloom, "Conservatives Are More Easily Disgusted than Liberals," *Cognition and Emotion* 23, no. 4 (2009): 714–25.

26. Yoel Inbar, David A. Pizarro, Joshua Knobe, and Paul Bloom, "Disgust Sensitivity Predicts Intuitive Disapproval of Gays," *Emotion* 9, no. 3 (June 2009): 435–39.

27. Yoel Inbar, David A. Pizarro, and Paul Bloom, "Disgusting Smells Cause Decreased Liking of Gay Men," *Emotion* 12, no. 1 (February 2012): 23–27.

28. William D. S. Killgore et al., "The Effects of 53 Hours of Sleep Deprivation on Moral Judgment," *Sleep* 30, no. 3 (2007): 345–52.

29. David A. Pizarro (website).

30. Jonathan Haidt, "The Emotional Dog and Its Rational Tail: A Social Intuitionist Approach to Moral Judgment," *Psychological Review* 108, no. 4 (October 2001): 814–934.

31. Eagleman, *Incognito.*

32. Ibid.

33. Walter Mischel, Ebbe B. Ebbesen, and Antonette Raskoff Zeiss, "Cognitive and

Attentional Mechanisms in Delay of Gratification," *Journal of Personality and Social Psychology* 21, no. 2 (February 1972): 204–18.

34. W. Mischel, Y. Shoda, and M. I. Rodriguez, "Delay of Gratification in Children," *Science* 244, no. 4907 (May 26, 1989): 933–38.

35. Yuichi Shoda, Walter Mischel, and Philip K. Peake, "Predicting Adolescent Cognitive and Self-Regulatory Competencies from Preschool Delay of Gratification: Identifying Diagnostic Conditions," *Developmental Psychology* 26, no. 6 (November 1990): 978–86.

36. Tanya R. Schlam et al., "Preschoolers' Delay of Gratification Predicts Their Body Mass 30 Years Later," *The Journal of Pediatrics* 162, no. 1 (January 2013): 90–93.

37. B. J. Casey et al., "Behavioral and Neural Correlates of Delay of Gratification 40 Years Later," *Proceedings of the National Academy of Sciences of the United States of America* 108, no. 36 (September 6, 2011), pp. 14998–15003.

38. Yu Gao et al., "Association of Poor Childhood Fear Conditioning and Adult Crime," *The American Journal of Psychiatry* 167, no. 1 (January 2010): 55–60.

39. Masahiko Haruno and Christopher D. Frith, "Activity in the Amygdala Elicited by Unfair Divisions Predicts Social Value Orientation," *Nature Neuroscience* 13, no. 2 (February 2010): 160–61.

40. Martin Reuter et al., "Investigating the Genetic Basis of Altruism: The Role of the COMT Val158Met Polymorphism," *Social Cognitive and Affective Neuroscience* 6, no. 5 (October 2011): 662–68.

41. Daniel M. Wegner, *The Illusion of Conscious Will* (Cambridge, MA: MIT Press / Bradford Books, 2002).

42. M. S. Gazzaniga, "Right Hemisphere Language Following Brain Bisection: A 20-Year Perspective," *American Psychologist* 38 (1983): 525–37.

43. Timothy Slater and Lydia Pozzato, "Focus on Psychopathy, Part One," *FBI Law Enforcement Bulletin*, July 2012.

44. Robert Hare, "Focus on Psychopathy, Part Two," *FBI Law Enforcement Bulletin*, July 2012.

45. Raymond Tallis, "The Anatomy of Violence by Adrian Raine—Review," *Guardian*, June 13, 2013, http://www.theguardian.com/books/2013/jun/13/anatomy-violence-adrian-raine -review (accessed June 28, 2015).

46. Ezequiel Morsella, Christine A. Godwin, Tiffany K. Jantz, Stephen C. Krieger, Adam Gazzaley, "Homing in on Consciousness in the Nervous System: An Action-Based Synthesis," *Behavioral and Brain Sciences*, June 21, 2015.

47. Beth Tagawa, "Consciousness Has Less Control than Believed, According to New Theory," ScienceDaily, June 23, 2015, http://www.sciencedaily.com/releases/2015/06/ 150623141911.htm (accessed June 28, 2015).

48. Julien Musolino, *The Soul Fallacy: What Science Shows We Gain from Letting Go of Our Soul Beliefs* (Amherst, NY: Prometheus Books, 2015).

49. Linda Zagzebski, "Foreknowledge and Free Will," *Stanford Encyclopedia of Philosophy*, Aug. 25, 2011, http://plato.stanford.edu/entries/free-will-foreknowledge (accessed June 9, 2015).

CHAPTER 10: BIBLICAL ARCHAEOLOGY

1. Eusebius of Caesarea, *Life of Constantine* 3.25–41.

2. Cf. Mk. 15:26; Matt. 27:37; Luke 28:38; John 19:19. John 19:20 actually states that the inscription "was written in Hebrew, in Latin, and in Greek."

3. Socrates Scholasticus, *Ecclesiastical History* 1.17.

4. Ibid.

5. Ron Wyatt, "Chariot Wheels in the Red Sea," *Wyatt Archaeological Research* 3 (1993), http://wyattmuseum.com/chariot-wheels-in-the-red-sea/2011–669.

6. Thomas A. Holland, "Jericho," in *The Oxford Encyclopedia of Archaeology in the Near East*, ed. Eric M. Meyers (Oxford: Oxford University Press, 1997), pp. 220–24.

7. Kathleen M. Kenyon, *Digging up Jericho: The Results of the Jericho Excavations, 1952–1956* (New York: Praeger, 1957), p. 229.

8. Piotr Bienkowski, *Jericho in the Late Bronze Age* (Warminster: Aris & Phillips, 1986), pp. 120–25.

9. H. J. Bruins and J. van der Plicht, "Tell es-Sultan (Jericho): Radiocarbon Results of Short-Lived Cereal and Multiyear Charcoal Samples from the End of the Middle Bronze Age," *Radiocarbon* 37 (1995): 213–20.

10. *Annals* 15.44 reads: "Consequently, to get rid of the report, Nero fastened the guilt and inflicted the most exquisite tortures on a class hated for their abominations, called Christians by the populace. Christus, from whom the name had its origin, suffered the extreme penalty during the reign of Tiberius at the hands of one of our procurators, Pontius Pilatus, and a most mischievous superstition, thus checked for the moment, again broke out not only in Judaea, the first source of the evil, but even in Rome, where all things hideous and shameful from every part of the world find their centre and become popular. Accordingly, an arrest was first made of all who pleaded guilty; then, upon their information, an immense multitude was convicted, not so much of the crime of firing the city, as of hatred against mankind. Mockery of every sort was added to their deaths."

11. Translation by Frederick Crombie in *Ante-Nicene Fathers, Vol. 4*, ed. Alexander Roberts, James Donaldson, and A. Cleveland Coxe (Buffalo, NY: Christian Literature Publishing Co., 1885).

12. Eusebius, *Historia Ecclesiastica* 1.11.7–8.

13. Cf. James 1:22-23, 25, 4:11.

14. Origen, *Against Celsus* 1.47. See p. 492 of Richard Carrier, "Origen, Eusebius, and the Accidental Interpolation in Josephus, *Jewish Antiquities* 20.200," *Journal of Early Christian Studies* 20, no. 4 (Winter 2012): 489–514. Remember, in the gospels, John the Baptist had a well-known, well-established ministry of his own, with his own disciples, who rivaled Jesus's disciples prior to John's death. The gospels state that John's disciples ultimately joined Jesus and his ministry. Cf. John 1:35: "John again was standing with two of *his* disciples"; Matt. 9:14: "Then the disciples of John came to him (Jesus), saying, 'Why do *we* and the Pharisees fast often, but *your* disciples do not fast?'"; Matt. 11:2: "When John heard in prison what the Messiah was doing, he sent word by *his disciples*"; Luke 7:20: "When the men had come to him, they said, 'John the Baptist has sent us to you to ask, "Are you the one who is to come, or are we to wait for another?"'"

15. Matt. 27:52–53.

16. Read more about the documentary *Exodus Decoded* at http://www.imdb.com/title/tt0847162/.

17. Read more about *The Lost Tomb of Jesus* at http://www.imdb.com/title/tt0974593/.

18. Read more about *Finding Atlantis* at http://channel.nationalgeographic.com/episode/finding-atlantis-4982/Overview.

19. See my article entitled, "A Critique of Simcha Jacobovici's *Secrets of Christianity: Nails of the Cross*," in *Bible and Interpretation*, May 2011, http://www.bibleinterp.com/articles/simcha358005.shtml.

20. See my blog post entitled, "If the Evidence Doesn't Fit, Photoshop It: Digital Image Manipulation in the Case of Simcha Jacobovici and James Tabor's Jonah Ossuary," *XKV8R*, Mar. 5, 2012, http://robertcargill.com/2012/03/05/if-the-evidence-doesnt-fit-photoshop-it/.

21. Simcha Jacobovici and Barrie Wilson, *The Lost Gospel: Decoding the Ancient Text that Reveals Jesus' Marriage to Mary the Magdalene* (New York: Pegasus, 2012).

22. Simcha Jacobovici, "Jesus and Paul in the Dead Sea Scrolls!" Simcha Jacobovici Television, March 30, 2016, http://www.simchajtv.com/jesus-in-the-dead-sea-scrolls-a-second-reference.

CHAPTER 11: THE CREDIBILITY OF THE EXODUS

1. For a full discussion, see Israel Finkelstein and Neil Asher Silberman, *The Bible Unearthed: Archaeology's New Vision of Ancient Israel and the Origin of Its Sacred Texts* (New York: Free Press, 2001), pp. 106–107.

2. See, for example, Gleason L. Archer, *Encyclopedia of Bible Difficulties* (Grand Rapids, MI: Zondervan, 1982), p. 191.

3. For example, David Rohl, *Pharaohs and Kings: A Biblical Quest* (New York: Crown Publishers, 1995).

4. Gerald Aardsma, "What's Wrong with the Conventional Dates for the Exodus?" *Biblical Chronologist*, http://www.biblicalchronologist.org/answers/wrongdates.php (accessed November 2015).

5. Gerald Aardsma, "Is there Evidence of the Exodus from Egypt?" *Bible Chronologist*, http://www.biblicalchronologist.org/answers/exodus_egypt.php (accessed November 2015).

6. William Shea, "Amenhotep II as Pharaoh of the Exodus," *Bible and Spade* 16, no. 2 (Spring 2003).

7. Charles H. Dyer, "The Date of the Exodus Reexamined," *Bibliotheca Sacra* 140 (1983): 225–43.

8. Finkelstein and Silberman, *Bible Unearthed*, pp. 81–83.

9. For example, Eric Lyons, "Philistines in the Time of Abraham—Fallacy or Fact?" *Apologetics Press*, 2004, http://www.apologeticspress.org/apcontent.aspx?category=6&article=671 (accessed November 2015).

10. Archer, *Encyclopedia of Bible Difficulties*, p. 95.

11. Gerald Aardsma, "The Chronology of Egypt in Relation to the Bible: 3000–1000 BC," *Biblical Chronologist* 2, no. 2 (1996): 7.

12. Donald Redford, *Canaan and Israel in Ancient Times* (Princeton, NJ: Princeton University Press, 1992), pp. 412–22.

13. Karl Butzer, *Early Hydraulic Civilization in Egypt* (Chicago: University of Chicago Press, 1976), pp. 82–85.

14. Rebecca Bradley, *Nomads in the Archaeological Record: Case Studies in the Northern Provinces of the Sudan* (Berlin: Akademie Verlag, 1992), pp. 128–37.

15. Nora Shalaby, "Wadi Maghara: A Copper and Turquoise Mine on the Periphery," *Current Research in Egyptology 2014: Proceedings of the Fifteenth Annual Symposium, University College London & King's College London 2014*, ed. Massimiliano S. Pinarello, Justin Yoo, Jason Lundock, and Carl Walsh (Oxford, UK: Oxbow Books, 2015), pp. 168–69.

16. Elizabeth Bloxam, "Miners and Mistresses: Middle Kingdom Mining on the Margins," *Journal of Social Archaeology* 6, no. 2 (June 2006): 297.

17. Rudolph Cohen, "Excavations at Kadesh-Barnea, 1976–1978," *Biblical Archaeologist* 44 (1981): 93–107.

18. Finkelstein and Silberman, *Bible Unearthed*, p. 114.

19. Redford, *Canaan and Israel in Ancient Times*, pp. 412–13.

20. Ibid., p. 422.

CHAPTER 12: PIOUS FRAUD AT NAZARETH

1. René Salm, *The Myth of Nazareth: The Invented Town of Jesus* (Cranford, NJ: American Atheist, 2008), p. 205.

2. Mark O'Keefe, "Israel's Evangelical Approach," *Washington Post*, January 6, 2002. See also: René Salm, *NazarethGate: Quack Archeology, Holy Hoaxes, and the Invented Town of Jesus* (Cranford, NJ: American Atheist, 2015), pp. 242–43.

3. Ibid.

4. See Salm, *Myth of Nazareth*, pp. 243–59.

5. Mishna Baba Bathra 2.9. Cf. Num. 5:1–3, 19:11; Lev. 5:3. See also Salm, *Myth of Nazareth*, pp. 219–20.

6. Thus Père P. Viaud wondered, regarding a tomb under the Church of the Annunciation, "Est-il de saint Joseph?" See *Nazareth et ses deux églises de L'Annonciation et de Saint-Joseph d'après les fouilles récentes* (Paris: Librarie A. Picard et Fils, 1910), p. 93.

7. C. Kopp, "Beiträge zur Geschichte Nazareths," *Journal of the Palestine Oriental Society* 19 (1939): 91–92.

8. Jack Finegan, *The Archaeology of the New Testament: The Life of Jesus and the Beginning of the Early Church* (Princeton, NJ: Princeton University Press, 1969), p. 27.

9. Salm, *Myth of Nazareth*, p. 250.

10. Salm, *Myth of Nazareth*, p. 170, and appendices 5, 6.

11. Per the work of H.-P. Kuhnen. E.g., Hans-Peter Kuhnen, *Norwest-Palästina in hellenistisch-römischer Zeit. Bauten und Gräber im Karmelgebiet* (Weinheim, Germany: VCH Verlag, 1986); *Palästina in griechisch-römischer Zeit* (München, Germany: C. H. Beck,

1990); "Grabbau und Bestattungssitten in Palästina zwischen Herodes und den Severern," in *Körpergräber des 1.–3. Jh. in der römischen Welt*, ed. A. Fabert et al., Kolloquium Frankfurt am Main (Frankfurt, Germany: Schriften des Archäologischen Museums Frankfurt am Main, 2004), pp. 57–76. Cf. Salm, *Myth of Nazareth*, pp. 158–64.

12. Stephen Pfann, Ross Voss, and Yehudah Rapuano, "Surveys and Excavations at the Nazareth Village Farm (1997–2002): Final Report," *Bulletin of the Anglo-Israel Archaeological Society* 25 (2007): 16–79.

13. René Salm, "A Response to 'Surveys and Excavations at the Nazareth Village Farm (1997–2002): Final Report,'" *Bulletin of the Anglo-Israel Archaeological Society* 26 (2008): 95–103 (reprinted in Salm, *NazarethGate*, pp. 48–61). In the same issue are Y. Rapuano's "Amendment" (pp. 113–35), S. Pfann's "A Reply to Salm" (pp. 105–108), Ken Dark's "A Reply to Salm" (pp. 109–11), and Dark's scathing review of Salm's *The Myth of Nazareth* (pp. 140–46).

14. See Salm, *NazarethGate*, pp. 245–313.

15. Bart Ehrman, *Did Jesus Exist?* (New York: HarperCollins, 2012), p. 195. See also Salm, *NazarethGate*, pp. 160, 165–66, 293–94.

16. S. Pfann et al., "Surveys and Excavations," p. 40. See Salm, *NazarethGate*, pp. 286–87.

17. For a digital scan and discussion of Alexandre's 2006 report, see Salm, *NazarethGate*, pp. 256–57. Discussed also at René Salm, "Archaeology, Bart Ehrman, and the Nazareth of 'Jesus,'" in *Bart Ehrman and the Quest of the Historical Jesus of Nazareth*, ed. F. Zindler and R. Price (Cranford NJ: American Atheist, 2013), pp. 346–47.

18. The numerous flaws in Berman's coin chapter are detailed in Salm, "Quackery at Mary's Well," chapter 11 in *NazarethGate*, pp. 295–307.

19. Two photos of this are reproduced at Salm, *NazarethGate*, plate 6.

20. Y. Alexandre, *Mary's Well, Nazareth: The Late Hellenistic to the Ottoman Periods* (Jerusalem: Israel Antiquities Authority, 2012), pp. 16, 18. Also cf. Salm, *NazarethGate*, pp. 310–11: "The Roman incipience of the water channels."

21. See the online IAA release of Dec. 21, 2009, "Residential Building from the Time of Jesus Exposed in Nazareth." Numerous articles appeared in the press substantially, echoing the IAA information.

22. Chapter 10 of my recent book *NazarethGate* devotes sixty-five pages to the site and reveals that no dwelling ever existed there.

23. Salm, *NazarethGate*, pp. 198–202.

24. Ibid., pp. 225–30.

25. Yuri Shahar, "The Underground Hideouts in Galilee and their Historical Meaning," in *The Bar Kokhba War Reconsidered*, ed. Peter Schafer (Tübingen, Germany: Mohr Siebeck, 2003), 217–40; Mordechai Aviam, "Secret Hideaway Complexes in the Galilee," chapter 12 in *Pagans and Christians in the Galilee: 25 Years of Archaeological Excavations and Surveys— Hellenistic to Byzantine Periods* (Rochester, NY: University of Rochester Press, 2004).

26. Ibid.

27. K. Dark, "Early Roman-Period Nazareth and the Sisters of Nazareth Convent," *Antiquaries Journal* 92 (September 2012), p. 38.

28. Salm, *NazarethGate*, p. 119.

29. Dark's publications on the SNC site (as of this writing) are tabulated in Salm,

NazarethGate, p. 79. They include a thirty-six-page report entitled "Archaeological Recording at the Sisters of Nazareth Convent in Nazareth, 2006" (published 2007), as well as three similar-length reports under the rubric "Nazareth Archaeological Project: A Preliminary Report" (published 2008, 2009, and 2010). All four publications bear the imprint Late Antiquity Research Group (London). In addition, Dark authored an *Antiquaries Journal* article on the SNC site (see note 27, above). Pending the appearance of a forthcoming monograph, he characterizes all five articles as "interim."

30. Dark, "Nazareth Archaeological Project," 2009, p. 11. (Reporting on the 2008 season.)

31. Salm, *NazarethGate*, chapter 6.

32. Dark extends this unusual dwelling-tomb sequence to other sites in Nazareth, proposing that they also were the sites of habitations before they became the locations of first century CE tombs! (See Salm, *NazarethGate*, pp. 75–76). This view, of course, is most convenient in explaining the existence of tombs below so many sites that the Christian tradition has claimed were domiciles in the time of Jesus.

33. Salm, *NazarethGate*, pp. 107–11.

34. Dark, "Nazareth Archaeological Project," 2009, p. 7. (Reporting on the 2008 season.)

35. Salm, *NazarethGate*, pp. 84–93.

36. Ibid., pp. 94–96.

37. An example is D. Trifon, "Did the Priestly Courses Transfer from Judaea to Galilee after the Bar Kokhba Revolt?" *Tarbits* 59 (1989), pp. 77–93 (Hebrew).

38. E. Schürer, *A History of the Jewish People in the Time of Christ*, 5 vols. (Edinburgh: T. & T. Clark, 1890; Peabody, MA: Hendrickson, 1998), I.ii.272.

39. The steps leading to this remarkable conclusion are detailed in Salm, *NazarethGate*, chapter 12.

40. Cahiers du Cercle Ernest Renan 252 (2010), pp. 35–64.

41. The photograph can be found in an article by S. Talmon entitled "The Calendar Reckoning of the Sect from the Judaean Desert," in *Aspects of the Dead Sea Scrolls*, ed. C. Rabin and Y. Yadin (Jerusalem: Magnes, 1958), p. 171.

42. M. Govaars, M. Spiro, and L. White. *Field O: The "Synagogue" Site*, vol. 9, The Joint Expedition to Caesarea Maritima: Excavation Reports (Boston: ASOR, 2009). See also: M. Govaars, *A Reconsideration of the Synagogue Site at Caesarea Maritima, Israel* (master's thesis, Madison, NJ: Drew University, 1983).

43. Though Josephus (War 2.14.4) describes a synagogue in Caesarea owned by a Hellenist Jew, it is certain (according to the work of M. Govaars) that the synagogue in question did not exist in the northwest quadrant of the city excavated by Avi-Yonah in 1956 and 1962. Nevertheless, it is precisely in that area that all three fragments of the "Caesarea inscription" have allegedly been found. How they could have been found in an area where no synagogue existed is thus a mystery.

44. Salm, *NazarethGate*, p. 369.

45. We learn this from G. E. Wright's letter of May 30, 1972, reproduced in Salm, *NazarethGate*, pp. 346–47.

46. See Wright's letter (details in previous note).

47. Salm, *NazarethGate*, pp. 360–67.

48. In 1962, Michael Avi-Yonah wrote that "Fragment B was found in area F." However, one years later he wrote that it was discovered along with "reused marble slabs" in "Area E." See Michael Avi-Yonah, "A List of Priestly Courses from Caesarea," *Israel Exploration Journal* 12 (1962): 137; Michael Avi-Yonah, "Notes and News: Caesarea," *Israel Exploration Journal* 13 (1963): 146. For discussion, see Salm, *NazarethGate*, p. 361.

CHAPTER 13: THE BETHLEHEM STAR

1. Aaron Adair, "The Star of Christ in the Light of Astronomy," *Zygon: Journal of Science & Religion* 47, no. 1 (March 2012): 7–29.

2. Adrienne Mayor, *The First Fossil Hunters: Paleontology in Greek and Roman Times* (Princeton, NJ: Princeton University Press, 2000).

3. Cf. Richard A Burridge, *What Are the Gospels? A Comparison with Greco-Roman Biography* (Grand Rapids, MI: Eerdmans, 2004).

4. Compare the work of Michael E. Vines, *The Problem of Markan Genre: The Gospel of Mark and the Jewish Novel* (Atlanta: Society of Biblical Literature, 2002), who argues that the Gospel of Mark bests fits into the category of ancient novels, specifically Jewish novels. Correct or not, the end of debate about the appropriate genre for the Gospels is not over, and the potentially biographical nature of the stories of Jesus hardly mean they are historically trustworthy; after all, fake gospels such as the other infancy narratives seem to fit the biography genre just as well as the canonical Gospels do, yet we don't even consider it plausible that they contain reliable information about Jesus.

5. David Hughes, "Astronomical Thoughts on the Star of Bethlehem," in *The Star of Bethlehem and the Magi: Interdisciplinary Perspectives from Experts on the Ancient Near East, the Greco-Roman World, and Modern Astronomy*, ed. Peter Barthel and George van Kooten (Boston: Brill, 2015), p. 105.

6. Cf. Colin R. Nicholl, *The Great Christ Comet: Revealing the True Star of Bethlehem* (Wheaton, IL: Crossway, 2015).

7. Michael Molnar, *The Star of Bethlehem: The Legacy of the Magi* (New Brunswick, NJ: Rutgers University Press, 1999).

8. John T. Ramsey and A. Lewis Licht, *The Comet of 44 B.C. and Caesar's Funeral Games* (Atlanta, GA: Scholars, 1997), 135–53.

9. John Ramsey, "Mithridates, the Banner of Ch'ih-yu, and the Comet Coin," *Harvard Studies in Classical Philology* 99 (1999): 197–253.

10. Origen, *Contra Celsum* 1.59.

11. David Hughes, *The Star of Bethlehem: An Astronomer's Confirmation* (New York: Walker, 1979), pp. 68, 96, 184–86; Roy Rosenberg, "The 'Star of the Messiah' Reconsidered," *Biblica* 53, no. 1 (1972): 105–109.

12. Cf. Azahiah ben Moses dei Rossi, *The Light of the Eyes*, trans. Joanna Weinberg (New Haven, CT: Yale University Press, 2001), pp. 548–50.

13. David E. Pingree, "Historical Horoscopes," *Journal of the American Oriental Society*

82, no. 4 (Oct.–Dec. 1962): 487–502; David E Pingree, "Astronomy and Astrology in India and Iran," *Isis* 54, no. 2 (1963): 229–46.

14. Ernest Martin, *The Star that Astonished the World* (Portland: ASK Publications, 1991).

15. Stephan Heilen, "The Star of Bethlehem and Greco-Roman Astrology, Especially Astrological Geography," in Barthel and van Kooten, *Star of Bethlehem and the Magi*, pp. 297–357 (esp. 330–32).

16. Aaron Adair, *The Star of Bethlehem: A Skeptical View* (Fareham, UK: Onus, 2013), p. 78.

17. John McGrew and Richard McFall, "A Scientific Inquiry into the Validity of Astrology," *Journal of Scientific Exploration* 4, no. 1 (1990): 75–83; Rob Nanninga, "The Astrotest: A Tough Match for Astrologers," *Correlation* 15, no. 2 (1996): 14–20.

18. Cf. Ptolemy, *Tetrabiblos* 1.2 (6–7); Firmicus, *Mathesis* 1.3.2.

19. Albert Schweitzer, *The Quest of the Historical Jesus*, trans. John Bowden (Minneapolis: Fortress, 2001), pp. 464–65.

20. Franz Boll, "Der Stern der Weisen," *Zeitschrift für die neutestamentliche Wissenschaft und die Kunde des Urchristentums* 18 (1917/1918): 40–48 (paraphrase comes from p. 48).

21. See Plato, *Timaeus* 41E-42A; *Republic* X.614–621; Pliny, *Natural History* 2.6.

22. Not only is this the view of all premodern interpreters of the tale, it is the one found among critical Bible scholars. This can be seen in the discussions of the Star of Bethlehem by Antonio Panaino (philologist and Iranian studies expert), Annette Merz (renowned historical Jesus scholar), and Stephan Heilen (philologist and ancient astrology expert) from the 2014 Groningen conference on the subject, all published in Peter Barthel and George van Kooten's *The Star of Bethlehem and the Magi: Interdisciplinary Perspectives from Experts on the Ancient Near East, the Greco-Roman World, and Modern Astronomy*. Merz in particular favorably cites my own discussion on the grammar and semantics of the Greek text.

23. Marcus Borg and John Dominic Crossan, *The First Christmas: What the Gospels Really Teach about Jesus' Birth* (New York: Harper One, 2007), p. 182.

24. Augustine, *Reply to Faustus* 2.6f. On Augustine's former interest in astrology, see Augustine, *Confessions* 4.1–3; 5.3; Leo Ferrari, "Astronomy and Augustine's Break with the Manichees," *Revue des Études Augustiniennes* 19 (1973): 263–76.

25. Barry Downing, *The Bible and Flying Saucers* (New York: Avon, 1970), p. 134.

26. Robert Newman, "The Star of Bethlehem: A Natural-Supernatural Hybrid?" *Interdisciplinary Biblical Research Institute* (2001): 1–16.

27. Paul Tobin, "The Bible and Modern Scholarship," in *The Christian Delusion: Why Faith Fails*, ed. John W. Loftus (Amherst, NY: Prometheus Books, 2010), pp. 148–80 (esp. 160–63).

28. Philippe Gignoux, "L'Inscription de Kartir à Sar Mashad," *Journal Asiatique* 256 (1968): 387–418.

29. Gerard Mussies, "Some Astrological Presuppositions of Matthew 2: Oriental, Classical and Rabbinical Parallels," in *Aspects of Religious Contact and Conflict in the Ancient World*, ed. Pieter Willem van der Horst (Ultrecht: Faculteit de Godgeleerdheid Universiteit Utrecht, 1995), pp. 25–44.

30. Tertullian, *Contra Marcion* 3.13.

31. War in Armenia: Tacitus, *Annals* 13; Dio, *Roman History* 62; Suetonius, *Nero* 57.

Diplomatic situation concerning Armenia: Josephus, *Antiquities of the Jews* 18.96–105; Tacitus, *Annals* 2.58; Suetonius, *Caligula* 41.3; *Vitellius* 2.4; Dio, *Roman History* 59.27.3–4. Also close to the time of Jesus' birth, a near-war because of a changing in control in Armenia: Dio, *Roman History* 55.18; Velleius Paterculus, *Compendium of Roman History* 2.100–102.

32. Robert A. Segal, *In Quest of the Hero* (Princeton, NJ: Princeton University Press, 1990); Richard Horsley, *The Liberation of Christmas: The Infancy Narratives in Social Context* (New York: Crossroad, 1989), pp. 162–63. See also Robert J. Miller, *Born Divine: The Births of Jesus & Other Sons of God* (Santa Rosa, CA: Polebridge, 2003).

33. Cf. Josephus, *Antiquities of the Jews* 2.9, §§ 2–3.

34. Virgil, *Aeneid* 2.687–711.

35. Servius, *In Virgilii Aeneidos* 1.382.

36. Annette Merz, "Matthew's Star, Luke's Census, Bethlehem, and the Quest of the Historical Jesus," in Barthel and van Kooten, *Star of Bethlehem and the Magi*, pp. 465–66. Merz also cites me on this point (p. 466n8), saying how two scholars noting the similarity between Matthew's story and the pious inventions in other infancy gospels "probably indicates that it forces itself on the historically minded as an illuminating comparison!"

37. This was in fact done by the quintessential ancient astronaut proponent, Erich von Daniken in his *Chariots of the Gods?* This is all the more humorous considering the Ali Baba story was likely invented in the eighteenth century by Antoine Galland.

38. I propose this as an avenue of future research, along with other pieces of supporting evidence, in Aaron Adair, "A Critical Look at the History of Interpreting the Star of Bethlehem in Scientific Literature and Biblical Studies," in Barthel and van Kooten, *Star of Bethlehem and the Magi*, pp. 74–79. This is something I intend to publish more on in the future.

CHAPTER 14: IF PRAYER FAILS,
WHY DO PEOPLE KEEP AT IT?

1. For information about Freedom From Religion's "Nothing Fails Like Prayer Award," see http://ffrf.org/news/news-releases/item/20595-ffrf-announces-%E2%80%98nothing -fails-like-prayer-award%E2%80%99-contest.

2. For details of the Pew Forum survey, see http://pewforum.org/docs/?DocID=179.

3. General Social Survey, GSS cumulative data file 1972–2006, 2008.

4. J. H. Ellens, "Communication Theory and Petitionary Prayer," *Journal of Psychology and Theology* 5 (Winter 1977): 48–54.

5. Bernard Spilka and Kevin Ladd, *The Psychology of Prayer: A Scientific Approach* (New York: Guilford Publications, 2012), Kindle edition.

6. R. J. Foster, *Prayer: Finding the Heart's True Home* (San Francisco, CA: HarperCollins Publishers, 1992).

7. Kevin L. Ladd and Bernard Spilka, "Inward, Outward, and Upward: Cognitive Aspects of Prayer," *Journal for the Scientific Study of Religion* 41 (September 2002): 475–84.

8. Thom Stark, *The Human Faces of God: What Scripture Reveals When It Gets God*

Wrong (and Why Inerrancy Tries To Hide It) (Eugene, OR: Wipf and Stock Publishers, 2010), http://humanfacesofgod.com/.

9. The Epistle of Ignatius to the Smyrnaeans, http://www.newadvent.org/fathers/0109.htm.

10. Massey Hamilton Shepherd Jr., "Smyrna in the Ignatian Letters: A Study in Church Order," *Journal of Religion* 20, no. 2 (April 1940): 152.

11. Francis Galton, *Inquiries into Human Faculty and Its Development* (New York: Macmillan, 1883), p. 282.

12. H. Benson et. al., "Study of the Therapeutic Effects of Intercessory Prayer (STEP) in Cardiac Bypass Patients: A Multicenter Randomized Trial of Uncertainty and Certainty of Receiving Intercessory Prayer," *American Heart Journal* 151, no. 4 (April 2006): 934–42, http://www.ncbi.nlm.nih.gov/pubmed/16569567/.

13. For more on the STEP study, see: http://www.templeton.org/newsroom/press _releases/060407step.html.

14. Leanne Roberts, Irshad Ahmed, and Andrew Davison, "Intercessory Prayer for the Alleviation of Ill Health," *Cochrane Database of Systematic Reviews*, April 15, 2009, onlinelibrary.wiley.com/doi/10.1002/14651858.CD000368.pub3/full.

15. Ibid.

16. Benedict Carey, "Long-Awaited Medical Study Questions the Power of Prayer," *New York Times*, March 31, 2006, http://www.nytimes.com/2006/03/31/health/31pray.html.

17. Richard Swinburne, "Response to a Statistical Study of the Effect of Petitionary Prayer," *Science and Theology News*, April 7, 2006, http://users.ox.ac.uk/~orie0087/framesetpdfs.shtml.

18. Carey, "Long-Awaited Medical Study."

19. Ken Collins, "Does Prayer Actually Work?" Ken Collins' Website, http://www .kencollins.com/explanations/why-14.htm.

20. Swinburne, "Response to a Statistical Study."

21. Spilka and Ladd, *Psychology of Prayer,* p. 169.

22. Ibid.

23. Newsweek Staff, "Newsweek Poll: 90% Believe in God," *Newsweek*, April 8, 2007, http://www.newsweek.com/newsweek-poll-90-believe-god-97611.

24. Adam Lee, "Nothing Fails Like Prayer," *Daylight Atheism*, http://www.patheos.com/ blogs/daylightatheism/essays/nothing-fails-like-prayer/.

25. Marian Wiggins, personal correspondence.

26. Pascal Boyer, *Religion Explained: The Evolutionary Origins of Religious Thought* (New York: Basic Books, 2001).

27. "Nothing Fails Like Prayer," The Thinking Atheist, February 9, 2012, http://www. thethinkingatheist.com/blog/28/Nothing-Fails-Like-Prayer.

28. Bill Maher, *Real Time with Bill Maher*, June 21, 2013.

29. Lawrence Krauss, "Join the Real World and Show Faith in Reasoned Debate," *A Celebration of Reason*, Global Atheist Convention, Melbourne, Australia, April 13–15, 2012, http://www.atheistconvention.org.au/2012/04/14/krauss-join-the-real-world-and-show -faith-in-reasoned-debate-smh/.

30. Julia Sweeney, "About *Letting Go of God*," Julia Sweeney, http://juliasweeney .com/letting-go-of-god/.

31. Lee, "Nothing Fails Like Prayer."

32. Michael Formica, "The Science, Psychology, and Metaphysics of Prayer," *Enlightened Living* (blog), *Psychology Today*, July 28, 2010, https://www.psychologytoday.com/blog/enlightened-living/201007/the-science-psychology-and-metaphysics-prayer.

CHAPTER 15: THE TURIN SHROUD

1. H. David Sox, *File on the Shroud* (London: Coronet Books, 1978), p. 39.

2. Thomas Humber, *The Sacred Shroud* (New York: Pocket Books, 1978), pp. 62–63; Joe Nickell, *Inquest on the Shroud of Turin: Latest Scientific Findings* (Amherst, NY: Prometheus Books, 1998), pp. 31–39.

3. Joe Nickell, *Relics of the Christ* (Lexington: University Press of Kentucky, 2007), pp. 77–95.

4. John Calvin, *Treatise on Relics* (1543; Amherst, NY: Prometheus Books, 2009), pp. 49–112.

5. Joe Nickell, "Testing Jesus: Alleged Relic vs. Radiocarbon Dating," *A Skeptical Eye* 1, no. 2 (Summer 2015): 3.

6. Joe Nickell, *The Science of Miracles* (Amherst, NY: Prometheus Books, 20), p. 102.

7. Nickell, *Inquest*, pp. 51–52.

8. Humber, *Sacred Shroud*, p. 78.

9. Ian Wilson, *The Shroud of Turin* (Garden City, NY: Image Books, 1979), pp. 106–24.

10. Nickell, *Science of Miracles*, pp. 111–15.

11. Pierre d'Arcis, "Draft Memorandum to Pope Clement VII, 1389," in Wilson, *The Shroud of Turin*, pp. 266–72.

12. Ibid.

13. Ibid.

14. Nickell, *Inquest*, pp. 17–27.

15. Nickell, *Science of Miracles*, p. 122.

16. Ibid.

17. Joe Nickell, "CNN's 'Finding Jesus': Disingenuous Look at Turin 'Shroud,'" *Investigative Briefs* (blog), *Center for Inquiry*, March 6, 2015, http://www.centerforinquiry.net/blogs/entry/cnns_finding_jesus_disingenuous_look_at_turin_shroud/ (accessed September 10, 2015).

18. Dr. Michael Baden, quoted in Reginald W. Rhein Jr., "The Shroud of Turin: Medical Examiners Disagree," *Medical World News* 21, no. 6 (December 26, 1980): 40.

19. Nickell, *Inquest*, pp. 62–63, 143.

20. Ibid., pp. 64–67, 143.

21. Ibid., pp. 142–43.

22. Robert K. Wilcox, *Shroud* (New York: Macmillan, 1977), p. 44.

23. Paul L. Kirk, *Crime Investigation*, 2nd ed. (New York: John Wiley & Sons, 1974), pp. 194–95.

24. Sox, *File*, pp. 90–93, 139–41; Wilson, *Shroud of Turin*, 72–77; Nickell, *Inquest*, pp. 109–14, 128–29.

25. Nickell, *Inquest*, pp. 58–59, 86, 114–18, 119–25; Walter C. McCrone, *Judgment Day for the Turin "Shroud"* (Chicago: Microscope, 1996); John F. Fischer, "A Summary Critique of Analyses of the 'Blood' on the Turin 'Shroud,'" appendix in Nickell, *Inquest*, pp. 155–58.

26. David Van Biema, "Science and the Shroud," *Time*, April 20, 1998, pp. 53–61.

27. McCrone, quoted in Nickell, *Inquest*, 124.

28. Nickell, *Inquest*, pp. 36–37.

29. P. E. Damon et al., "Radiocarbon Dating of the Shroud of the Turin" *Nature* 337, no. 6208 (February 16, 1989): 611–15.

30. Nickell, *Science of Miracles*, pp. 123–24.

31. Ian Wilson, *The Blood and the Shroud* (New York: Free Press, 1998), pp. 219–23.

32. Leoncio Garza-Valdez. *The DNA of God?* (New York: Doubleday, 1999), p. 37; Thomas J. Pickett, "Can Contamination Save the Shroud of Turin?" *Skeptical Briefs* 6, no. 2 (June 1996): 3.

33. Nickell, *Relics*, pp. 135–36.

34. Ian Wilson, *The Shroud* (New York: Bantam Books, 2010), pp. 100–103; Joe Nickell, "UPDATED: Another Easter for the Turin 'Shroud,'" *Investigative Briefs* (blog), *Center for Inquiry*, March 28, 2013, http://www.centerforinquiry.net/blogs/entry/another_easter_for_the _turin_shroud/ (accessed March 29, 2013).

35. Mary Whanger and Alan Whanger, *The Shroud of Turin: An Adventure of Discovery* (Franklin, TN: Providence House, 1998); Wilson, *Blood and the Shroud*, p. 242.

36. Joe Nickell, "Claims of Invalid 'Shroud' Radiocarbon Date Cut from Whole Cloth," *Skeptical Inquirer* 29, no. 3 (May/June 2005): 14–16; Marco Bella, Luigi Garlaschelli, and Roberto Samperi, "There Is No Mass Spectrometry Evidence that the C14 Sample from the Shroud of Turin Comes from a 'Medieval Invisible Mending,'" *Thermochimica Acta* 617 (October 10, 2015): 160–61.

37. Nickell, *Science of Miracles*, pp. 133–34.

38. Mark Guscin, *The Oviedo Cloth* (Cambridge, UK: Lutterworth, 1998), pp. 33–34.

39. Mary Jo Anderson, "Scientists: Relic Authenticates Shroud of Turin," *World Net Daily*, October 4, 2000, http://www.wnd.com/2000/10/4279/ (accessed Feb. 14, 2013).

40. Guscin, *Oviedo Cloth*, p. 110.

41. Wilson, *Shroud of Turin*, p. 89.

42. Raymond N. Rogers. "Shroud Not Hoax, Not Miracle," letter to the editor, *Skeptical Inquirer* 28, no. 4 (July/August 2004): 69.

43. Guscin, *Oviedo Cloth*, pp. 17–18.

44. Ibid., 77–88.

45. Nickell, *Relics of the Christ*, pp. 154–56, 177–79.

46. Nickell, *Inquest*, p. 124.

47. Ibid., plates 6–9.

48. Walter McCrone, speaking at a 1980 meeting of the British Society for the Turin Shroud, quoted in Nickell, *Inquest*, 99.

49. See, for example, Mark Antonacci, *The Resurrection of the Shroud* (New York: M. Evans, 2000), pp. 73–76.

50. Ibid., p. 73.

51. *The Mysterious Man of the Shroud*, CBS documentary, aired April 1, 1997.

52. Nickell, *Inquest*, pp. 77–81, 95–106, 138–40.

53. Luigi Garlaschelli, quoted in Massimo Polidoro, "The Shroud of Turin Duplicated," *Skeptical Inquirer* 34, no. 1 (January/February 2010): 18.

54. Steven D. Schafersman, "Science, the Public, and the Shroud of Turin," *Skeptical Inquirer* 6 (Spring 1982): 51.

55. Lynn Picknett and Clive Prince, *Turin Shroud: In Whose Image? The Truth behind the Centuries-Long Conspiracy of Silence* (New York: HarperCollins, 1994), pp. 107–28.

56. H. W. Janson, *History of Art* (New York: Harry N. Abrams, 1963), pp. 267, 286; J. Leroy Davidson, and Phillipa Gerry, eds., *The New Standard Encyclopedia of Art* (New York: Garden City, 1939), pp. 226–27.

57. Calvin, *Treatise*, p. 40.

58. Nickell, *Inquest*, p. 26.

59. Nickell, *Science of Miracles*, p. 132.

60. Ibid., pp. 131–32.

61. Joe Nickell, response to Raymond N. Rogers, "Shroud Not Hoax, Not Miracle."

62. Ulysse Chevalier, 1900, quoted in Humber, *The Sacred Shroud*, p. 106.

63. B.C., "Pope Francis and the Turin Shroud: Making Sense of a Mystery" *Erasmus* (blog), *Economist*, March 31, 2013, http://www.economist.com/blogs/erasmus/2013/03/pope-francis-and-Turin-shroud/ (accessed April 2, 2013).